Partial Least Squares Structural Equation Modeling (PLS-SEM) Applications in Economics and Finance

Partial Least Squares Structural Equation Modeling (PLS-SEM) Applications in Economics and Finance

Editors

María del Carmen Valls Martínez
Pedro Antonio Martín Cervantes

MDPI • Basel • Beijing • Wuhan • Barcelona • Belgrade • Manchester • Tokyo • Cluj • Tianjin

Editors
María del Carmen Valls Martínez
University of Almería
Spain

Pedro Antonio Martín Cervantes
University of Almería
Spain

Editorial Office
MDPI
St. Alban-Anlage 66
4052 Basel, Switzerland

This is a reprint of articles from the Special Issue published online in the open access journal *Mathematics* (ISSN 2227-7390) (available at: https://www.mdpi.com/journal/mathematics/special_issues/PLS-SEM).

For citation purposes, cite each article independently as indicated on the article page online and as indicated below:

LastName, A.A.; LastName, B.B.; LastName, C.C. Article Title. *Journal Name* **Year**, *Volume Number*, Page Range.

ISBN 978-3-0365-2620-1 (Hbk)
ISBN 978-3-0365-2621-8 (PDF)

© 2021 by the authors. Articles in this book are Open Access and distributed under the Creative Commons Attribution (CC BY) license, which allows users to download, copy and build upon published articles, as long as the author and publisher are properly credited, which ensures maximum dissemination and a wider impact of our publications.

The book as a whole is distributed by MDPI under the terms and conditions of the Creative Commons license CC BY-NC-ND.

Contents

Preface to "Partial Least Squares Structural Equation Modeling (PLS-SEM) Applications in
Economics and Finance" vii

Luis Miguel López-Bonilla, Borja Sanz-Altamira and Jesús Manuel López-Bonilla
Self-Consciousness in Online Shopping Behavior
Reprinted from: *Mathematics* 2021, 9, 729, doi:10.3390/ math9070729 1

Marvello Yang, Abdullah Al Mamun, Muhammad Mohiuddin, Sayed Samer Ali Al-Shami
and Noor Raihani Zainol
Predicting Stock Market Investment Intention and Behavior among Malaysian Working Adults
Using Partial Least Squares Structural Equation Modeling
Reprinted from: *Mathematics* 2021, 9, 873, doi:10.3390/math9080873 15

Ricardo E. Buitrago R., María Inés Barbosa Camargo and Favio Cala Vitery
Emerging Economies' Institutional Quality and International Competitiveness:
A PLS-SEM Approach
Reprinted from: *Mathematics* 2021, 9, 928, doi:10.3390/math9090928 31

Jorge Luis García Alcaraz, Flor Adriana Martínez Hernández, Jesús Everardo Olguín Tiznado,
Arturo Realyvásquez Vargas, Emilio Jiménez Macías and Carlos Javierre Lardies
Effect of Quality Lean Manufacturing Tools on Commercial Benefits Gained by
Mexican Maquiladoras
Reprinted from: *Mathematics* 2021, 9, 971, doi:10.3390/math9090971 49

Andreea-Ionela Puiu, Anca Monica Ardeleanu, Camelia Cojocaru and Anca Bratu
Exploring the Effect of Status Quo, Innovativeness, and Involvement Tendencies on Luxury
Fashion Innovations: The Mediation Role of Status Consumption
Reprinted from: *Mathematics* 2021, 9, 1051, doi:10.3390/math9091051 65

Alberto Peralta and Luis Rubalcaba
How Governance Paradigms and Other Drivers Affect Public Managers' Use of Innovation
Practices. A PLS-SEM Analysis and Model
Reprinted from: *Mathematics* 2021, 9, 1055, doi:10.3390/math9091055 83

María Dolores Benítez-Márquez, Guillermo Bermúdez-González, Eva María Sánchez-Teba
and Elena Cruz-Ruiz
Exploring the Antecedents of Cruisers' Destination Loyalty: Cognitive Destination Image and
Cruisers' Satisfaction
Reprinted from: *Mathematics* 2021, 9, 1218, doi:10.3390/math9111218 111

Alicia Ramírez-Orellana, María del Carmen Valls Martínez and Mayra Soledad Grasso
Using Higher-Order Constructs to Estimate Health-Disease Status: The Effect of Health System
Performance and Sustainability
Reprinted from: *Mathematics* 2021, 9, 1228, doi:10.3390/math9111228 135

Rubén Jesús Pérez-López, Jesús Everardo Olguín-Tiznado, Jorge Luis García-Alcaraz,
María Mojarro-Magaña, Claudia Camargo-Wilson and Juan Andrés López-Barreras
Integrating and Controlling ICT Implementation in the Supply Chain: The SME Experience
from Baja California
Reprinted from: *Mathematics* 2021, 9, 1234, doi:10.3390/math9111234 159

Mariano Méndez-Suárez
Marketing Mix Modeling Using PLS-SEM, Bootstrapping the Model Coefficients
Reprinted from: *Mathematics* **2021**, *9*, 1832, doi:10.3390/math9151832 . . . . **173**

**Rafael Sancho-Zamora, Isidro Peña-García, Santiago Gutiérrez-Broncano and
Felipe Hernández-Perlines**
Moderating Effect of Proactivity on Firm Absorptive Capacity and Performance:
Empirical Evidence from Spanish Firms
Reprinted from: *Mathematics* **2021**, *9*, 2099, doi:10.3390/math9172099 . . . . **185**

**María del Carmen Valls Martínez, Pedro Antonio Martín-Cervantes,
Ana María Sánchez Pérez and María del Carmen Martínez Victoria**
Learning Mathematics of Financial Operations during the COVID-19 Era: An Assessment with
Partial Least Squares Structural Equation Modeling
Reprinted from: *Mathematics* **2021**, *9*, 2120, doi:10.3390/math9172120 . . . . **201**

Bin Wang, Qiuxia Zheng, Ao Sun, Jie Bao and Dianting Wu
Spatio-Temporal Patterns of CO_2 Emissions and Influencing Factors in China Using ESDA and
PLS-SEM
Reprinted from: *Mathematics* **2021**, *9*, 2711, doi:10.3390/math9212711 . . . . **221**

**Fernando Gimeno-Arias, José Manuel Santos-Jaén, Mercedes Palacios-Manzano and
Héctor Horacio Garza-Sánchez**
Using PLS-SEM to Analyze the Effect of CSR on Corporate Performance: The Mediating
Role of Human Resources Management and Customer Satisfaction. An Empirical Study
in the Spanish Food and Beverage Manufacturing Sector
Reprinted from: *Mathematics* **2021**, *9*, 2973, doi:10.3390/math9222973 . . . . **245**

Preface to "Partial Least Squares Structural Equation Modeling (PLS-SEM) Applications in Economics and Finance"

Partial least squares structural equation modeling (PLS-SEM) represents a new-generation statistical data analysis technique that, despite its recent creation, is exponentially gaining popularity in academia. It has attracted the interest of scholars for its use in various methodologies, becoming a dynamic and constantly evolving technique.

In today's world, managers of companies and public administrations, academics, and researchers have at their disposal a large amount of data to analyze for decision making and the discovery of new findings. This requires the establishment of a theoretical framework, the use of sciences such as mathematics and statistics, as well as experience and intuition. In this regard, PLS-SEM involves a multivariate data analysis technique that combines the methodologies of regression and linear analysis, offering advantages for its application, such as the non-requirement of normality or the possibility of obtaining reliable results with small sample sizes.

It is widely employed in the field of social sciences, where it is necessary to handle unobservable or latent variables. It allows jointly analyzing the relationships between observable and latent variables (evaluation of the measurement model) and the relationships between latent variables (evaluation of the structural model). Likewise, it is a widespread approach in management since it allows the study of complex models with innumerable indicators for each latent variable and numerous relationships among latent variables.

The present book contains the 14 articles accepted for publication among the 32 manuscripts in total that were submitted to the Special Issue "Partial Least Squares Structural Equation Modeling (PLS-SEM): Applications in Economics and Finance" of the MDPI journal Mathematics.

The 14 articles, which appear in the present book in the order in which they were published in Volume 9 (2021) of the journal, cover a wide range of topics connected to the theory and application of PLS-SEM methodology. These topics include, among others, prediction of stock market investment intention, institutional quality and international competitiveness, governance paradigms and public innovation, information and communication technologies in the supply chain, influence of the ability to absorb information from the environment and proactivity on the company's results, quality management, effects of the corporate social responsibility on financial performance, resource management for the improvement of the healthcare system, self-consciousness in online shopping behavior, status quo as a predictor of brand loyalty and propensity to innovate, the application of maximum entropy bootstrapping to time series.

It is expected that the book will be worthwhile and helpful for those working in the area of PLS-SEM, whatever your field of application (economics, finance, marketing, education or other). Applications of higher order constructs, mediating variables, multigroup analysis and the latest advances in applied methodology are found in this book.

As the Guest Editors of the Special Issue, we are grateful to the authors of the papers for their quality contributions, to the reviewers for their valuable comments toward improving the submitted works and to the administrative staff of the MDPI publications for their support in completing this project. Special thanks are due to the Managing Editor of the Special Issue, Dr. Syna Mu, for his professionalism, human qualities, excellent collaboration and valuable assistance.

María del Carmen Valls Martínez, Pedro Antonio Martín Cervantes
Editors

 *mathematics*

Article
Self-Consciousness in Online Shopping Behavior

Luis Miguel López-Bonilla *, Borja Sanz-Altamira and Jesús Manuel López-Bonilla *

Department of Business Administration and Marketing, University of Seville, 41018 Seville, Spain; borja@us.es
* Correspondence: luismi@us.es (L.M.L.-B.); lopezbon@us.es (J.M.L.-B.)

Abstract: Self-consciousness can be considered as the internal disposition to direct attention to oneself. This dispositional tendency can be focused on private aspects of the self, but also on public characteristics of the individual. We examine self-consciousness in online consumer behavior. This concept has been poorly investigated in consumer research. The main objective of this paper is to analyze the influence of the dimensions of self-consciousness in consumer adoption of online shopping. This study is based on a sample of 725 Spanish undergraduates. Findings indicated that public self-consciousness is a significant predictor of the adoption of online shopping, and inversely affects perceived ease of use and usefulness. These results may have important implications in the segmentation of users of self-service technologies.

Keywords: self-consciousness; e-commerce; consumer behavior; Technology Acceptance Model

Citation: López-Bonilla, L.M.; Sanz-Altamira, B.; López-Bonilla, J.M. Self-Consciousness in Online Shopping Behavior. *Mathematics* **2021**, *9*, 729. https://doi.org/10.3390/math9070729

Academic Editor: María del Carmen Valls Martínez

Received: 26 February 2021
Accepted: 26 March 2021
Published: 28 March 2021

Publisher's Note: MDPI stays neutral with regard to jurisdictional claims in published maps and institutional affiliations.

Copyright: © 2021 by the authors. Licensee MDPI, Basel, Switzerland. This article is an open access article distributed under the terms and conditions of the Creative Commons Attribution (CC BY) license (https://creativecommons.org/licenses/by/4.0/).

1. Introduction

The internet has been a relevant distribution channel and promotion tool for organizations for several years. However, as Arce-Urriza and Cebollada-Calvo [1] indicate, although online shopping sales are still low compared to offline sales, their growth rates are much higher, which suggests that the proportion of online sales will be even higher in the future. This paper focuses on the study of self-consciousness and its influence on the adoption of online shopping.

The self-consciousness of the person discusses the predisposition or willingness to direct one's attention inside or outside of oneself. Fierro [2] points out that this concept is one of the integral phenomena of the "self" system. As Jiménez [3] explains, the philosophical, anthropological and psychological fields show the importance of this self-ownership, that is, reflexivity: the ability to use oneself as an object of attention. There is a dichotomy regarding objects of conscious attention, so that the attention of the person considered, at any time, is absolutely directed to give importance to an internal point of view of oneself, or to external events of the subject, without the possibility of attention being focused on both aspects at the same time.

Duval and Wicklund [4] proposed the objective self-awareness theory, and this originated the scientific research on self-consciousness which investigates the differential effects of focusing attention on oneself, but only reacting to a temporary state of self-awareness, without considering the individual differences. This omission provoked the necessary impulse for the development of a widely used measurement instrument, the Fenigstein, Scheier, and Buss [5] Self-consciousness Scales. Using this scale of measurement and the formulation of the model of Carver and Scheier [6], self-consciousness is considered a transient state induced by concrete stimuli, but also a feature that denotes the tendency of people to be self-focused. This scale allows individual differences to be established in accordance with the degree of self-consciousness.

The Fenigstein et al. [5] Self-Consciousness Scale identifies three aspects to assess individual differences in self-consciousness: Private Self-Consciousness, Public Self-Consciousness, and Social Anxiety. The first two factors measure psychological tendencies centered on oneself. Private self-consciousness is related to the tendency to be introspective about one's thoughts and feelings. Meanwhile, public self-consciousness refers to attending to oneself as a social object, being concerned with how others view the self, and the impression that one makes on others. Social anxiety refers to discomfort in situations that involve being in the same place with other people.

The construct of self-consciousness has been poorly studied in consumer behavior literature. The present work aims to expand this knowledge by analyzing the effects caused by the dimensions of self-consciousness on the process of consumer adoption of online shopping. Understanding the antecedents of online shopping behavior is essential in order to develop and improve the use of Information and communications technologies (ICT).

The electronic ticket has been chosen as a product to be adopted through online purchasing. The use of electronic tickets has been widely spread among passenger transport companies in the last decade, but we must bear in mind that it is a type of sale that the consumer can adopt in its entirety or only partially; that is, it is an alternative system to the traditional method of buying tickets. Therefore, as López-Bonilla and López-Bonilla [7] indicate, consumers can buy an electronic ticket autonomously through digital means, offered by transport service providers or intermediary agents, without having to interact with the employees of any of these organizations. However, the purchase of the electronic ticket can be made through the staff of the transport company or the intermediary company.

The structure of the study continues with a review of the literature on self-consciousness within the framework of consumer behavior. The objectives and hypotheses of the study are presented below, proposing a research model based on the acceptance of ICT. Next, the study methodology is described, as well as the results obtained from the empirical analysis and the final considerations follow.

2. Literature Review

Many studies deal with the relationship of self-consciousness and personality of the individual, particularly the extensive review conducted by Scandell and Scandell [8]. But Jimenez [3] warns that all these investigations have given little attention to the healthy personality variables, and consequently he dedicates a study to it, proving that the profile of individuals who tend to concentrate their attention on the private aspects of themselves, with less experience of personal wellbeing and social adaptation, have a poor assessment of themselves. Moreover, people who focus their attention on the public aspects of themselves are less up-to-date and tend to behave more in accordance with situational behavioral clues than with personal ones.

Abe, Bagozzi and Sadarangani [9] warn that self-consciousness has been little studied in the field of consumer behavior. This is verified in the present review of the literature, although there is a certain favorable trend toward greater applicability of self-consciousness in the consumer field in the last decade. In this sense, most of the previous works have focused on public self-consciousness [10–13]. In contrast, there are two studies that have analyzed the private self-consciousness dimension exclusively [13–15]. Finally, few studies have examined the three dimensions or even two of them [9,16–18].

Table 1 contains a summary of the published works that have dealt with self-consciousness in the field of consumer behavior. This table details in chronological order the authors of the studies carried out, specifying the sample used and the content analyzed in relation to self-consciousness: PUSC is public self-consciousness, PRSC is private self-consciousness, and SA is social anxiety.

Table 1. Self-consciousness research on consumer behavior.

Authors	Content	Sample
Burnkrant and Page (1982)	PUSC	Unspecified
Solomon and Schopler (1982)	PUSC	104 US individuals
Gould and Barak (1989)	PUSC	679 US individuals
Gould (1993)	PRSC	337 US individuals
Bushman (1993)	PUSC	160 university students and 160 non-university students
Abe et al. (1996)	PUSC, PRSC, SA	402 Japanese university students and 233 US university students
Marquis (1998)	PUSC, PRSC	250 Canadian individuals
Marquis and Filiatrault (2000)	PUSC	159 Canadian university students
Marquis and Filiatrault (2002)	PUSC, PRSC	159 Canadian university students
Dabholkar and Bagozzi (2002)	PUSC, SA	392 US university students
Marquis and Filiatrault (2003)	PUSC	159 Canadian university students
Xu, Summers and Belleau (2004)	PUSC	690 US female respondents
Xu (2008)	PUSC	96 US university students
Sun, Horn and Merrit (2009)	PUSC	21,974 individuals from 25 countries
Workman and Lee (2011)	PUSC	400 US university students
Workman and Lee (2013)	PRSC	400 US university students
Tolbert, Kohli and Suri (2014)	PUSC, PRSC	198 US consumers
López-Bonilla and López-Bonilla (2015)	PUSC, PRSC, SA	819 Spanish university students
Lennon, Kim, Lee and Johnson (2017)	PUSC	411 US students
Shah and Amjad (2017)	PUSC, PRSC, SA	388 Pakistan consumers

Source: own elaboration.

Table 1 suggests that most papers have only focused on public self-consciousness. Therefore, in a study on image management, Burnkrant and Page [10] noted that high public self-consciousness subjects are more sensitive to social situations and are more inclined to act on those situations. They recommend using public self-consciousness as a segmentation variable for socially consumed products, and also as a means of identifying subjects who choose products for their ability to cause an external impression. In this sense, Solomon and Schopler [19] found that women high in public self-consciousness are more fashion-aware. Likewise, Gould and Barak [14] related public self-consciousness with various psychological and demographic variables of consumer behavior, observing their influence on different aspects of the consumer's daily behavior, such as the purchase of products that are socially consumed and interest in fashion. Bushman [11] revealed that high public self-consciousness subjects prefer branded to unbranded products in order to improve their public image. Workman and Lee [20] compared consumer vanity and public self-consciousness among fashion change agents and fashion followers.

In a study on the use of information technological self-services, Dabholkar and Bagozzi [12] conclude that high public self-consciousness strengthens the relationships between reliability and fun with attitudes, as well as between attitudes and intentions in the proposed model.

Likewise, some authors have related public self-consciousness with responsible consumption, such as Sun et al. [13], who noted that public self-consciousness moderates the relationship between social factors and the intention to follow a healthy diet. Other authors have related public self-consciousness with some negative aspects of consumers, such as Xu [21], who proved the influence of public self-consciousness on compulsive buying through materialism, finding that young consumers high in public self-consciousness are more likely to be materialistic. In contrast, Xu, Summers and Belleau [22] analyzed the relationship between public self-consciousness and the desire to buy controversial products, such as those made with alligator skin, thinking that these products can help them improve their look, but they found that this factor does not have a significant influence. Recently, Lennon, Kim, Lee and Johnson [23] examined the effects of public self-consciousness on buying behavior on Black Friday, noting that it moderates the relationship between anxiety and bad behavior on Black Friday in the case of women.

On the contrary, there are just two papers that analyzed private self-consciousness as the only dimension exclusively. Gould [24] studied the combined effect of private self-consciousness and self-monitoring on the consumer's self-concept, finding that high private self-consciousness subjects and low self-monitors show greater discrepancies between their real self-concept and their ideal self-concept. Workman and Lee [15] related private self-consciousness to these dimensions of consumer vanity: appearance view, appearance concern, achievement view, and achievement concern.

On the other hand, few studies examine the three dimensions or even two of them. Abe et al. [9] analyzed self-consciousness in the transcultural context, comparing Americans living in a relatively independent culture with Japanese people living in an interdependent culture. They concluded that Americans experience greater levels of private self-consciousness than Japanese people but lower levels of social anxiety, and both show similar levels of public self-consciousness.

In addition, two authors especially study public self-consciousness, but also private self-consciousness. Marquis [25], and Marquis and Filiatrault [16–18] analyze consumer reactions in situations where there is an additional delay in entering a movie theater. Marquis [25] found that people high in private self-consciousness consider the antecedents and the consequences of the waiting situation in a more cautious and reflective way. Marquis and Filiatrault [16,17] noted that high public self-consciousness subjects show a more negative evaluation of the experience of delay and have a more negative perception of service, and this significantly results in a negative word-of-mouth communication to express their dissatisfaction. Likewise, Marquis and Filiatrault [18] observe that people high in public self-consciousness tend to focus their attention on the time of delay; that is, they are quickly oriented toward the environment that surrounds them, blaming the company, and evaluating the service negatively.

Finally, some more recent studies can be cited. Tolbert and Kohli and Suli [26] examined public and private self-consciousness in the context of customer loyalty. These authors predict that the relative price of an offer in an exclusive trade moderates the effects of public and private self-consciousness on the perceptions of value. López-Bonilla and López-Bonilla [27] studied the adoption of the electronic ticket from a self-service perspective through self-consciousness profiles obtained from three dimensions. They found that a consumer who is high in private self-consciousness but low or neutral in public self-consciousness and social anxiety is the consumer profile that best explains the adoption of the electronic ticket. Shah and Amjad [28] analyzed self-consciousness with its three dimensions and found that self-consciousness moderates the relationship between moral intensity and neutralization techniques in the context of ethical consumer decisions.

3. Approach and Objectives

The purpose of this paper is to examine the influence of the dimensions of self-consciousness in consumer adoption of online shopping. As Farias, Kovacs and Silva [29] pointed out, the internet has modified the behavior of individuals, especially on the topic of online shopping behavior. There are many studies conducted on the behavior of individuals and the use of the internet. However, this broad field of study on the use of the internet has caused some controversy. On the one hand, the internet is considered as a means that causes isolation, and on the other hand, the network is a medium that offers opportunities to experience disinhibition and social competence [30]. In particular, the internet has been described as one of the most impersonal and socially distant forms of communication media [31] because people spend a lot of time online, limiting the possibilities of acquiring and maintaining the necessary skills to interact socially [32].

In contrast to the above arguments, the internet is a social communication medium that complements and spreads traditional face-to-face behaviors [33]. This is what has been called the social network theory, which has been supported by many studies that suggest that communication over the internet has enough potential to foster satisfying, long-lasting and healthy relationships [34–36].

As Suler [37] suggests, internet users say and do things that they would not normally say or do in a face-to-face interaction. They feel looser and less self-conscious, and they express themselves more openly. This phenomenon has been called the online disinhibition effect, and it has been related to personality variables. In some cases, it results in a slight deviation of the individual's behavior, while in others it can cause relevant changes.

Morahan-Martin and Schumacher [38] published a paper related to the present study. In it, they propose that anonymity and the lack of face-to-face communication on the internet can reduce public and private self-consciousness, as well as social anxiety, while facilitating prosocial behavior at the same time. These authors find differences between solitary and non-solitary people regarding the use of the internet. Thus, they confirm that the social behavior of solitary people on the internet is related to the effect of disinhibition. These authors suggest that the internet provides an ideal social environment for lonely people, since it allows them to interact with others more easily. The internet not only offers a wide social network, but also provides diverse models of social interaction that can be particularly attractive for solitary people. Anonymity and lack of physical presence allow internet users greater control over social interaction. Therefore, the internet facilitates disinhibition and provides a space to practice and develop social skills.

According to these approaches, the internet influences the personal characteristics of the individual's self-consciousness, and these characteristics of the subject affect the adoption of online purchasing. Consequently, this work suggests that individuals who are more inclined to adopt online shopping may be conditioned by their self-consciousness. Therefore, we can assume that those individuals high in social disinhibition through the internet have a greater predisposition to adopt online shopping.

There are several models to measure the adoption of technology, among which the Technology Acceptance Model (TAM) stands out as it is highly effective in predicting the use of any technology [39]. This model has been widely applied in practice [40,41] and it explains the reasons for the users' technology adoption [42]. Although the TAM was initially designed to be applied to organizations, it has been used by many researchers to explain the adoption of various technological advances, such as the internet and electronic commerce [43]. The TAM is simple [44], but it has become a robust, powerful and parsimonious model for predicting user adoption of technology [45]. Davis, Bagozzi, and Warshaw [46] note that the TAM model is a specific adaptation of the Theory of Reasoned Action by Fishbein and Ajzen [47] to model user adoption of information systems. These authors state that the objective of the TAM is to provide an explanation of the determinants of information systems acceptance, which can explain the behavior of users over a wide range of computer technologies and user groups.

In the TAM, the attitude toward the use of an information system is founded on two main determinants: perceived usefulness and perceived ease of use. Furthermore, the TAM postulates that these two variables relative to a particular technology shapes the attitude toward its use and behavioral intention to make use of that technology. Perceived usefulness refers to the degree to which people believe that using a particular system would enhance their performance [46]. The second determinant, perceived ease of use, is defined as the extent to which a person believes that using a particular system would be free of effort [48]. Hence, as Izquierdo, Martínez and Jiménez [49] point out, the TAM assumes that online purchases that the consumer considers easy to use are very likely to lead to an increase in perceived usefulness and, in addition, the consumer seeks value from this online shopping process that other alternative channels cannot provide. Customers seek products' features, price and functionality online [50].

Davis, Bagozzi and Warshaw [46] and Venkatesh and Davis [45] exclude the construct attitudes in their later model. However, some controversies have been generated about including or excluding attitudes in the TAM [51–55]. In this sense, López-Bonilla and López-Bonilla [53,54] assure that attitude is a necessary construct in the TAM, especially in voluntary scenarios, in which users have greater autonomy.

As shown in Figure 1, external variables directly affect these two constructs: perceived usefulness and perceived ease of use. These variables can be linked to the characteristics of the technological system to be used and/or to the individual differences to use it. Therefore, this research considers the three dimensions of self-consciousness as external variables. From this theoretical perspective, our work aims to achieve three objectives. First, as a general and theoretical objective, we try to contribute to the broadening of knowledge of self-consciousness in the field of consumer behavior. Secondly, we intend to know how the dimensions of self-consciousness influence the adoption of online shopping. And thirdly, we analyze the effect of online disinhibition through self-consciousness in this environment of consumer adoption of online shopping. In this sense, it would be expected that individuals characterized by greater private self-consciousness and/or lower public self-consciousness and/or greater social anxiety will be more likely to adopt online shopping, given that these personal characteristics are more related to the disinhibition effect online.

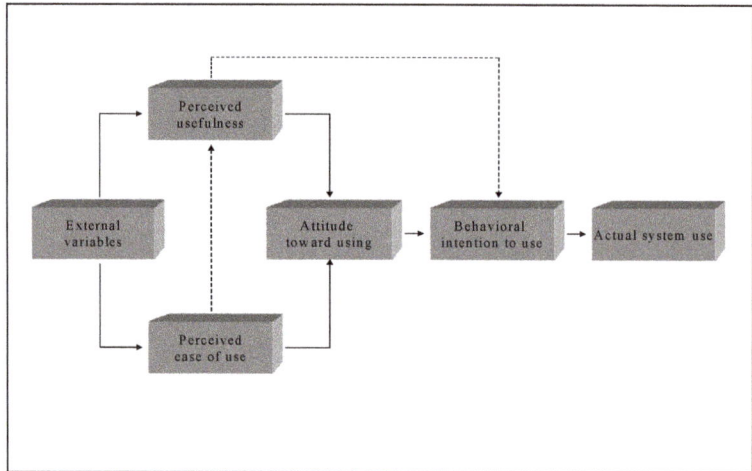

Figure 1. Technology Acceptance Model (TAM). Source: [48].

From the above theoretical approaches, the following hypotheses are tested:

Hypothesis 1 (H1). *Public self-consciousness influences the perceived ease of use.*

Hypothesis 2 (H2). *Public self-consciousness influences the perceived usefulness.*

Hypothesis 3 (H3). *Private self-consciousness influences the perceived ease of use.*

Hypothesis 4 (H4). *Private self-consciousness influences the perceived usefulness.*

Hypothesis 5 (H5). *Social anxiety influences the perceived ease of use.*

Hypothesis 6 (H6). *Social anxiety influences the perceived usefulness.*

Hypothesis 7 (H7). *Perceived usefulness influences the perceived ease of use.*

Hypothesis 8 (H8). *Perceived ease of use influences the attitude toward using technology.*

Hypothesis 9 (H9). *Perceived usefulness influences the attitude toward using technology.*

Hypothesis 10 (H10). *Perceived usefulness influences the intention to use technology.*

Hypothesis 11 (H11). *Attitude toward using technology influences the intention to use technology.*

4. Materials and Methods

The present study is based on a sample for convenience. The data were collected using a personal survey completed by 724 Spanish university students. Online shopping is very common among young students. This sample consists of 459 women and 265 men, all of whom buy products online.

A structured questionnaire was administered in the Faculty of Economics and Business Administration and in the Faculty of Tourism and Finance of the University of Seville. A self-administered survey was conducted. The survey was carried out at the beginning of the classes after having requested the collaboration of the professors. When distributing the questionnaire, explicit instructions were given to complete it. We use the PLS (Partial Least Squares) method to examine the relationships of the theoretical model to be tested. PLS is a Variance-Based Structural Equation Model technique that has become very popular in management and social sciences [56,57]. As indicated by Joreskog and Wold [58], PLS is primarily intended for causal–predictive analysis in situations of high complexity but low theoretical information.

The variables of the proposed model are based on the TAM and on the measurement scales shown below (see Appendix A). The perceived usefulness and perceived ease of use are based on the measures employed by Davis [48], Davis et al. [46] and Dabholkar [59–61], using two scales with four items each in our work. The variable attitude toward online shopping is based especially on Fishbein and Ajzen [47], and Ajzen and Fishbein's [62] studies, using a scale with four items. On the other hand, the measure of online purchase intention is gathered from Ajzen and Fishbein [62], and Dabholkar [59] studies, using a three-item scale. All these measures are semantic differential scales with seven response options. The semantic differential has been frequently used in research concerning attitudes in social psychology and related fields, such as marketing [63].

Regarding the external variables of the model, the three dimensions of self-consciousness (private and public self-consciousness and social anxiety) are integrated independently. For this purpose, we use a self-consciousness scale, an updated version by Scheier and Carver [64] based on 22 items regarding the three dimensions: nine items concerning private self-consciousness, seven regarding public self-consciousness, and six concerning social anxiety.

5. Results

We study the measurement scales of the constructs included in the model to carry out the statistical analysis of the data. Therefore, in order to check the convergent validity and the discriminant validity, we analyze the relationships between the variables and their items. As Hair, Anderson, Tatham, and Black [65] claim, it is necessary to fulfill these criteria in order to accept the model. Three metric tests are applied to check the convergent validity: reliability of the indicators, composite reliability and average variance extracted (AVE) from the latent variables.

Regarding the reliability of the indicators, all their values with respect to the variables that represent the original model of the TAM are higher than the recommended values. However, the same does not occur with the three dimensions of self-consciousness. Therefore, the results obtained recommended to remove several items of each dimension, finally leaving three indicators of private self-consciousness, two indicators of public self-consciousness, and three indicators of social anxiety.

The composite reliability and the average variance extracted (AVE) are the other two measures of convergent validity. They are detailed in Table 2 with the refined indicators of self-consciousness. Following the recommendations of Fornell and Larcker [66], with regards to the first measure, it is observed that the values exceed the recommended minimum of 0.7 for all variables of the model. Likewise, the values of the average variance extracted (AVE) exceed 0.5, which is the recommended value. These results prove the convergent validity of the measurement model.

Table 2. Compound reliability and variance extracted from latent variables.

Variable	AVE	Compound Reliability
ATT	0.5138	0.7602
PEU	0.7181	0.8355
PU	0.5411	0.7792
ITU	0.773	0.9316
PRSC	0.7354	0.9174
PUSC	0.8285	0.9354
SA	0.8008	0.9414

Note. ATT = Attitude; PEU = Perceived ease of use; PU = Perceived usefulness; ITU = Intention to use; PRSC = Private self-consciousness; PUSC = Public self-consciousness; SA = Social anxiety.

The discriminant validity completes the analysis of the measurement model. As Chin [67] indicates, discriminant validity consists of verifying that variables relate more strongly to their own factor than to another factor. Table 3 shows these data, thereby testing the discriminant validity of the measurement scales used.

Table 3. Discriminant validity of latent variables.

Variable	ATT	PEU	PU	ITU	PRSC	PUSC	SA
ATT	0.96519						
PEU	0.4165	0.95781					
PU	0.5175	0.5013	0.97025				
ITU	0.5206	0.4183	0.5684	0.96716			
PRSC	−0.0578	−0.0958	−0.0833	−0.0538	0.87189		
PUSC	−0.1228	−0.1423	−0.1613	−0.1137	0.3436	0.91401	
SA	−0.0721	−0.1085	−0.1034	−0.0379	0.2277	0.2919	0.88272

Note. ATT = Attitude; PEU = Perceived ease of use; PU = Perceived usefulness; ITU = Intention to use; PRSC = Private self-consciousness; PUSC = Public self-consciousness; SA = Social anxiety.

Once the measurement scales used have been analyzed, the structural analysis of the model concludes with the empirical study. Figure 2 details these results. It is observed that all relationships of the TAM original model have been checked. However, private self-consciousness and social anxiety do not affect any of the two assumptions of the model. Only public self-consciousness significantly influences perceived ease of use and perceived usefulness. It should be pointed out that the effect of public self-consciousness is negative in both cases. This means that a decrease in this characteristic of the individual significantly increases perceived ease of use and perceived usefulness in terms of their attitude and intention toward online shopping. These results can be understood in the sense that higher public self-consciousness makes individuals give more importance to the opinion or judgment of other people and less importance to their own beliefs about the service, such as usefulness and perceived ease of use.

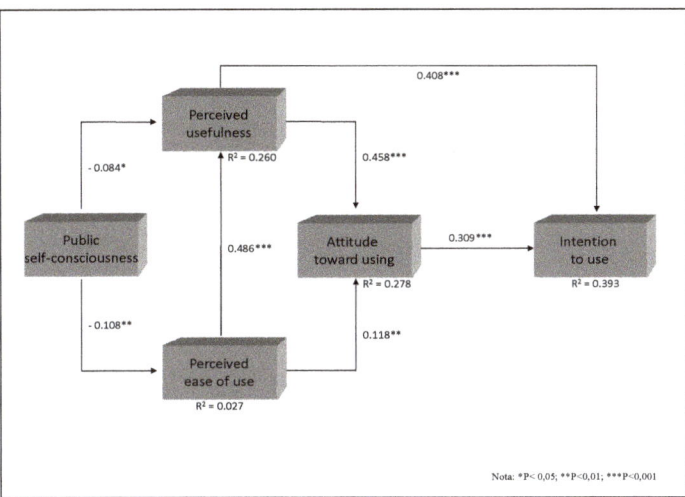

Figure 2. Significant results of the proposed model.

6. Discussion and Conclusions

This paper analyzes the construct of self-consciousness, which contains three dimensions: public self-consciousness, private self-consciousness and social anxiety. These three factors are related to psychological tendencies based on oneself and one's disposition towards the social environment that surrounds them.

Self-consciousness is a relevant psychological characteristic in the behavior of people, but it has been little studied in the field of consumer behavior. Our general objective is to cover this lack in electronic commerce. As the first specific objective, an empirical study was carried out to contrast the influence of the three dimensions of self-consciousness in consumer adoption of online shopping. The results obtained indicate that self-consciousness affects consumer adoption of online shopping. According to Shah and Amjad [28], private self-consciousness obtained the highest scores, closely followed by public self-consciousness and, at a greater distance, by social anxiety. Our results indicate a stronger relationship between public self-consciousness and self-insecurity than in the Gould and Barak [14] study. In addition, usefulness and ease of use can avoid uncertainty in the use of technology. However, as in the study by Sun, Horn and Merrit [13], the fact of trying to avoid uncertainty leads to less public self-consciousness.

In particular, public self-consciousness directly influences perceived usefulness and perceived ease of use. Xu [21] warns that the public self-consciousness of young consumers can influence their compulsive buying decisions. In this sense, buyers with high public self-consciousness do not look so much for the usefulness of the purchased product. Therefore, we can conclude that public self-consciousness has a negative effect on cognitive factors such as usefulness and ease of use. Furthermore, private self-consciousness and social anxiety do not have a direct impact on these two constructs. This means that private self-consciousness and social anxiety do not imply an intention to accept online shopping. Therefore, hypotheses H1 and H2 have been accepted, in addition to the basic hypotheses posed by the Technology Acceptance Model (TAM): hypotheses H7, H8, H9 and H10.

The second specific objective of this study was that the adoption of online shopping would have a greater interest for those individuals to whom the use of this technology would give a higher degree of social disinhibition. The results seem to support this idea to a limited extent. There is evidence that anonymity and lack of face-to-face communication on the internet can influence the adoption of online shopping. These results are in line with the studies of Morahan-Martin and Schumacher [38], but in the consumer field since the decrease in public self-consciousness has an effect on the adoption of online shopping. In this way, low public self-consciousness subjects are more likely to use the internet for purchasing. The internet can be considered as a form of social communication complementary to face-to-face interaction, which can reduce individuals' social limitations, increasing their disinhibition in online shopping.

One of the advantages of this study compared to others on self-consciousness in the consumer field is that it empirically analyzes the three dimensions of self-consciousness at the same time. Most of the previous work on the subject has focused on just one dimension, with few on two of the dimensions, and even fewer on the three dimensions. The analysis of the three dimensions together, comparing the relative influence of each of them, is closer to reality. Previous studies that have focused on one or two components of self-consciousness entail a restriction of information and a bias in that self-consciousness must be considered conceptually as a combination of the three dimensions.

Regarding the implications of this work, it is necessary to understand consumer buying behavior in order to design an adequate marketing strategy to attract the consumer to purchase products offered by organizations. The electronic ticket is a basic or standardized product, hence consumers do not buy it for the social impact they can cause, as Burnkrant and Page [10] state, but quite the opposite. In line with Bushman [11], individuals high in public self-consciousness may reject products that they consider more basic and cheaper than those offered in other sales channels. Therefore, passenger transport companies can adopt marketing strategies to market their tickets online focusing on the internal motivations of individuals rather than the public image that can be conveyed with the purchase of these products.

The main limitation of this paper is the use of a homogeneous sample with individuals of similar ages and educational levels. Although it is not possible to generalize these results to the whole population, a homogeneous sample is useful and advisable to verify theoretical models based on the study of behavior [68]. Another limitation of this study is the fact that it is based on a single product, even if it is one of the best-selling products on the internet. On the other hand, a widely recognized model has been used to explain the adoption of technology, the Technology Acceptance Model (TAM), although it is based on a more utilitarian perspective. However, future research can analyze online shopping from a more hedonistic approach and compare it with the utilitarian perspective. It is also possible to study self-consciousness profiles that identify the three dimensions in the population in relation to the use of social networks, given its relation to the psychosocial characteristics of people.

Furthermore, it is possible to study how different types of self-consciousness can react by inhibiting or preventing the co-creation of value from the dominant logic of service [69,70]. Other relationships within the Technology Acceptance Model (TAM) can be analyzed in future studies, such as observing the influence of ease of use on the intention to buy online. Likewise, the direct relationships of self-consciousness with attitudes can be analyzed and, as suggested by Marquis and Filiatrault [18], with the intention of use. In addition, the moderating effect of self-consciousness in the relationships between the variables considered can be contrasted, as proposed by previous studies [12,13]. It is possible to contemplate other relationships in the model such as the influence of perceived ease of use on the intention to use technology. This model can be studied in different e-channels and e-channel touchpoints in the lens of Wagner, Schramm-Klein and Steinmann [71].

Author Contributions: For research articles with several authors, a short paragraph specifying their individual contributions must be provided. Conceptualization, J.M.L.-B., L.M.L.-B., B.S.-A.; methodology, J.M.L.-B., L.M.L.-B.; validation, J.M.L.-B., L.M.L.-B.; formal analysis, J.M.L.-B., L.M.L.-B.; investigation, J.M.L.-B., L.M.L.-B., B.S.-A.; resources, J.M.L.-B., L.M.L.-B., B.S.-A.; writing—original draft preparation, J.M.L.-B., L.M.L.-B., B.S.-A.; writing—review and editing, J.M.L.-B., B.S.-A.; visualization, J.M.L.-B., L.M.L.-B., B.S.-A.; supervision, J.M.L.-B., L.M.L.-B., B.S.-A. All authors have read and agreed to the published version of the manuscript.

Funding: This research received no external funding.

Conflicts of Interest: The authors declare no conflict of interest.

Appendix A

Scales	Items
ease of use	Using the self-service system through internet to book air tickets . . . - be/not be complicated - be/not be confusing - take a lot of/little effort - require a lot of/little work
usefulness	Using the self-service through internet to book air tickets . . . - means I will get/not get what I ordered - is something I expect/do not expect to work well - will/will not result in errors - will be/not be reliable
attitudes	How do you describe your feelings towards using the self-service through Internet to buy air tickets? - extremely good/extremely bad - extremely pleasant/extremely unpleasant - extremely harmful/extremely beneficial - extremely favorable/extremely unfavorable
intention to use	Would you intend to use the self-service through Internet to buy air tickets? - extremely likely/extremely unlikely - extremely possible/extremely impossible - definitively I will use/definitively I will not use
Self-consciousness	Private Self-consciousness - always trying to figure himself out - concerned about his style of doing things - thinks a lot about himself - never takes a hard look at himself - generally pays attention to inner feelings - constantly thinks about reasons for doing things - steps back to examine himself from a distance - quickly notices changes in own mood - knows his cognitive process while solving a problem - percentage of variances explained Public Self-consciousness - cares about presenting himself to others - self-conscious about the way he looks - worries about making a good impression - before leaving the house, checks how he looks - concerned about what others think of him - usually aware of his appearance - percentage of variances explained Social Anxiety - takes time to get over shyness in new situations - hard for him to work when someone is watching - is embarrassed very easily - finds it easy to talk to strangers - feels nervous when speaking in front of a group - large groups make him nervous

References

1. Arce-Urriza, M.; Cebollada-Calvo, J.J. Elección del canal de compra y estrategia multicanal: Internet vs. tradicional. Aplicación a la compra en una cadena de supermercados. *Cuad. Econ. Dir. Empresa* **2013**, *16*, 108–122. [CrossRef]
2. Fierro, A. *Manual de Psicología de la Personalidad*; Ediciones Piados Ibérica: Barcelona, Spain, 1996.
3. Jiménez, J.A. Autoconciencia, personalidad sana y sistema autorreferente. *Ann. Psicol.* **1999**, *15*, 169–177.
4. Duval, S.; Wicklund, R.A. *A Theory of Effects of Objective Self-Awareness*; Academic: New York, NY, USA, 1972.
5. Fenigstein, A.; Scheier, M.F.; Buss, A.H. Public and private self-consciousness: Assessment and theory. *J. Consult. Clin. Psychol.* **1975**, *43*, 522–527. [CrossRef]
6. Carver, C.S.; Scheier, M.F. *Attention and Self-Regulation: A Control-Theory Approach to Human Behavior*; Springer: New York, NY, USA, 1981.
7. López-Bonilla, J.M.; López-Bonilla, L.M. Self-service technology versus traditional service: Examining cognitive factors in the purchase of the airline ticket. *J. Travel Tour. Mark.* **2013**, *30*, 497–508. [CrossRef]
8. Scandell, D.J.; Scandell, D. The personality correlates of public and private self-consciousness from a five-factor perspective. *J. Soc. Behav. Pers.* **1998**, *13*, 579–592.
9. Abe, S.; Bagozzi, R.P.; Sadarangani, P. An investigation of construct validity and generalizability of the self-concept: Self-consciousness in Japan and Unites States. *J. Int. Consum. Mark.* **1996**, *8*, 97–123. [CrossRef]
10. Burnkrant, R.E.; Page, T.J. On the management of self-images in social situations: The role of public self-consciousness. *Adv. Consum. Res.* **1982**, *9*, 452–455.
11. Bushman, B.J. What's in a name? The moderating role of public self-consciousness on the relation between brand label and brand preference. *J. Appl. Psychol.* **1993**, *78*, 857–861. [CrossRef]
12. Dabholkar, P.A.; Bagozzi, R.P. An attitudinal model of technology-based self-service: Moderating effects of consumer traits and situational factors. *J. Acad. Mark. Sci.* **2002**, *30*, 184–201. [CrossRef]
13. Sun, T.; Horn, M.; Merrit, D. Impacts of cultural dimensions on healthy diet through public self-consciousness. *J. Consum. Mark.* **2009**, *26*, 241–250. [CrossRef]
14. Gould, S.J.; Barak, J. Public self-consciousness and consumption behaviour. *J. Soc. Psychol.* **1989**, *128*, 393–400. [CrossRef]
15. Workman, J.E.; Lee, S.H. Relationships among consumer vanity, gender, brand sensitivity, grand consciousness and private self-consciousness. *Int. J. Consum. Stud.* **2013**, *37*, 206–213. [CrossRef]
16. Marquis, M.; Filiatrault, P. Cognitive and affective reactions when facing an additional delay while waiting in line: A matter of self-consciousness disposition. *Soc. Behav. Pers.* **2000**, *28*, 355–376. [CrossRef]
17. Marquis, M.; Filiatrault, P. Understanding complaining responses through consumers' self-consciousness disposition. *Psychol. Mark.* **2002**, *19*, 267–292. [CrossRef]
18. Marquis, M.; Filiatrault, P. Public self-consciousness disposition effect on reactions to waiting in line. *J. Consum. Behav.* **2003**, *2*, 212–231. [CrossRef]
19. Solomon, M.R.; Schopler, J. Self-consciousness and clothing. *Pers. Soc. Psychol. Bull.* **1982**, *8*, 508–514. [CrossRef]
20. Workman, J.E.; Lee, S.H. Vanity and public self-consciousness: A comparison of fashion consumer groups and gender. *Int. J. Consum. Stud.* **2011**, *35*, 307–315. [CrossRef]
21. Xu, Y. The influence of public self-consciousness and materialism on young consumers' compulsive buying. *Young Consum.* **2008**, *9*, 37–48. [CrossRef]
22. Xu, Y.; Summers, T.A.; Belleau, B.D. Who buys American alligator? Predicting purchase intention of a controversial product. *J. Bus. Res.* **2004**, *27*, 1189–1198. [CrossRef]
23. Lennon, S.J.; Kim, M.; Lee, J.; Johnson, K.K.P. Effects of emotions, sex, self-control, and public self-consciousness on Black Friday misbehaviour. *J. Glob. Fash. Mark.* **2017**, *8*, 163–179. [CrossRef]
24. Gould, S.J. Assessing self-concept discrepancy in consumer behaviour: The joint effect of private self-consciousness and self-monitoring. *Adv. Consum. Res.* **1993**, *20*, 419–424.
25. Marquis, M. Self-consciousness disposition sheds lights of consumers' reactions to waiting. *Adv. Consum. Res.* **1998**, *25*, 544–550.
26. Tolbert, S.L.; Kohli, C.; Suri, R. Who pays the price for loyalty? The role of self-consciousness. *J. Prod. Brand Manag.* **2014**, *23*, 362–371. [CrossRef]
27. López-Bonilla, J.M.; López-Bonilla, L.M. Self-consciousness profiles in the acceptance of airline e-ticketing services. *Anatolia* **2015**, *26*, 447–458. [CrossRef]
28. Shah, S.A.M.; Amjad, S. Consumer ethical decision making: Linking moral intensity, self-consciousness and neutralization techniques. *Australas. Account. Bus. Finance J.* **2017**, *11*, 99–130. [CrossRef]
29. Farias, S.A.; Kovacs, M.H.; Silva, J.M. On-line consumer behavior: The flow theory perspective. *Rev. Bras. Gest. Negocios* **2008**, *10*, 27–44. [CrossRef]
30. Saunders, P.L.; Chester, A. Shyness and the Internet: Social problem or panacea? *Comput. Hum. Behav.* **2008**, *24*, 2649–2658. [CrossRef]
31. Matheson, K.; Zanna, M.P. The impact of computer-mediated communication on self-awareness. *Comput. Hum. Behav.* **1998**, *4*, 221–233. [CrossRef]
32. Henderson, L.; Zimbardo, P. *Encyclopedia of Mental Health*; Academic Press: San Diego, CA, USA, 1999.

33. Birnie, S.; Horvath, P. Psychological predictors of internet social communication. *J. Comput. Mediat. Commun.* **2002**, *7*, 13–27. [CrossRef]
34. Bargh, J.A.; McKenna, K.Y.A.; Fitzsimmons, G.M. Can you see the real me? Activation and expression of the "true self" on the internet. *J. Soc. Issues* **2002**, *58*, 33–48. [CrossRef]
35. Cornwell, B.; Lundgren, D.C. Love on the internet: Involvement and misrepresentation in romantic relationships in cyberspace vs. realspace. *Comput. Hum. Behav.* **2001**, *17*, 197–211. [CrossRef]
36. Mckenna, K.Y.A.; Green, A.S.; Gleason, M.E.J. Relationship formation on the internet: What's big attraction? *J. Soc. Issues* **2002**, *58*, 9–32. [CrossRef]
37. Suler, J. The online disinhibition effect. *CyberPsychol. Behav.* **2004**, *7*, 321–326. [CrossRef] [PubMed]
38. Morahan-Martin, J.; Schumacher, P. Loneliness and social uses of the Internet. *Comput. Hum. Behav.* **2003**, *19*, 659–671. [CrossRef]
39. Lorenzo, C.; Alarcón, M.C.; Gómez, M.A. Adopción de redes sociales virtuales: Ampliación del modelo de aceptación tecnológica integrando confianza y riesgo percibido. *Cuad. Econ. Dir. Empresa* **2011**, *14*, 194–205. [CrossRef]
40. Tasai, C.H. Integrating Social Capital Theory, Social Cognitive Theory, and the Technology Acceptance Model to explore a behavioral model of telehealth systems. *Int. J. Environ. Res. Public Health* **2014**, *11*, 4905–4925. [CrossRef] [PubMed]
41. Ijaz, M.F.; Rhee, J. Constituents and consequences of online-shopping in sustainable e-business: An experimental study of online-shopping malls. *Sustainability* **2018**, *10*, 3756. [CrossRef]
42. Santana-Mancilla, P.C.; Anido-Rifón, L.E. The technology acceptance of TV platform for the elderly living alone or in public nursing homes. *Int. J. Environ. Res. Public Health* **2017**, *14*, 617. [CrossRef]
43. Belanche, D.; Casaló, L.V.; Flavián, C. Integrating trust and personal values into Technology Acceptance Model: The case of e-government services adoption. *Cuad. Econ. Dir. Empresa* **2012**, *15*, 192–204. [CrossRef]
44. Kim, J. Platform adoption factors in the internet industry. *Sustainability* **2018**, *10*, 3185. [CrossRef]
45. Venkatesh, V.; Davis, F.D. A theoretical extension of the Technology Acceptance Model: Four longitudinal field studies. *Manag. Sci.* **2000**, *46*, 186–204. [CrossRef]
46. Davis, F.D.; Bagozzi, R.P.; Warshaw, P.R. User acceptance of computer technology: A comparison of two theoretical models. *Manage. Sci.* **1989**, *35*, 983–1003. [CrossRef]
47. Fishbein, M.; Ajzen, I. *Belief, Attitude, Intention and Behaviour: An Introduction to Theory and Research*; Addison-Wesley Publishing: Menlo Park, CA, USA, 1975.
48. Davis, F.D. Perceived usefulness, perceived ease of use, and user acceptance of Information Technology. *MIS Q.* **1989**, *13*, 319–340. [CrossRef]
49. Izquierdo, A.; Martínez, M.P.; Jiménez, A.I. El papel de la conveniencia y de la norma subjetiva en la intención de compra por Internet (B2C): Una aplicación en la industria hotelera. *Rev. Bras. Gest. Negocios* **2011**, *13*, 137–158.
50. Moslehpour, M.; Pham, V.K.; Wong, W.K.; Bilgiçli, I. e-Purchase intention of Taiwanese consumers: Sustainable mediation of perceived usefulness and perceived ease of use. *Sustainability* **2018**, *10*, 234. [CrossRef]
51. Teo, T. Is there an attitude problem? Reconsidering the role of attitude in the TAM. *Br. J. Educ. Technol.* **2009**, *40*, 1139–1141. [CrossRef]
52. Nistor, N.; Heymann, J.O. Reconsidering the role of attitude in the TAM: An answer to Teo (2009a). *Br. J. Educ. Technol.* **2010**, *41*, E142–E145. [CrossRef]
53. López-Bonilla, L.M.; López-Bonilla, J.M. The role of attitudes in the TAM: A theoretically unnecessary construct? *Br. J. Educ. Technol.* **2011**, *42*, E160–E162. [CrossRef]
54. López-Bonilla, L.M.; López-Bonilla, J.M. Explaining the discrepancy in the mediating role of attitudes in the TAM. *Br. J. Educ. Technol.* **2017**, *48*, 940–949. [CrossRef]
55. Ursavas, O.F. Reconsidering the role of attitude in the TAM: An answer to Teo (2009) and Nistor and Heymann (2010), and López-Bonilla and López-Bonilla (2011). *Br. J. Educ. Technol.* **2013**, *44*, E22–E25. [CrossRef]
56. Nitz, C.; Roldán, J.L.; Cepeda, G. Mediation analysis in Partial Least Squares path modeling: Helping researchers discuss more sophisticated models. *Ind. Manage. Data Syst.* **2016**, *116*, 1849–1864. [CrossRef]
57. Richter, N.F.; Cepeda, G.; Roldán, J.L.; Ringle, C. Editorial: European management research using Partial Least Squares Structural Equation Modeling (PLS-SEM). *Eur. Manag. J.* **2016**, *34*, 589–597. [CrossRef]
58. Jöreskog, K.G.; Wold, H. The ML and PLS techniques for modelling with latent variables: Historical and competitive aspects. In *Systems under Indirect Observation*; Jöreskog, K.G., Wold, H., Eds.; North-Holland: Amsterdam, The Netherlands, 1982; Part 1; pp. 263–270.
59. Dabholkar, P.A. Decision-Making in Consumer Trial of Technology-Based Self-Service Options: An Attitude-Based Choice Model. Ph.D. Thesis, Georgia State University, Atlanta, GA, USA, 1991.
60. Dabholkar, P.A. Incorporating choice into an attitudinal framework: Analyzing models of mental comparison processes. *J. Consum. Res.* **1994**, *21*, 100–118. [CrossRef]
61. Dabholkar, P.A. Consumer Evaluations of New Technology-based Self-service options: An Investigation of Alternative Models of Service Quality. *Int. J. Res. Mark.* **1996**, *13*, 29–51. [CrossRef]
62. Ajzen, I.; Fishbein., M. *Understanding Attitudes and Predicting Social Behaviour*; Prentice-Hall: Englewood Cliffs, NJ, USA, 1980.
63. Stoklasa, J.; Talášek, T.; Stoklasová, J. Semantic differential for the twenty-first century: Scale relevance and uncertainty entering the semantic space. *Qual. Quant.* **2019**, *53*, 435–448. [CrossRef]

64. Scheier, M.F.; Carver, C.S. The self-consciousness scale: A revised version for use with general populations. *J. Appl. Soc. Psychol.* **1985**, *15*, 687–699. [CrossRef]
65. Hair, J.F.; Anderson, R.E.; Tatham, R.L.; Black, W.C. *Análisis Multivariante*; Pearson Education: Madrid, Spain, 2004.
66. Fornell, C.; Larcker, D. Evaluating structural equation models with unobservable variables and measurement error. *J. Mark. Res.* **1981**, *18*, 39–50. [CrossRef]
67. Chin, W.W. The Partial Least Squares Approach to Structural Equation Modeling. In *Modern Methods for Business Research*; Marcoulides, G.A., Ed.; Lawrence Erlbaum Associates: Hillsdale, NJ, USA, 1998; pp. 295–336.
68. Calder, B.J.; Phillips, L.W.; Tybout, A.M. Designing research for application. *J. Consum. Res.* **1981**, *8*, 197–207. [CrossRef]
69. Vargo, S.L.; Lusch, R.F. Evolving to a New Dominant Logic. *J. Mark.* **2004**, *68*, 1–17. [CrossRef]
70. Vargo, S.L.; Lusch, R.F. Service-Dominant Logic: Continuing the evolution. *J. Acad. Mark. Sci.* **2008**, *36*, 1–10. [CrossRef]
71. Gerhard, W.; Schramm-Klein, H.; Steinmann, S. Online retailing across e-channels and e-channel touchpoints: Empirical studies of consumer behavior in the multichannel e-commerce environment. *J. Bus. Res.* **2020**, *107*, 256–270. [CrossRef]

Article

Predicting Stock Market Investment Intention and Behavior among Malaysian Working Adults Using Partial Least Squares Structural Equation Modeling

Marvello Yang [1], Abdullah Al Mamun [2,*], Muhammad Mohiuddin [3], Sayed Samer Ali Al-Shami [4] and Noor Raihani Zainol [5]

1. Department of Management, Faculty Economic and Business, Widya Dharma University Pontianak, Kota Pontianak, Kalimantan Barat 78243, Indonesia; marvelo_yang@widyadharma.ac.id
2. Faculty of Business and Management, UCSI University, Cheras, Kuala Lumpur 56000, Malaysia
3. Faculty of Business Administration, Laval University, Quebec City, QC 2375, Canada; muhammad.mohiuddin@fsa.ulaval.ca
4. Institute of Technology Management and Entrepreneurship, Universiti Teknikal Malaysia Melaka, Durian Tunggal, Melaka 76100, Malaysia; samerali@utem.edu.my
5. Faculty of Entrepreneurship and Business, Universiti Malaysia Kelantan, Pengkalan Chepa, Kota Bharu 16100, Malaysia; raihani@umk.edu.my
* Correspondence: abdullaham@ucsiuniversity.edu.my or mamun7793@gmail.com; Tel.: +6-013-300-3630

Citation: Yang, M.; Mamun, A.A.; Mohiuddin, M.; Al-Shami, S.S.A.; Zainol, N.R. Predicting Stock Market Investment Intention and Behavior among Malaysian Working Adults Using Partial Least Squares Structural Equation Modeling. *Mathematics* 2021, 9, 873. https://doi.org/10.3390/math9080873

Academic Editors: María del Carmen Valls Martínez and Pedro Antonio Martín Cervantes

Received: 12 March 2021
Accepted: 13 April 2021
Published: 15 April 2021

Publisher's Note: MDPI stays neutral with regard to jurisdictional claims in published maps and institutional affiliations.

Copyright: © 2021 by the authors. Licensee MDPI, Basel, Switzerland. This article is an open access article distributed under the terms and conditions of the Creative Commons Attribution (CC BY) license (https://creativecommons.org/licenses/by/4.0/).

Abstract: The purpose of this study was to investigate the effects of risk tolerance, financial well-being, financial literacy, overconfidence bias, herding behavior, and social interaction on stock market investment intention and stock market participation among working adults in Malaysia. Adopting the cross-sectional design, this study collected quantitative data from a total of 349 respondents in an online survey via Google form link across various social media platforms. This study used the partial least squares structural equation modeling (PLS-SEM) approach to test the hypotheses. This study revealed the significant positive effects of risk tolerance, herding behavior, and social interaction on stock market investment intention. Stock market investment intention also had a significant effect on stock market participation. Stock market investment intention was also found to successfully mediate the relationships of risk tolerance and overconfidence bias with stock market participation. When it comes to stock market investment, the government and related authorities should focus on developing programs and policies that provide a financial safety net for investors and promote investment-related social platforms. This study linked risk tolerance, financial well-being, financial literacy, overconfidence bias, herding behavior, social interaction, stock market investment intention, and stock market participation. This is one of the few early attempts to address issues in light of the stock market investment participation among the working adults in a developing country.

Keywords: risk tolerance; financial well-being; financial literacy; overconfidence bias; herding behavior; social interaction; investment intention; stock market participation

1. Introduction

The stock market becomes a crucial key for the development of the financial system in developing countries. The stock market provides opportunities for both new or experienced investors to increase their wealth [1]. However, the stock market is risky given the unpredictable price of the market, and this makes stock trading attractive to aggressive or high-risk appetite investors, as they pursue the old advice of buying low and selling high. In Malaysia, the Chinese have the highest average wealth of RM 128,325, 76% and 47% higher than Malays and Indians, respectively [2]. This evidence indicates the high correlation between ethnicity and investment behavior in relation to future financial decision-making [3]. During the Covid-19 outbreak, many countries suffered an economic recession. Back on the 27 March 2020, the Malaysian government announced a national economic stimulus package, specifically known as the Prihatin Rakyat Economic Stimulus Package 2020 (PRIHATIN Package), allocating RM 250 billion to ease the financial burden of Malaysian citizens and another RM 10 billion to ease the financial burden of small and medium-sized enterprises (SMEs). Due to the uncertainties caused by the Covid-19 outbreak, major shareholders of some listed companies even disposed their shares during the first few trading days when the movement control order (MCO) was enforced [4].

Over the years, the Malaysian government has tried to promote stock market investment among Malaysians who otherwise would generally deposit their money into various savings schemes that provide very little returns. Therefore, it is crucial to understand human behavior and decision-making from a financial perspective and examine factors that influence stock market investment intention among Malaysian working adults. In addition, investors need to develop a positive vision, foresight, patience, and drive [5]. Many factors influence investment stock market participation, such as cognitive and emotional weaknesses, risk tolerance, financial well-being, financial literacy, overconfidence bias, demographic characteristics, herding behavior, social interaction, income level, and investment intention [6]. There is a need to determine how these factors can influence stock market participation among the working adults in Malaysia.

According to the theory of planned behavior (TPB), behavioral achievement can be obtained through behavioral intention. Hence, when a behavior or situation affords a person complete control over behavioral performance, behavioral intention alone should be sufficient to predict the behavior [7]. Therefore, this study attempted to examine the relationships of risk tolerance, financial well-being, financial literacy, overconfidence bias, herding behavior, social interaction, stock market investment intention, and stock market participation within the Malaysian context.

2. Literature Review

2.1. Theoretical Foundation

TPB concerns one's intention to perform a given behavior. It is assumed that behavioral intention captures motivational factors that influence a particular behavior [7]. Ahmad and Shah [6] investigated the effect of behavioral biases on individual investors' decision-making and their performance in different cultures or environments. Behavioral finance, however, assumes that investment decisions are often irrational due to imperfect information. Investment decisions can be actively driven by the current and expected macroeconomic environments. Information related to macroeconomic variables may predict the variation in trading decisions. The significance of macroeconomic expectations in decision-making has been echoed in financial literature [8]. According to the consumption-based capital asset pricing model and other models, it was admitted that stock market participation could be much lower than predicted [9]. When individuals have a better attitude towards a particular action, they are more likely to perform the action. Therefore, it is reasonable that, in an investment context, an investor is more favorable towards investment [10]. Moreover, attempts of conducting analysis related to investment performance and start investing their resources, including evaluating their financial position, also show one's level of stock market investment intention. With the application of TPB, this study investigated the mediating effect of stock market investment intention on stock market participation (SMP) in Malaysia. Investor's attitudes can be classified as traditional mode investor, causal investor, long-term investor, and well-informed investor [11]. Traditional investors are relatively young investors who invest a very small percentage in the stock market. Causal investors invest 10 to 20 percent of their disposable income, and a third of investor's attitudes are long-term investors, who consider capital gain and give high importance to the past movement of stock. This kind of investor being highly focused on capital gain, their less active role in the recovery phase is understandable and self-explanatory. The last is the biggest in number and constitutes almost 65 per cent of the total respondents with very alert and informed investors who study the market very carefully, read expert opinion, risk factors of stock, tracks past movement very often, and invests for the medium term.

Moreover, Garg and Singh [12] pointed out that understanding the factors that contribute to the acquisition of financial literacy among working adults can help in making policy interventions targeted at working adults to enhance their financial wellbeing. In addition, working adult's intention to invest in peer to peer lending is strongly influenced by their knowledge regarding mechanism, risk, and return [13]. Therefore, there are various factors that can affect a working adult's intention to invest on stock market in developing countries such as Indonesia, Malaysia, and India.

2.2. Hypothesis Development

2.2.1. Risk Tolerance

Financial risk tolerance happens when one is committed to accept the uncertainties of the investment decision. Risk tolerance is said to have a substantial effect on one's investment behavior [14]. Low-risk tolerance investors have the tendency to take up investment without understanding the financial risk involved [15]. The level of risk tolerance affects investor's investment behavior. High-risk tolerance investors trade in stocks of higher value. Lim, Soutar, and Lee [16] examined the differences in investment decisions and behaviors between investors of high uncertainty avoidance and low uncertainty avoidance. Investors with low uncertainty avoidance in the study displayed the following characteristics: (1) They were more flexible; (2) they accepted uncertainties without a great deal of discomfort; (3) they took risks easily; (4) they showed a greater tolerance for others' opinions and behaviors. Therefore, they showed high risk tolerance and uncertainty or vagueness attitudes towards their investment on whether it will give them a profit or loss. An investor invests in volatile investments in order to get higher profits than average. Based on the above discussion, the following hypothesis was proposed for testing:

Hypothesis 1 (H1). *Risk tolerance has a significant positive effect on stock market investment intention among Malaysian working adults.*

2.2.2. Financial Well-Being

Financially-informed individuals make wiser decisions for their family members, and they are in a better position to in terms of financial sustainability [17]. Khalil and Akhtar [18] stated that investors with positive investment behaviors can experience positive impact on their financial well-being. Financial well-being associated individuals move from distinctive perspective with psychological adjustment, physical health and life satisfaction of the investor life cycle [19]. Kamakia, Mwangi, and Mwangi [20] identified financial well-being may impact the individual's evaluation and intention related to retirement investment. Based on the above discussion, the following hypothesis was suggested for testing:

Hypothesis 2 (H2). *Financial well-being has a significant positive effect on stock market investment intention among Malaysian working adults.*

2.2.3. Financial Literacy

Financial literacy has become a phenomenon of interest for financial decisions [21]. Financial literacy is considered as means to expedite financial well-being. Financial literacy enables informed judgments and effective decision-making on the usage of investment. Financial literacy not only helps investors to build a settled way of thinking for their investment decisions but also makes them confident to perform rational and well-calculated judgements [22]. Moreover, financial literacy can help individuals with day-to-day financial tasks and to deal with financial emergency [12]. Considering the various financial behaviors, Kamakia, Mwangi, and Mwangi [20] acknowledged the need for motivation and confidence to apply financial knowledge in one's decision-making. Financial literacy affects one's financial decisions in many aspects, such as wealth management, stock holding, and insurance demand [23]. Based on the findings of previous studies, this study proposed the following hypothesis:

Hypothesis 3 (H3). *Financial literacy has a significant positive effect on stock market investment intention among Malaysian working adults.*

2.2.4. Overconfidence Bias

Overconfidence is basically heuristic bias, in which investors rely on ostensibly, to reduce the risk of losses in unpredictable situations. Overconfidence leads stock market investors towards understating investment-related risks and overstating their stock market knowledge and trading excessively, which ultimately affect their behaviors [24]. When investors are overconfident, they are more likely to take higher risk [25]. When individual investors use heuristics, their technical knowledge and reasoning faculties are impaired, leading to errors in judgement [6]. Bakar, Ng, and Yi [26] noted the significant positive effect of overconfidence bias on investors' decision-making. Investors with overconfidence bias tend to focus more on profitability, usage of debt financing, and preference for short-term external investment in the cost of a long-term project [27]. Based on the above discussion, this study proposed the following hypothesis:

Hypothesis 4 (H4). *Overconfidence bias has a significant positive effect on stock market investment intention among Malaysian working adults.*

2.2.5. Herding Behavior

Individuals with herding behavior base their investment decisions on the crowd actions of buying and selling, which create speculative bubbles and subsequently, make the stock market inefficient [26]. Herding behavior affects investors for two reasons: (1) To protect themselves from losses; (2) to reward themselves with maximum profit [25]. Herding behavior happens when investors pick stocks for investment and do not avoid stocks. With that, herding behavior may drive the industry market values away from the fundamentals [28]. Based on the above discussion, the following hypothesis was suggested:

Hypothesis 5 (H5). *Herding behavior has a significant positive effect on stock market investment intention among Malaysian working adults.*

2.2.6. Social Interaction

Social influence refers to one's perception towards other individuals on the target behavior and whether they expect others to perform that behavior. Social influence considers users' perceptions of how other users perceive about a certain product or services. A feedback mechanism is used—for instance, by receiving recognition in the forms of "likes" and comments and conforming to the perceived expectations of other users [29]. The spreading of investment success stories within the social networks may partly explain the fluctuations of stock market. Internet and social interaction increase stock market participation, and the usage of modern communication devices may crowd out the informational effect of social interaction [9]. Studies have revealed the positive effect of social interaction and media on trading decisions. Furthermore, among the social factors, social interaction yields major impact on trading decisions [30]. Moreover, Wu, Huang, Chen, Davison, and Hua [31] demonstrated the positive impact of social interaction on customers' investment intention. Therefore, based on the above discussion, the following hypothesis was proposed for testing:

Hypothesis 6 (H6). *Social interaction has a significant positive effect on stock market investment intention among Malaysian working adults.*

2.2.7. Stock Market Investment Intention

When it comes to decision-making, investors have to choose a certain course of action among various alternatives in the world of uncertainties [32]. Identifying the relative importance of the determinants of behavioral intention is one essential step in studying the behavioral intention of individual investors [33]. Moreover, intention is assumed to capture motivational factors that influence a particular behavior and indicate one's willingness to try or how much effort one exerts to perform the behavior [7]. Investment intention was found to significantly predict behavior in relation to stock market [34]. Therefore, based on the above discussion, the following hypothesis was suggested:

Hypothesis 7 (H7). *Stock market investment intention has a significant positive effect on stock market participation among Malaysian working adults.*

2.2.8. Mediating Effect of Stock Market Investment Intention

Investment intention can be predicted by several predictors, such as risk tolerance, herding behavior, and financial literacy, towards obtaining stock market participation [33]. One's financial behavior is termed as investment intention—short- and long-term investment intentions are intended to reflect behavioral intentions [35]. Hence, the behavior of an investor, investment experience, and social interaction significantly influence stock market investment intention and subsequently, stock market participation [36]. According to [7], one's intention can predict future behavior because intention is a preliminary step to the subsequent pattern of behavior. Intention is an attitudinal construct based on intrinsic values and plays an important role in predicting one's future behavior. Consequently, intention indicates the direction of one's possible behavior in the future [37]. Based on the above discussion, the following hypothesis was proposed for testing:

Hypothesis 8 (H8). *Stock market investment intention mediates the relationships of risk tolerance, financial well-being, financial literacy, overconfidence bias, herding behavior, and social interaction with stock market participation among Malaysian working adults.*

All association hypothesized and tested, presented in Figure 1 below.

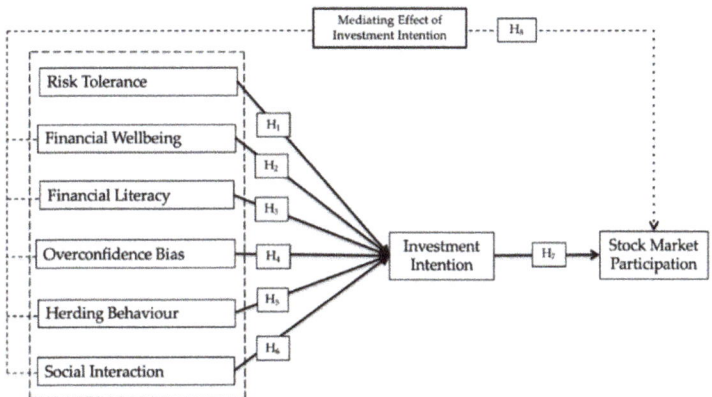

Figure 1. Research Framework.

3. Research Methodology

3.1. Population and Sample

Adopting the cross-sectional design, this study collected quantitative data in an online survey via a Google form link across various social media platforms, including Facebook, WhatsApp, and Instagram. An online survey was one of the convenient sampling methods, which has been widely adopted by researchers to collect data during the lockdown due to the Covid-19 outbreak. In order to obtain the minimum sample size, this study used G*Power 3.1 (Heinrich Heine University, Düsseldorf, Germany) (source: https://webpower.psychstat.org/models/kurtosis/ accessed on 23 November 2020). With the power of 0.95, effect size of 0.15, and 4 predictors, the calculated minimum sample size for this study was 74. In order to avoid any complications due to the small sample size, this study aimed to collect data from more than 300 respondents. The online survey was conducted from September 2020 to October 2020, resulting in a total of 349 respondents. The collected data were analyzed to test the effects of selected constructs on stock market investment intention and stock market participation. This study used partial least squares used structural equation modelling (PLS-SEM) to analyze the data.

3.2. Measures of Constructs

Risk tolerance in this study was generally defined as the willingness to accept the maximum amount of uncertainties when one makes a financial decision [38]. Five items were used to measure risk tolerance, which were adopted from Pak and Mahmood [39] and Sarwar and Afaf [15]. Meanwhile, this study viewed financial well-being as a composite concept contributing to one's assessment of financial status [17]. Five items were also used to measure financial well-being, which were adopted from Lee, Lee, and Kim [17]. Financial literacy in this study was viewed as the required knowledge that enables individuals to make financial decisions in their best interest. Five items were used to measure financial literacy, which were adopted from Sarwar and Afaf [15] and Raut [22]. Overconfidence bias happens when one overstates knowledge, skills, and capabilities, understates the risk, and even ignore the actual facts [40]. Five items were also used to measure overconfidence bias in this study, which were adopted from Sarwar and Afaf [15]. Herding behavior in this study was defined as the tendency of investors to imitate activities of other investors, disregarding their own personal information and expectations. In order to measure herding behavior, this study adopted five items from Sarwar and Afaf [15]. Social interaction played an important role in transmitting relevant information to potential investors, which may be affected by other information channels [9]. In order to measure social interaction, this study adopted 5 items from Wu et al. [31]. Stock market investment intention in this study referred to the indication of one's willingness to perform a specific behavior. This study adopted 5 items from Akhtar and Das [29]. For the measurement of stock market participation, this study adopted 5 items from Khan, Tan, and Chong [8] and Akhtar and Das [29]. All items adapted in this study presented in Appendix A. Finally, in this study, a 5-point Likert scale with the endpoints of "strongly disagree" (1) and "strongly agree" (5) was used, and its purpose was to determine how significant the relationships of the selected constructs were with stock market investment intention and stock market participation.

3.3. Multivariate Normality

The study obtained the multivariate normality using Web Power [41]. The calculated Mardia's multivariate skewness and kurtosis coefficient and p-values revealed that the data had non-normality issue as the p-values were less than 0.05 [42].

3.4. Data Analysis Method

The partial least square structural equation modeling (PLS-SEM) was employed to estimate complex cause–effect relationship models with latent variables [43]. Contrasting covariance-based approaches to structural equation modeling were suitable to assess higher-order constructs and complex conceptual model with mediation effects [41]. Since the study sample had exceeded 100 (n = 349), the PLS-SEM technique via SmartPLS was suitable for this study to test the causal–effect relationships proposed in this study model.

4. Data Analysis

4.1. Demographic Characteristics of Respondents

Table 1 shows the demographic characteristics of the respondents in this study. As for the gender aspect, the survey was dominated by female respondents (51.0%), and the remaining 49.0% were male respondents. In addition, most of respondents in this study were 18 to 30 years old (84.8%), followed by those between 31 to 40 years old (10.0%). The remaining respondents were between 41 to 50 years old (2.6%) and above 50 years old (2.6%). Furthermore, Chinese (92.8%) represented the majority of respondents in this study, followed by Indians (4.0%) and Malays (3.2%). As for their education level, the survey was dominated by bachelor degree holders (51.9%), followed by diploma holders (21.2%), secondary school certificates (19.8%), and lastly, Master's degree or doctoral degree holders (7.2%). About 47.0% of the respondents in this study were students, while 41.8% of the respondents were employed. About 8.9% of the respondents were self-employed, and the remaining respondents (2.3%) were unemployed. Based on the data on annual income, 58.2% of the respondents reported an annual income of below RM 24,000; 28.9% of the respondents reported an annual income ranging from RM 24,000 to RM 48,000; 8.6% of the respondents reported an annual income ranging from RM 48,001 to RM 72,000; 2.3% of the respondents reported an annual income ranging from RM 72,001 to RM 96,000. The remaining respondents reported an annual income of above RM 96,000 (2.0%).

Table 1. Demographic Characteristics.

	n	%		n	%
Gender			Education		
Female	178	51.0	Secondary school certificate	69	19.8
Male	171	49.0	Diploma/technical school certificate	74	21.2
Total	349	100.0	Bachelor degree or equivalent	181	51.9
			Master and/or Doctoral degree	25	7.2
Age Group			Total	349	100.0
18 to 30	296	84.8			
31 to 40	35	10.0	Occupation		
41 to 50	9	2.6	Employed	146	41.8
Above 50	9	2.6	Self-employed	31	8.9
Total	349	100.0	Student	164	47.0
			Un-employed	8	2.3
Ethnicity			Total	349	100.0
Chinese	324	92.8			
Indian	14	4.0	Annual Income (RM)		
Malay	11	3.2	Below RM24,000	203	58.2
Total	349	100.0	RM24,000 to RM48,000	101	28.9
			RM48,001 to RM72,000	30	8.6
			RM72,001 to RM96,000	8	2.3
			Above RM96,000	7	2.0
			Total	349	100.0

4.2. Reliability and Validity

The measurement model was the first assessment in SEM that included the evaluation of construct reliability, indicator reliability, convergent validity, and discriminant validity of the outlined constructs. Construct reliability can be assessed in terms of composite reliability (CR) and Cronbach's alpha (CA). A CR of greater than 0.07 indicates adequate construct reliability [44]. Table 2 presents the results of the measurement model, which showed CR values of greater than 0.07; thus, confirming adequate construct reliability. Indicator reliability in this study was assessed in terms of CA, in which CA must be higher than 0.06. The results indicated CA of constructs were all acceptable. Convergent validity was assessed using average variance extracted (AVE). The criterion was that the values of AVE must be higher than 0.50 [45]. The results revealed that all constructs recorded substantial AVE values and achieved convergent validity. The values of CR, CA, and AVE are tabulated in Table 2.

Table 2. Reliability and Validity.

Variables	No. Items	Mean	SD	CA	DG rho	CR	AVE	VIF
RT	5	3.328	1.121	0.894	0.905	0.922	0.703	3.717
FW	5	3.523	1.014	0.890	0.891	0.919	0.694	4.750
FL	4	3.270	1.068	0.861	0.863	0.906	0.706	4.778
OB	5	3.150	1.100	0.920	0.921	0.940	0.758	3.817
HB	5	3.427	1.080	0.918	0.919	0.938	0.753	3.572
SI	5	3.406	1.096	0.928	0.928	0.946	0.777	4.038
INT	4	4.474	1.588	0.930	0.930	0.950	0.825	1.000
SMP	5	3.344	1.134	0.916	0.918	0.937	0.748	-

Note: RT: Risk Tolerance, FW: Financial Wellbeing; FL: Financial Literacy; OB: Overconfidence Bias; HB: Herding Behavior; SI: Social Interaction; INT: Investment Intention; SMP: Stock Market Participation; SD: Standard Deviation; CA: Cronbach's Alpha; DG rho: Dillon-Goldstein's rho; CR: Composite Reliability; AVE: Average Variance Extracted; VIF: Variance Inflation Factors. Source: Author's data analysis.

The assessment of discriminant validity involves 2 types of methods, which were used in this study: Fornell–Larcker criterion and cross-loadings. Initial discriminant validity of the construct was tested with another method for measuring discriminant validity to assess the cross-loadings of indicators [44]. Fornell–Larcker criterion was used to assess the discriminant validity of constructs, which involved comparing the square root values of AVE of each construct with the correlation between constructs. On the other hand, the method of cross-loadings suggests that the outer loadings of constructs should be greater than the loadings of corresponding constructs. With that, adequate discriminant validity of all constructs can be validated. The results of the Fornell–Larcker criterion are presented in Table 3, while the results of cross-loadings were tabulated in Table 4. This study confirmed adequate discriminant validity for all constructs, as all constructs loading were higher than other constructs. Finally, the variance inflation factor (VIF) values were less than 5, which indicates the absence of multicollinearity. Following the recommendation by Kock [46], this study tested full collinearity diagnostics of all independent variables. All the study constructs regressed on the common variable and the VIF value less than 5 indicates the absence of bias from the single-source data. Full collinearity analysis shows no issue of single-source bias.

Table 3. Discriminant Validity.

	RT	FW	FL	OB	HB	SI	INT	SMP
RT	0.838							
FW	0.806	0.833						
FL	0.777	0.784	0.841					
OB	0.748	0.716	0.840	0.871				
HB	0.748	0.808	0.763	0.707	0.868			
SI	0.772	0.824	0.769	0.740	0.782	0.881		
INT	0.792	0.775	0.723	0.725	0.739	0.811	0.909	
SMP	0.825	0.801	0.800	0.810	0.755	0.844	0.879	0.865

Note: RT: Risk Tolerance, FW: Financial Wellbeing; FL: Financial Literacy; OB: Overconfidence Bias; HB: Herding Behavior; SI: Social Interaction; INT: Investment Intention; SMP: Stock Market Participation. Source: Author's data analysis.

Table 4. Loadings and Cross-Loading.

Code	RT	FW	FL	OB	HB	SI	INT	SMP
RT—Item 1	0.831	0.658	0.605	0.610	0.604	0.663	0.669	0.683
RT—Item 2	0.881	0.736	0.679	0.670	0.681	0.728	0.749	0.767
RT—Item 3	0.869	0.677	0.621	0.596	0.590	0.629	0.699	0.689
RT—Item 4	0.855	0.697	0.699	0.621	0.665	0.650	0.667	0.696
RT—Item 5	0.750	0.606	0.675	0.661	0.601	0.549	0.499	0.611
FW—Item 1	0.661	0.807	0.666	0.606	0.670	0.695	0.643	0.645
FW—Item 2	0.664	0.810	0.711	0.654	0.676	0.648	0.618	0.687
FW—Item 3	0.650	0.845	0.599	0.544	0.663	0.713	0.657	0.664
FW—Item 4	0.717	0.868	0.665	0.638	0.700	0.721	0.696	0.702
FW—Item 5	0.664	0.833	0.629	0.538	0.654	0.652	0.608	0.638
FL—Item 1	0.642	0.569	0.798	0.758	0.600	0.612	0.585	0.655
FL—Item 2	0.633	0.709	0.837	0.659	0.622	0.683	0.630	0.679
FL—Item 3	0.674	0.695	0.878	0.701	0.655	0.657	0.633	0.703
FL—Item 4	0.665	0.658	0.848	0.710	0.689	0.631	0.582	0.648
OB—Item 1	0.718	0.767	0.715	0.622	0.875	0.721	0.679	0.701
OB—Item 2	0.676	0.745	0.676	0.584	0.848	0.691	0.650	0.677
OB—Item 3	0.625	0.661	0.635	0.627	0.870	0.645	0.612	0.629
OB—Item 4	0.614	0.661	0.645	0.635	0.877	0.675	0.625	0.644
OB—Item 5	0.604	0.660	0.633	0.597	0.867	0.657	0.636	0.621
HB—Item 1	0.667	0.636	0.757	0.883	0.662	0.661	0.648	0.709
HB—Item 2	0.703	0.628	0.742	0.882	0.640	0.653	0.640	0.714
HB—Item 3	0.629	0.644	0.720	0.856	0.591	0.663	0.641	0.724
HB—Item 4	0.627	0.605	0.756	0.883	0.608	0.624	0.629	0.685
HB—Item 5	0.629	0.603	0.680	0.850	0.572	0.621	0.598	0.693
SI—Item 1	0.722	0.751	0.736	0.686	0.721	0.886	0.706	0.761
SI—Item 2	0.694	0.746	0.698	0.663	0.731	0.893	0.706	0.744
SI—Item 3	0.668	0.720	0.668	0.624	0.678	0.887	0.711	0.749
SI—Item 4	0.635	0.668	0.629	0.657	0.630	0.860	0.715	0.722
SI—Item 5	0.682	0.748	0.659	0.633	0.689	0.881	0.734	0.741
INT—Item 1	0.728	0.696	0.657	0.698	0.647	0.754	0.910	0.821
INT—Item 2	0.714	0.688	0.689	0.692	0.686	0.732	0.906	0.813
INT—Item 3	0.718	0.720	0.629	0.582	0.669	0.737	0.920	0.796
INT—Item 4	0.718	0.712	0.653	0.661	0.686	0.724	0.898	0.763
SMP—Item 1	0.709	0.685	0.744	0.759	0.656	0.732	0.752	0.883
SMP—Item 2	0.794	0.775	0.709	0.661	0.723	0.784	0.826	0.898
SMP—Item 3	0.703	0.749	0.652	0.613	0.662	0.744	0.775	0.865
SMP—Item 4	0.646	0.609	0.653	0.749	0.582	0.683	0.727	0.829
SMP—Item 5	0.707	0.637	0.703	0.734	0.638	0.701	0.716	0.849

Note: RT: Risk Tolerance, FW: Financial Wellbeing; FL: Financial Literacy; OB: Overconfidence Bias; HB: Herding Behavior; SI: Social Interaction; INT: Investment Intention; SMP: Stock Market Participation. The Italic values in the matrix above are the item loadings and others are cross-loadings. Source: Author's data analysis.

4.3. Path Analysis

The results on the structural model in Table 5 revealed factors that affect stock market investment intention. This study demonstrated the significant effects of risk tolerance, herding behavior, and social interaction on stock market investment intention. This study also revealed the significant effect of stock market investment intention on stock market participation. On the contrary, financial well-being, financial literacy, and overconfidence bias were found to contribute significant effects on stock market investment intention.

Based on the effect size (f^2), all constructs in this study exhibited a small effect size on stock market investment intention, which ranged from 0.000 to 0.124. According to Hair et al. [44], the blindfolding procedure showed how the values of constructs were well-observed by reconstructing the estimates of the parameters. In addition, the blindfolding procedure can be applied only on endogenous constructs with reflective indicators. The predictive relevance of the model in this study was calculated collectively with Q^2 for all factors at the individual level (single factor). The results of predictive relevance Q^2 were also presented in Table 5. The results of the blindfolding procedure revealed substantial predictive relevance of the model at 0.603%, which confirmed the integration of the predictors in relation to stock market investment intention.

Table 5. Path Coefficients.

Hypo		Beta	CI-Min	CI-Max	t	p	r^2	f^2	Q^2	Decision
	Factors Effecting Intention to Invest in Stock Market									
H_1	RT → INT	0.296	0.154	0.466	3.167	0.001		0.091		Accept
H_2	FW → INT	0.110	−0.034	0.257	1.278	0.101		0.010		Reject
H_3	FL → INT	−0.062	−0.196	0.074	0.751	0.227	0.735	0.003	0.603	Reject
H_4	OB → INT	0.092	−0.028	0.220	1.173	0.121		0.022		Reject
H_5	HB → INT	0.147	0.016	0.270	1.882	0.030		0.009		Accept
H_6	SI → INT	0.360	0.185	0.511	3.731	0.000		0.124		Accept
	Factor Effecting the Stock Market Participation									
H_7	INT → SMP	0.879	0.848	0.906	48.390	0.000	0.722	3.403	0.573	Accept

Note: RT: Risk Tolerance, FW: Financial Wellbeing; FL: Financial Literacy; OB: Overconfidence Bias; HB: Herding Behavior; SI: Social Interaction; INT: Investment Intention; SMP: Stock Market Participation. Source: Author's data analysis.

4.4. Mediating Effects

In this study, stock market investment intention exhibited a partial mediating effect on the relationship between the predictors and stock market participation. The coefficient of risk tolerance in relation to stock market participation recorded 0.260 (p-value = 0.001). This indicates the mediating effect of stock market investment intention on the relationship between risk tolerance and stock market participation. Besides that, stock market investment intention was also found to significantly mediate the relationship between overconfidence bias and stock market participation (coefficient of 0.129, p-value = 0.031). On the contrary, stock market investment intention did not exhibit any mediating effect on the individual relationships of financial well-being (p-value = 0.098), financial literacy (p-value = 0.226), herding behavior (p-value = 0.122), and social interaction (p-value = 0.122) with stock market participation (p-value > 0.05). Table 6 presents the results on the mediating effect of stock market investment intention on the proposed relationships.

Table 6. Mediating Effects.

Associations	Beta	CI-Min	CI-Max	t	p	Decision
RT → INT → SMP	0.260	0.136	0.407	3.165	0.001	Accept
FW → INT → SMP	0.096	−0.031	0.218	1.296	0.098	Reject
FL → INT → SMP	−0.055	−0.171	0.066	0.751	0.226	Reject
OB → INT → SMP	0.129	0.014	0.238	1.873	0.031	Accept
HB → INT → SMP	0.081	−0.025	0.192	1.168	0.122	Reject
SI → INT → SMP	0.081	−0.025	0.192	1.168	0.122	Reject

Note: RT: Risk Tolerance, FW: Financial Wellbeing; FL: Financial Literacy; OB: Overconfidence Bias; HB: Herding Behaviors; SI: Social Interaction; INT: Investment Intention; SMP: Stock Market Participation. Source: Author's data analysis.

4.5. Multiple Group Analysis

Multiple group analyses were applied to determine the differences between the model based on gender and education. Table 7 demonstrates the path values for the 2 groups and the differences within the groups in terms of p-values.

The results of the 2 groups based on the gender of the sample demonstrated a significant difference in the relationship between herding behaviors and social interaction on investment intention and the effect of investment intention on stock market participation. The effect of herding behaviors on investment intention was high among the female working adults, whereas the effect of social interaction and investment intention on stock market participation was significantly higher among the male working adults than that of others.

Table 7. Multi-group Analysis.

	Female		Male		Difference		
	Beta	p-Value	Beta	p-Value	Beta	p-Value	Decision
RT → INT	0.198	0.094	0.382	0.000	0.184	0.167	No Difference
FW → INT	0.116	0.229	0.050	0.277	−0.066	0.364	No Difference
FL → INT	−0.089	0.265	−0.082	0.209	0.006	0.491	No Difference
OB → INT	0.230	0.011	0.074	0.248	−0.156	0.144	No Difference
HB → INT	0.252	0.022	−0.005	0.474	−0.257	0.041	Sig. Difference
SI → INT	0.223	0.051	0.533	0.000	0.310	0.032	Sig. Difference
INT → SMP	0.845	0.000	0.912	0.000	0.066	0.023	Sig. Difference
	High School/Diploma		Bachelor Degree and Above		Difference		
	Beta	p-Value	Beta	p-value	Beta	p-Value	Decision
RT → INT	0.394	0.001	0.239	0.026	0.154	0.194	No Difference
FW → INT	−0.029	0.394	0.149	0.109	−0.178	0.134	No Difference
FL → INT	−0.119	0.173	−0.031	0.388	−0.088	0.296	No Difference
OB → INT	0.040	0.340	0.237	0.010	−0.198	0.082	No Difference
HB → INT	0.097	0.157	0.100	0.173	−0.003	0.477	No Difference
SI → INT	0.560	0.000	0.245	0.032	0.315	0.041	Sig. Difference
INT → SMP	0.901	0.000	0.856	0.000	0.045	0.094	No Difference

Note: RT: Risk Tolerance, FW: Financial Wellbeing; FL: Financial Literacy; OB: Overconfidence Bias; HB: Herding Behaviors; SI: Social Interaction; INT: Investment Intention; SMP: Stock Market Participation. Source: Author's data analysis.

The results of the 2 groups based on education show that the effect of social interaction on investment intention among the working adults with high school and/or diploma was significantly higher than the working adults with a bachelor's degree and above. The findings highlight that social interaction plays a much bigger role in investment decisions among the less educated working adults in Malaysia.

5. Discussion

This study aimed to investigate the mediating effect of stock market investment intention on the relationships of selected constructs with stock market participation within the Malaysian context. This study verified the significant positive effect of risk tolerance on stock market investment intention, which is compatible with the findings of a study by Fauzi, Husniyah, and Amim [14]. The study specifically stated that risk-takers tend to be involved in stock market investment. The perception towards one's capability to control stock investment decisions is also important due to the influence of the individual's confidence in the stock investment market. Besides that, the relationship between herding behavior and stock market investment intention was also found statistically significant. This particular finding is consistent with the finding of a prior study [47]. Kumari et al. [47] identified herding behavior as a specific investment-related behavior that an investor assumes in combating the volatility of the stock market. With the lack of knowledge and restricted information, many investors concurrently replicate the actions of other investors. Meanwhile, social interaction in this study was found to exhibit a significant effect on stock market investment intention. This particular finding supports the finding of a study by Wu et al. [31], which stated that there is a major impact of social interaction on stock market investment intention. Inexperienced investors can acquire both higher utilitarian and hedonic values from social values, and stock market investment intention relies more on hedonic values. Meanwhile, experienced investors place greater emphasis on utilitarian values. In addition, Shanmugham and Ramya [30] revealed that social interaction (e.g., social media and information from close friends) promotes stock market investment intention and subsequently increases stock market participation.

Furthermore, this study demonstrated the significant effect of stock market investment intention on stock market participation. Sarwar and Afaf [15] stated that stock market investment intention plays a crucial role in the relationship between risk tolerance and herding behavior. However, this study demonstrated insignificant effects of financial well-being, financial literacy, and overconfidence bias on stock market investment intention. The findings of this study are not in line with the findings of previous studies [12,26]. These studies highlighted that understanding the factors that contribute to or detract from the acquisition of financial literacy among working adults can help in making policy interventions targeted at working adults for a higher level of financial well-being.

In addition, this study evidenced that the contradictory findings may be caused by the differences in demographic characteristics. The current study argued that female working adults generally possess higher financial literacy than male working adults. In most cases, a higher educational level was also found to be a significant indicator of higher financial knowledge, financial attitude, financial behavior, and financial literacy. Moreover, employment status, annual income, and financial socialization were also found to influence financial knowledge, financial well-being, overconfidence bias, and financial literacy at the individual level. In addition, this study argued the propensity of the Chinese to have a better understanding of investment for asset prospects in the future compared to other ethnic groups in Malaysia.

Adding to that, this study investigated the mediating effect of stock market investment intention. This study proved that stock market investment intention mediates the effects of risk tolerance and overconfidence bias on stock market participation within the context of developing countries. On the other hand, this study proved that the intention to invest among the majority of Malaysian working adults does not mediate the effects of certain factors on stock market participation, such as financial well-being, financial literacy, herding behavior, and social interaction.

Nonetheless, this study argued the different rates of stock market participation across countries, specifically between developing and developed countries, which generally increases with wealth. Moreover, at a higher wealth level, working adults in developing countries, such as Malaysia, do not hold stock, as they make a rational choice of not holding part of their assets in the stock market given their lack of financial literacy and inability to imitate the activities of other investors (of disregarding their own personal information and expectations). Meanwhile, working adults in developed countries have a better understanding of investing part of their assets in the stock market for higher income.

Majority of respondents in this study are employees with an annual income below RM 24,000, and the respondents in this study are dominated by young women with higher education graduates. These findings indicated that respondents in this study need a better understanding of stock market investment to gain their income and self-confidence. In addition, lack of investment knowledge and intentions become barriers for them to participate in the stock market. Perkins and Jones [48] stress the importance of demographics and hypothesized that different demographics generally have a different outlook on finance and spending. These factors translate into market dynamics that can be leveraged while selecting a portfolio. They emphasize creating a consistent and habitual savings and investing plan. Therefore, based on the statistical correlation in this study, financial wellbeing, financial literacy, and overconfidence bias had no significant relationship on stock market investment among Malaysian working adults.

The majority of Malaysian working adults in this study is dominated by female worker around 18 to 30 years old, with 84.8%. These findings showed that the participation of women is more sensitive towards stock market investment in Malaysia. The participants of Chinese (92.8%) who hold a bachelor's degree tend to have a better understanding of the stock market. Consequently, these results pointed out that Chinese employees had a greater intention to gain their knowledge, investment, and income. Moreover, based on received data from respondents, 58.2% of respondents in this study have an annual income below Rm.24.000. This indicated that Malaysia's stock market investment has big potential to motivate Malaysian working adults with annual income under RM. 24.000 to invest their money or disposable income in the stock market to gain their income in SMI. Working adults in developing countries tend to score low on financial knowledge, financial attitude, financial behavior, and financial literacy. Therefore, an individual's perception and financial opportunity recognition are important to obtain Malaysian working adults' intentions in stock market investment. This finding is also supported in the study by Garg and Singh [12], which stated that educational status, employment status, and family background play important factors to determine if high financial knowledge, financial attitude, financial behavior, and financial literacy of Participants.

6. Theory and Practical Implications

One of the major contributions of this study came from the development of stock market investment intention among working adults in relation to their investment decision-making processes from different behavioral perspectives. This study argued the importance of risk tolerance and overconfidence bias as predictors of working adult's participation in the stock market through stock market investment intention. Practitioners need to develop a better understanding of the significant factors that affect stock market investment intention among working adults. Based on the findings of this study, it is suggested that increasing such investing capabilities can increase the willingness to invest in the stock market. Besides that, financial advisors may consider conducting financial training to equip working adults with knowledge before making any investment, and this may also encourage them to be more involved in their investment decisions. The nature of such training is important, as it provides information and a sense of connection to build involvement. Moreover, the findings of this study on stock market participation and the mediating effect of stock market investment intention within the context of working adults in developing countries have also extended the current literature on behavioral finance and behavioral theories, such as TPB. The practical contribution of this study is that the capital market and security authority have to put more events and workshops in place for Malaysian working adults. Moreover, the capital market and security authority (CMSA) need to motivate and provide adequate training, seminars, and awareness among working adults on the potential benefits of investing in the stock market in Malaysia. Thus that the individual has a better understanding of the stock market in the future. Other factors that might influence financial wellbeing, financial wellbeing, and overconfidence bias on stock market participation, such as income, mindset, culture, and gender, should be investigated. As a consequence, this will give more comprehensive results and influence more individuals to participate in the stock market.

7. Conclusions and Future Research

Stock market investment intention among working adults is crucial for their future assets. Creating a campaign or investment program to socialize stock investment is likely to provide a better understanding of the perspectives of Malaysian working adults towards stock market investment and subsequently contribute to the growing number of investors. The findings of this study can help firms that intend to maximize their funding sources from the stock market. Although the current study has successfully provided significant insights on the factors that influence stock market investment intention among working adults and the applicability of TPB within the context of developing countries, this study encountered several limitations. Firstly, this study exclusively focused on Malaysian working adults. Therefore, the generalization of the findings needs to be carefully considered. Secondly, for a better understanding, longitudinal data should be considered in future research. It is recommended for future research to incorporate individual characteristics, such as annual income, ethnicity, education level, as potential moderating variables in relation to stock market investment intention. This study adopted a convenient sampling method because it was easy to obtain a big sample and be time-efficient. However, this sampling method has weaknesses, such as the contribution of the findings can only be applicable to the group of the target respondents.

Author Contributions: M.Y., M.M., S.S.A.A.-S., and N.R.Z. focused on conceptualization, methodology, resources, and writing—original draft preparation; A.A.M. focused on conceptualization, methodology, supervision, and writing—review and editing. All authors have read and agreed to the published version of the manuscript.

Funding: This research received no external funding.

Institutional Review Board Statement: Local ethics committees (UCSI University) ruled that no formal ethics approval was required for such research. This study has been performed in accordance with the Declaration of Helsinki.

Informed Consent Statement: Written informed consent for participation was obtained from respondents who participated in the survey. For the respondents who participated the survey online (using google form), they were asked to read the ethical statement posted on the top of the form (*There is no compensation for responding nor is there any known risk. In order to ensure that all information will remain confidential, please do not include your name. Participation is strictly voluntary and you may refuse to participate at any time*) and proceed only if they agree.

Data Availability Statement: The data presented in this study are available on request from the corresponding author.

Conflicts of Interest: The authors declare no conflict of interest.

Appendix A

Table A1. Survey Instrument.

Code	Question
RT—Item 1	I consider myself as a high-risk taker
RT—Item 2	If I unexpectedly received some easy money, I would surely invest a certain amount of money in stocks
RT—Item 3	I would prefer to invest in stocks rather than to keep money in a bank account
RT—Item 4	I consider risk in investments as an opportunity
RT—Item 5	In the investment process, if it happens, I would not mind losing some money
FW—Item 1	I am securing my financial future
FW—Item 2	I am behind with my finances
FW—Item 3	My finances control my life
FW—Item 4	I am just getting by financially
FW—Item 5	I am concerned that the money I have or will save won't last
FL—Item 1	I have complete knowledge of stock exchange
FL—Item 2	I check financial statements of company of past 5 years before investing
FL—Item 3	I consider the financial position of a company before investing
FL—Item 4	Considering a long-term period (e.g., 10–20 years) stocks normally give the highest return
OB—Item 1	I feel confident to evaluate securities prices in my investment portfolio myself
OB—Item 2	My past profitable investments were mainly due to my specific investment skills
OB—Item 3	My ability to predict future prices is better
OB—Item 4	My investments decisions can mostly earn higher than average return in the market
OB—Item 5	I believe that my skills and knowledge of the market help me to outperform the market
HB—Item 1	My volume of investment also depends on others opinion (broker, financial consultant)
HB—Item 2	I am confident about accuracy of my investment decisions
HB—Item 3	I believe that information from friends has high reliability
HB—Item 4	I believe that information from colleagues has high reliability
HB—Item 5	I believe that information from relatives has high reliability.
SI—Item 1	I maintain close social relationships with my friends (investors).
SI—Item 2	I spend a lot of time interacting with my friends (investors).
SI—Item 3	I have frequent communication with my friends (investors).
SI—Item 4	I am a very active person in investment related conversation.
SI—Item 5	I really enjoy talking to people (investors).
INT—Item 1	I will invest in stock market frequently
INT—Item 2	I will encourage my friend and family to invest in stock market
INT—Item 3	I will invest in stock market in near future
INT—Item 4	I believe that the Stock Exchange is an attractive investment channel
SMP—Item 1	I have a portfolio that focuses on multiple asset classes (i.e., stocks, bonds, cash, real estate, etc.).
SMP—Item 2	I invest in stocks about which I think will definitely grow in future.
SMP—Item 3	I invest in stocks in which I can get the profit as soon as possible.
SMP—Item 4	I often buy and sell stock/shares.
SMP—Item 5	I manage my portfolio for maximum gross return rather than tax and cost efficiency.

References

1. Kalam, K. The Effects of Macroeconomic Variables on Stock Market Returns: Evidence the Effects of Macroeconomic Variables on Stock Market Returns: Malaysia's Stock Market Return Performance Evidence from. *J. World Bus.* **2020**, *55*. Available online: https://www.researchgate.net/publication/344158504_The_Effects_of_Macroeconomic_Variables_on_Stock_Market_Returns_Evidence_from_Malaysia\T1\textquoterights_Stock_Market_Return_Performance (accessed on 27 November 2020).
2. Khalid, M.A. NEP to NEM: Who Cares? Wealth Distribution in Malaysia. *Pros. Perkem* **2011**, *4*, 400–409.
3. Rahman, M.; Albaity, M.; Isa, C.R. Behavioural propensities and financial risk tolerance: The moderating effect of ethnicity. *Int. J. Emerg. Mark.* **2019**, *15*, 728–745. [CrossRef]
4. Lee, K.Y.-M.; Jais, M.; Chan, C.-W. Impact of Covid-19: Evidence from Malaysian Stock Market. *Int. J. Bus. Society* **2020**, *21*, 607–628.
5. Ansari, M.L.; Moid, M.S. Factors Affecting Investment Behaviour among Young Professionals. *Int. J. Tech. Res. Appl.* **2013**, *1*, 27–32.
6. Ahmad, M.; Shah, S.Z.A. Overconfidence heuristic-driven bias in investment decision-making and performance: Mediating effects of risk perception and moderating effects of financial literacy. *J. Econ. Adm. Sci.* **2020**. [CrossRef]
7. Ajzen, I. The Theory of Planned Behavior. *Organ. Behav. Hum. Decis. Process.* **1991**, *50*, 179–211. [CrossRef]
8. Khan, M.T.I.; Tan, S.H.; Chong, L.L. Active trading and retail investors in Malaysia. *Int. J. Emerg. Mark.* **2017**, *12*, 708–726. [CrossRef]
9. Liang, P.; Guo, S. Social interaction, Internet access and stock market participation-An empirical study in China. *J. Comp. Econ.* **2015**, *43*, 883–901. [CrossRef]
10. Nugraha, B.A.; Rahadi, R.A. Analysis of young generations toward stock investment intention: A preliminary study in an emerging market. *J. Account. Investig.* **2021**, *22*, 80–103. [CrossRef]
11. Shah, M.; Verma, A. Analysis of Investment Behaviour during Recovery Phase among Youth Investors of Indian Stock Market. *Vis. J. Bus. Perspect.* **2011**, *15*, 1–9. [CrossRef]
12. Garg, N.; Singh, S. Financial literacy among youth. *Int. J. Soc. Econ.* **2018**, *45*, 173–186. [CrossRef]
13. Ichwan, I.; Rahmatina, A.K. Why are youth intent on investing through peer to peer lending? Evidence from Indonesia. *J. Islamic Monet. Econ. Financ.* **2019**, *5*, 741–762. [CrossRef]
14. Fauzi, A.A.W.; Husniyah, S.M.F.; Amim, O.M. *Financial Risk Tolerance as a Predictor for Malaysian Employees' Gold Investment Behavior*; Springer International Publishing: Berlin/Heidelberg, Germany, 2017. [CrossRef]
15. Sarwar, A.; Afaf, G. A comparison between psychological and economic factors affecting individual investor's decision-making behavior. *Cogent Bus. Manag.* **2016**, *3*. [CrossRef]
16. Lim, K.L.; Soutar, G.N.; Lee, J.A. Factors affecting investment intentions: A consumer behaviour perspective. *J. Financ. Serv. Mark.* **2013**, *18*, 301–315. [CrossRef]
17. Lee, J.M.; Lee, J.; Kim, K.T. Consumer Financial Well-Being: Knowledge Is Not Enough. *J. Fam. Econ. Issues* **2019**, *41*, 218–228. [CrossRef]
18. Khalil, U.R.W.; Akhtar, H.; Shah, Y.Z.A. Framing effect and financial wellbeing: Role of Investment Behaviors as Mediator. *Rev. Econ. Dev. Stud.* **2019**, *5*, 343–352.
19. Koekemoer, D.Z.; Ferreira, S. A conceptual model of financial well-being for South African investors a conceptual model of financial well-being for South African investors. *Cogent Bus. Manag.* **2019**, *6*. [CrossRef]
20. Kamakia, M.G.; Mwangi, C.I.; Mwangi, M. Financial Literacy and Financial Wellbeing of Public Sector Employees: A Critical Literature Review. *Eur. Sci. J. ESJ* **2017**, *13*, 233–245. [CrossRef]
21. Sadiq, M.N.; Ased, R.; Khan, A. Impact of Personality Traits on Investment Intention: The Mediating Role of Risk Behaviour and the Moderating Role of Financial Literacy Literature Review Personality Traits. *J. Financ. Econ. Res.* **2019**, *4*, 1–18. [CrossRef]
22. Raut, R.K. Past behaviour, financial literacy and investment decision-making process of individual investors. *Int. J. Emerg. Mark.* **2020**, *15*, 1243–1263. [CrossRef]
23. Niu, G.; Zhou, Y.; Gan, H. Financial literacy and retirement preparation in China. *Pac. Basin Financ. J.* **2020**, *59*. [CrossRef]
24. Sabir, S.A.; Mohammad, H.B.; Shahar, H.B.K. The role of overconfidence and past investment experience in herding behaviour with a moderating effect of financial literacy: Evidence from Pakistan stock exchange. *Asian Econ. Financ. Rev.* **2019**, *9*, 480–490. [CrossRef]
25. Qasim, M.; Hussain, R.Y.; Mehboob, I.; Arshad, M. Impact of herding behavior and overconfidence bias on investors' decision-making in Pakistan. *Accounting* **2019**, *5*, 81–90. [CrossRef]
26. Bakar, S.; Ng, A.; Yi, C. The Impact of Psychological Factors on Investors' Decision Making in Malaysian Stock Market: A Case of Klang Valley and Pahang. *Procedia Econ. Financ.* **2016**, *35*, 319–328. [CrossRef]
27. Sadaqat, S.H.S.; Xinping, X.; Khan, M.A.; Harjan, S.A. Investor and Manager Overconfidence Bias and Firm Value: Micro-level Evidence from the Pakistan Equity Market. *Int. J. Econ. Financ. Issues* **2018**, *8*, 190–199.
28. Spyrou, S. Herding in financial markets: A review of the literature. *Rev. Behav. Financ.* **2013**, *5*, 175–194. [CrossRef]
29. Akhtar, F.; Das, N. Predictors of investment intention in Indian stock markets: Extending the theory of planned behaviour. *Int. J. Bank Mark.* **2019**, *37*, 97–119. [CrossRef]
30. Shanmugham, R.; Ramya, K. Impact of Social Factors on Individual Investors' Trading Behaviour. *Procedia Econ. Financ.* **2012**, *2*, 237–246. [CrossRef]

31. Wu, W.; Huang, V.; Chen, X.; Davison, R.M.; Hua, Z. Social value and online social shopping intention: The moderating role of experience. *Inf. Technol. People* **2018**, *31*, 688–711. [CrossRef]
32. Kumar, S.; Goyal, N. Behavioural biases in investment decision making—A systematic literature review. *Qual. Res. Financ. Mark.* **2014**, *7*, 88–108. [CrossRef]
33. Phan, K.C.; Zhou, J. Factors Influencing Individual Investor Behavior: An Empirical Study of the Vietnamese Stock Market. *Am. J. Bus. Manag.* **2014**, *3*, 77–94.
34. Sivaramakrishnan, S.; Srivastava, M.A.R. Attitudinal factors, financial literacy, and stock market participation. *Int. J. Bank Mark.* **2017**, *35*. [CrossRef]
35. Sashikala, V.; Chitramani, P. The Impact of Behavioural Factors on Investment Intention of Equity Investors. *Asian J. Manag.* **2018**, *9*, 183–189. [CrossRef]
36. Sondari, M.C. Using Theory of Planned Behavior in Predicting Intention to Invest: Case of Indonesia. *Int. Acad. Res. J. Bus. Technol.* **2015**, *1*, 137–141.
37. Vuk, K.; Pifar, A.; Aleksić, D. Should I, Would I, Could I: Trust and Risk Influences on Intention to Invest. *Dyn. Relatsh. Manag. J.* **2017**, *6*, 61–67. [CrossRef]
38. Grable, J.E.; Joo, S.H. Environmental and biopsychosocial factors associated with financial risk tolerance. *J. Financ. Couns. Plan.* **2004**, *15*, 73–82.
39. Pak, O.; Mahmood, M. Impact of personality on risk tolerance and investment decisions: A study on potential investors of Kazakhstan. *Int. J. Commer. Manag.* **2015**, *25*, 370–384. [CrossRef]
40. Kukacka, J.; Barunik, J. Behavioural breaks in the heterogeneous agent model: The impact of herding, overconfidence, and market sentiment. *Phys. A Stat. Mech. Appl.* **2013**, *392*, 5920–5938. [CrossRef]
41. Peng, D.X.; Lai, F. Using partial least squares in operations management research: A practical guideline and summary of past research. *J. Oper. Manag.* **2012**, *30*, 467–480. [CrossRef]
42. Cain, M.K.; Zhang, Z.; Yuan, K.-H. Univariate and multivariate skewness and kurtosis for measuring nonnormality: Prevalence, influence, and estimation. *Behav. Res. Methods* **2017**, *49*, 1716–1735. [CrossRef] [PubMed]
43. Carrion, C.G.; Cegarra-Navarro, J.G.; Cillo, V. Tips to use partial least squares structural equation modelling (PLS-SEM) in knowledge management. *J. Knowl. Manag.* **2019**, *23*, 67–89. [CrossRef]
44. Hair, J.F., Jr.; Black, W.C.; Black Babin, B.J.; Andreson, R.E. *Multivariate Data Analysis*, 7th ed.; Pearson: Bloomington, MN, USA, 2014.
45. Fornell, C.; Larcker, D.F. Evaluating Structural Equation Models with Unobservable Variables and Measurement Error. *J. Mark. Res. This* **1981**, *18*, 39–50. [CrossRef]
46. Kock, N. Common method bias in PLS-SEM: A full collinearity assessment approach. *Int. J. e-Collab.* **2015**, *11*, 1–10. [CrossRef]
47. Kumari, S.; Chandra, B.; Pattanayak, J.K.; Kumari, S.; Chandra, B.; Pattanayak, J.K. Personality traits and motivation of individual investors towards herding behaviour in Indian stock market. *Kybernetes* **2019**. [CrossRef]
48. Perkins, J.; Jones, K. *Wanted: The Next Generation of Investors*; Young Money Media, LLC: Orlando, FL, USA, 2007.

Article

Emerging Economies' Institutional Quality and International Competitiveness: A PLS-SEM Approach

Ricardo E. Buitrago R. [1,2,*], María Inés Barbosa Camargo [3] and Favio Cala Vitery [1]

[1] Facultad de Ciencias Naturales e Ingeniería, Universidad Jorge Tadeo Lozano, Bogotá 11001, Colombia; favio.cala@utadeo.edu.co
[2] Escuela de Administración, Universidad del Rosario, Bogotá 11001, Colombia
[3] Facultad de Ciencias Económicas y Sociales, Universidad de La Salle, Bogotá 11001, Colombia; mibarbosa@unisalle.edu.co
* Correspondence: ricardo.buitrago@urosario.edu.co

Abstract: The home country's institutional framework determines the capacity to compete in the global arena. This paper discusses the linkage between institutional quality (IQ) and international competitiveness (IC). We measured institutions' quality in emerging economies through the use of selected indicators between 2007–2017. To evaluate the proposed IQ constructs and their relationship with IC, we applied partial least squares – structural equation modeling (PLS-SEM) analysis. The model outcomes suggest that political and lack of systemic conditions have a significant and negative effect on international competitiveness, while science, technology, engineering and mathematics (STEM) resource conditions have a significant and positive effect.

Keywords: institutional quality; international competitiveness; emerging economies; PLS-SEM

Citation: Buitrago R., R.E.; Barbosa Camargo, M.I.; Cala Vitery, F. Emerging Economies' Institutional Quality and International Competitiveness: A PLS-SEM Approach. *Mathematics* **2021**, *9*, 928. https://doi.org/10.3390/math9090928

Academic Editors: María del Carmen Valls Martínez and Pedro Antonio Martín Cervantes

Received: 28 February 2021
Accepted: 7 April 2021
Published: 22 April 2021

Publisher's Note: MDPI stays neutral with regard to jurisdictional claims in published maps and institutional affiliations.

Copyright: © 2021 by the authors. Licensee MDPI, Basel, Switzerland. This article is an open access article distributed under the terms and conditions of the Creative Commons Attribution (CC BY) license (https://creativecommons.org/licenses/by/4.0/).

1. Introduction

This study is aimed to empirically explore the role of home country institutional quality on international competitiveness [1–6]. Past studies have used traditional econometric models and variables to measure institutions' effect on international competitiveness [2]. To fill in gaps and expand previous studies, this paper analyzes the influence of different institutional conditions on emerging economies' competitiveness. This paper selects several quantitative proxies to determine the institutional quality and its relationships in the process of international competition. We follow the partial least squares-structural equation modeling (PLS-SEM) method to conduct this analysis.

There are various measures of the concept of International Competitiveness. One is proposed by Sachs, focused on macro indicators defined as "the set of institutions and economic policies supportive of high rates of economic growth in the medium term." Another, proposed by Porter, focused on microeconomic indicators to measure the "set of institutions, market structures, and economic policies supportive of high current levels of prosperity" [7]. A third approach looks at "the capability of firms engaged in value-added activities in a specific industry in a particular country to sustain this value-added over long periods in spite of international competition" [8] (p. 139). The last approach, proposed by the OECD (Organization for Economic Cooperation and Development), argues that "competitiveness is the degree to which a nation can, under free trade and fair market conditions, produce goods and services, which meet the test of international markets, while simultaneously maintaining and expanding the real income of its people over the long-term" [9].

Over the last decade, authors, reviewers, and editors have universally accepted PLS-SEM as a multivariate analysis method. A search in specialized data bases for the term "partial least squares path modeling" reveals that it has assisted researchers in empirically validating their theoretical project developments in various disciplines, such as accounting, family business, management information systems, operations management, supply chain, and many others [10–14].

According to the literature review, our paper is the first approach to study the interplay between institutional quality and international competitiveness in emerging economies using PLS-SEM. It also extends the use of PLS-SEM to the field of international business and international political economy by the use and combination of alternative data sources to explain the proposed constructs [2,15,16].

This paper is structured as follows. Section 2 briefly describes the literature review and hypothesis development. Section 3 details the methodological structure. Section 4 presents the results and discussion. Sections 5 and 6 present the conclusions, contributions, limitations, and future research directions.

2. Literature Review and Hypothesis Development on Institutional Quality and International Competitiveness

The modern economy institutions must be taken into account when thinking about economic growth and prosperity. North [15] argues that consistent, dependable institutions are necessary for the modern economic system's overall functioning. Institutions provide a defined legal system, a structured judicial system to enforce property rights and settle disputes, and a contracting and trading system that reduces firms' transaction costs [15,16].

While some institutions are more mature than others, the majority of them are underdeveloped in emerging economies. Lack of institutional development in the country has been examined in the literature to be a cause of macroeconomic volatility and can be accounted for by the adverse effects on economic growth and prosperity [17–21].

North's work [22] has been the basis for further studies that has influenced literature in growth, internationalization, and competitiveness. Another noteworthy contribution was the origin of the "institutional framework" construct that emerged in literature featured in the works of Acemoglu [17–19,23–25], which is understood to be the basis of economic transformation.

The institutional framework is determined by the quality of the institutions, both inclusive and extractive. Inclusive economic institutions create inclusive markets, while "extractive economic institutions are designed to extract incomes and wealth from one subset of society to benefit a different subset" [19].

On the other hand, the academic debate on international competitiveness focuses on the lack of a generally accepted theory on the roots of international competitiveness [26]. Summarizing the academic approaches to competitiveness:

- Technology and production capacity are more important for economic growth than cost competitiveness [27].
- International competitiveness boils down to the discussion on international trade [28].
- International competitiveness is a matter of export performance with technological capacities [8,29–35].
- International competitiveness is based on regulations and policy frameworks [36–43].

Graham and Naim [44] identified three types of institutional functions. The first is the development of rules and laws. Institutions that fall into this category are legislative, ministries, municipal councils, and related agencies. The second category of the institutional role is the application and award of rules and laws. The institutions involved here are tribunals, boards, control, and regulatory bodies. The third institutional role is the supply of public services. These are the institutions that guarantee the provision of different types of public goods and services.

There are many explanations for institutional quality that could be classified into three categories for analysis [44]:

- Resource conditions: related to the quantity, quality, and allocation of available resources.
- Political conditions: related to co-optation, corruption, and politicization in the allocation of resources.

- Systemic conditions: related to the clarity in setting long-term goals, the concentration of power in economic agents, and external state intervention.

Thus, we wanted to understand what the various institutional quality dimensions encourage international competitiveness and deter it. Due to the firm's interaction with a wide range of stakeholders, including political and social actors, they are dependent on the institutional environment in which they operate. Regulatory and normative pressures exist in a business environment, which causes firms' particular behavior [45,46]. Factors like government stability, political parties, predictability of the legal system, and contractual enforcement determine economic outcomes and internationalization [47–51]. The above arguments lead to our first hypothesis.

Hypothesis 1 (H1). *A lower degree of political conditions has a negative effect on international competitiveness.*

Porter [41] identifies the nation's competitive advantage due to the quality of endogenous variables like demand conditions, complementary industries, strategy, structure, and rivalry. The country's competitiveness is determined by resource allocation, including human capital, that helps create economic development.

The pace of economic growth is highly dependent on innovation [52]. Economic progress is made possible through technological innovation and development. New or improved technology can be developed through invention and innovation and foreign technology absorption. Allowing for such technological advances requires adequate institutions and policies to support them. It means that an economy's competitiveness relies on how well government policy can support it [53]. The nature and pace of economic growth depend on the degree of institutions and systemic factors that support technological advancements [54,55].

Technology and human capital are interdependent, inseparable, and essential. A large part of technological progress is a result of investing in human capital. In the absence of skilled workers, machines, tools, scientific instruments, the legal system, financial system, and most modern society would not function. To develop more technology, it is necessary to create and maintain skilled employees. To better utilize technology and human capital, society needs technical and business skills [56,57]. Hence, we propose the next hypothesis.

Hypothesis 1 (H2). *Science, technology, engineering, and mathematics (STEM) resources enhance international competitiveness.*

Individual property rights and property-based capitalism are vital elements to entrepreneurship. As private property becomes less prevalent or concentrated in a small elite's hands, it becomes more extractive and undermines broader economic growth [17,58]. Political restraint leads to a pattern of captured democracy in which the game's rules favor the elite [59].

A country's legal infrastructure's capacity to resolve disputes and enforce contracts motivates firms to rely on it [60]. For Kramer [61], rules are based on the ability to predict institutional action. "At the country-level, trust in country's laws is reflected in confidence in their country's legal system" [62,63]. Based on the specific application, rule-based trust is expected to reduce transaction costs and guide organizational strategic choices [64,65]. We, thus, hypothesize that:

Hypothesis 1 (H3). *Lack of structural systemic conditions have a negative effect on international competitiveness.*

3. Methodology

The problem intended to analyze is the institutional framework and how it affects international competitiveness. International competitiveness is affected when a country's "rules of the game" generate present and future uncertainty and question the economy's perceived potential productive capacity.

The aim is to analyze the period of 2007–2017 in 48 emerging economies given the changes in these regions' institutional conditions during that period (see Table 1). The selected countries are classified as emerging economies because they are moving from an informal institutional system to a more formal structure with rules of the game that are transparent and apply equally to all participants in the market. Besides, they often experience faster economic growth as measured by gross domestic product (GDP) and improvement in infrastructure and market conditions. However, there is still a higher risk due to political instability, domestic infrastructure problems, currency volatility, and limited equity opportunities.

Table 1. Countries included in the study.

Region	Countries
Latin America and the Caribbean	Argentina, Brazil, Chile, Colombia, Jamaica, Mexico, Peru, and Venezuela
Europe	Bulgaria, Croatia, Czech Republic, Estonia, Greece, Hungary, Latvia, Lithuania, Poland, Romania, Slovenia, Serbia, and Ukraine
Asia	Bangladesh, China, India, Indonesia, Kazakhstan, Malaysia, Pakistan, the Philippines, Russia, Sri Lanka, Thailand, and Vietnam
Africa	Kenya, Nigeria, Namibia, South Africa, Uganda, and Zambia
MENA	Egypt, Jordan, Kuwait, Morocco, Qatar, Tunisia, Turkey, and the United Arab Emirates

Source: Author's elaboration.

We propose three latent variables: political (POL), resources (RES), and systemic conditions (SYS), to measure institutional quality and its impact in a fourth latent variable named international competitiveness (IC). Figure 1 shows the basic model.

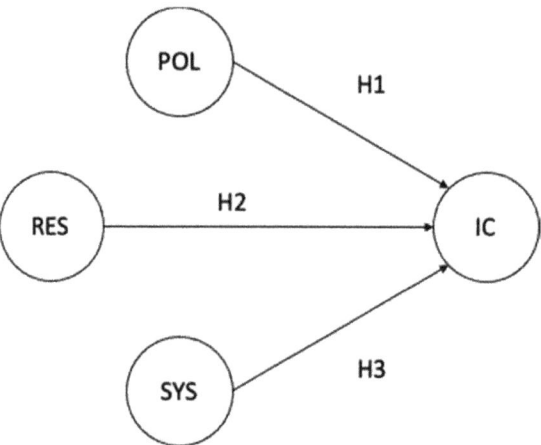

Figure 1. Institutional Quality and International Competitiveness-Basic Model. Source: Author's elaboration.

3.1. Sources and Measures

To test the proposed hypotheses, alternative reliable secondary data sources were utilized [2]. We collected indicators from the Fragile States Index (FSI) [66], from the Global Competitiveness Index (GCI) [67], from the International Country Risk Guide (ICRG) [68], and from the Index of Economic Freedom (IEF) [69]. Table 2 summarizes the structure and scales of each source.

Table 2. Data sources and scales.

Index	Categories/Pillars	Indicators	Scale High	Scale Low
GCI	Basic requirements	Institutions, infrastructure, macroeconomic environment, and health and primary education.	7	1
GCI	Efficiency enhancers	Goods, labor, and financial markets, higher education and training, and technological readiness.	7	1
GCI	Innovation and sophistication	Business sophistication and innovation.	7	1
FSI	Cohesion	Security apparatus, factionalized elites, and group grievance.	1	10
FSI	Economic	Economic decline, uneven economic development, and human flight and brain drain.	1	10
FSI	Political	State legitimacy, public services, and human rights and rule of law	1	10
FSI	Social and cross-cutting indicators	Demographic pressures, refugees and IDPs (Internal Displaced People), and external intervention	1	10
EFI	Rule of law	Property rights, government integrity, judicial effectiveness	100	0
EFI	Government Size	Government spending, tax burden, fiscal health	100	0
EFI	Regulatory efficiency	Business freedom, labor freedom, monetary freedom	100	0
EFI	Open markets	Trade freedom, investment freedom, financial freedom	100	0
ICRG	Government stability	Government unity, legislative strength, popular support	12	0
ICRG	Socioeconomic conditions	Unemployment, consumer confidence, poverty	12	0
ICRG	Investment profile	Contract viability/expropriation, profits repatriation, payment delays	12	0
ICRG	Internal conflict	Civil war/coup threat, terrorism/political violence, civil disorder	12	0
ICRG	External conflict	War, cross-border conflict, foreign pressures	12	0
ICRG	Corruption	Special payments and bribes	6	0
ICRG	Military in politics	Domination of society and/or governance by military forces	6	0
ICRG	Religious tensions	Domination of society and/or governance by a single religious group	6	0
ICRG	Law and order	Strength and impartiality of the legal system, observance of the law	6	0
ICRG	Ethnic tensions	Tension within a country attributable to racial, nationality, or language divisions	6	0
ICRG	Democratic accountability	Government's responsiveness to its people	6	0
ICRG	Bureaucracy quality	Institutional strength to govern without drastic changes in policy or interruptions in government services	6	0

Source: Author's elaboration based on the respective source.

A country's productive structure results from its level of social capital and the quality of its institutions. Research has shown that the complexity and the diversity of products a nation exports are a reliable indicator of the resources available in the economy. Complex products require a great deal of tacit knowledge and entail more distributed knowledge than those produced with a product based on resource richness or low labor costs [31,70–72]. In a world where economic power is indicative of political power, economies characterized by narrow resource endowment are more susceptible to capture due to economic and political corruption. Hence, we selected the economic complexity index (ECI) (http://atlas.cid.harvard.edu, accessed on 12 January 2021), developed by Hausmann and Hidalgo [73] as the proxy to measure international competitiveness (IC).

3.2. Constructs and Indicators

From the mentioned sources, we selected specific indicators related to the meaning of the proposed constructs. In Table 3, we describe each construct's composition. Table A1 shows the descriptions of the indicators.

Table 3. Indicators and constructs.

Indicator		Description	Construct		Source
efi_pr	X_1	Property rights	POL	Y_1	Index of Economic Freedom
gci_dpf	X_2	Diversion of public funds			Global Competitiveness Index
gci_ipp	X_3	Intellectual property protection			Global Competitiveness Index
icrg_corr	X_4	Corruption			International Country Risk Guide
icrg_lwo	X_5	Law and order			International Country Risk Guide
gci_art	X_6	Availability of research and training services	RES	Y_2	Global Competitiveness Index
gci_ftf	X_7	FDI (Foreign Direct Investment) and technology transfer			Global Competitiveness Index
gci_qes	X_8	Quality of the education system			Global Competitiveness Index
gci_qms	X_9	Quality of math and science education			Global Competitiveness Index
gci_qri	X_{10}	Quality of scientific research institutions			Global Competitiveness Index
gci_uic	X_{11}	University-industry collaboration in R&D			Global Competitiveness Index
fsi_bd	X_{12}	Human flight and brain drain	SYS	Y_3	Fragile States Index
fsi_fe	X_{13}	Factionalized elites			Fragile States Index
fsi_gg	X_{14}	Group grievance			Fragile States Index
fsi_sl	X_{15}	State legitimacy			Fragile States Index
eci	X_{16}	Economic complexity	IC	Y_4	Economic Complexity Index

Source: Author's elaboration.

3.3. Method

The study opted for structural equation modeling (SEM) because of its ability to model all paths at once. We choose Partial Least Square (PLS-SEM) instead of covariance-based (CB-SEM) for the following reasons: (1) PLS has minimal restrictions on measurement scales, sample size, and residual distributions, (2) PLS analysis does not assume that the variables are truly independent, leading to more reliable results, and (3) PLS is robust against data skewness and omitting an independent variable [11,74–81].

The literature regarding international business research shows the increasing complexity in the research problems and models observed due to the contemporary interaction between established theories and data availability [82–84]. PLS-SEM is regarded as one of the most innovative approaches in international fields that are very difficult to understand. The method proves particularly valuable for exploratory purposes and is considered proper to explain intricate relationships, like those arising from institutions and global competition [85,86].

Data were assessed using SmartPLS [87] to help determine the relationship between the latent variables POL, RES, and SYS as indicators of institutional quality and their impact on international competitiveness (IC).

Variables have been modeled as reflective constructs since the indicators are expected to covary with each other. The indicators share the same theme in the reflective model. Therefore, indicators must have the same antecedents and consequences [88,89].

Model Specification

Our model consists of 16 indicators ($X_1, X_2, X_3, \ldots, X_{16}$) and four latent variables (Y_1, Y_2, Y_3, Y_4). Latent variables Y_1, Y_2, Y_3 influence Y_4, and the measurement model is specified as follows:

$$X_1 = Y_1 C_1 + \varepsilon_1$$
$$X_2 = Y_1 C_2 + \varepsilon_2$$
$$X_3 = Y_1 C_3 + \varepsilon_3$$
$$X_4 = Y_1 C_4 + \varepsilon_4$$
$$X_5 = Y_1 C_5 + \varepsilon_5$$
$$X_6 = Y_2 C_6 + \varepsilon_6$$
$$X_7 = Y_2 C_7 + \varepsilon_7$$
$$X_8 = Y_2 C_8 + \varepsilon_8$$
$$X_9 = Y_2 C_9 + \varepsilon_9$$
$$X_{10} = Y_2 C_{10} + \varepsilon_{10}$$
$$X_{11} = Y_2 C_{11} + \varepsilon_{11}$$
$$X_{12} = Y_3 C_{12} + \varepsilon_{12}$$
$$X_{13} = Y_3 C_{13} + \varepsilon_{13}$$
$$X_{14} = Y_3 C_{13} + \varepsilon_{13}$$
$$X_{15} = Y_3 C_{15} + \varepsilon_{15}$$
$$X_{16} = Y_4 C_{16} + \varepsilon_{16}$$

In our model, X's are the indicators, Y's are the latent variables, C's are the loadings that relate latent variables to indicators, and ε's are the residuals of indicators that are unexplained. All indicators are considered reflective in our measurement model because each is assumed to affect the corresponding latent variable. As a result, all endogenous variables are observed.

The measurement model can be generally written as follows:

$$X = C'Y + \varepsilon \tag{1}$$

In the measurement (outer) model, X is a J by 1 vector of all indicators, Y is a P by 1 vector of all latent variables, C is a P by the J matrix of loadings relating P latent variables to J indicators, and ε is a J by 1 vector of the residuals of all indicators. In our model, J and P are equal to 16 (indicators) and 4 (latent variables), respectively.

The proposed structural (inner) model expresses the relationships among latent variables and can be expressed as follows:

$$Y_4 = Y_1\beta_1 + Y_2\beta_2 + Y_3\beta_3 + \zeta_4$$

where β's are path coefficients relating a latent variable to other latent variables and ζ's are the residuals of the latent variable left unexplained by the corresponding exogenous latent variables. In the model, Y_1, Y_2, and Y_3 are exogenous, whereas Y_4 is endogenous.

The above model can be expressed as:

$$Y = B'Y + \zeta \qquad (2)$$

In the structural model, B is a P-by-P matrix of path coefficients relating P latent variables among themselves, and ζ is a P by 1 vector of the residuals of all latent variables.

The weighted relation for the proposed model is as follows:

$$Y_1 = X_1w_1 + X_2w_2 + X_3w_3 + X_4w_4 + X_5w_5$$

$$Y_2 = X_6w_6 + X_7w_7 + X_8w_8 + X_9w_9 + X_{10}w_{10} + X_{11}w_{11}$$

$$Y_3 = X_{12}w_{12} + X_{13}w_{13} + X_{14}w_{14} + X_{15}w_{15}$$

$$Y_4 = X_{16}w_{16}$$

In the weighted relation model, W is a J by the P matrix of weights assigned to J indicators, which, in turn, lead to P latent variables. This can be rewritten compactly as:

$$Y = W'X \qquad (3)$$

In sum, generalized, structured component analysis involves three sub-models taking the general forms as follows:

Measurement model	$X = C'Y + \varepsilon$
Structural model	$Y = B'Y + \zeta$
Weighted model	$Y = W'X$

where:

X is a J by 1 vector of indicators
Y is a P by 1 vector of latent variables
C is a P by J matrix of loadings
B is a P by P matrix of path coefficients
W is a J by P matrix of component weights
ε is a J by 1 vector of the residuals of indicators
ζ is a P by 1 vector of the residuals of latent variables

3.4. Assessment of the Measurement Model

PLS bootstrapping with 10,000 samples [11,12,85] was used to assess the statistical significance of the model. The results of the PLS-SEM analysis are shown in Figure 2. The model tested their reliability and validity and measured the level of consistency of their scores. The indicators are all highly correlated with their intended constructs. The construct indicators were nearly all above the cutoff score of 0.708, proving that all of them represented the construct [77,79–81,89].

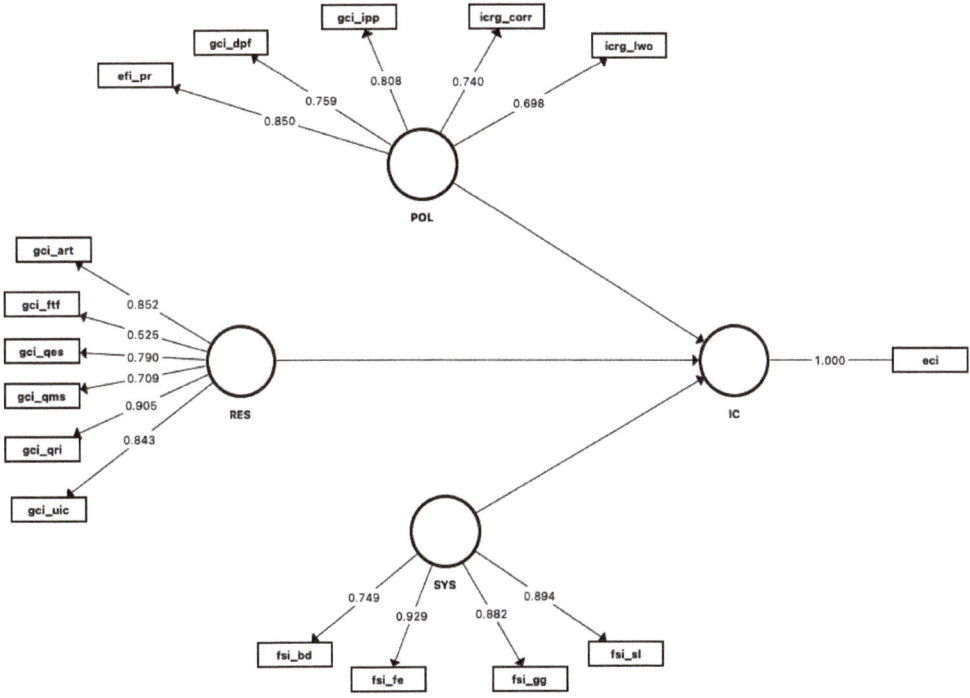

Figure 2. Indicator loadings. Source: Results from SmartPLS software 3.3.3.

To assess internal consistency, Cronbach's alpha and Heterotrait-Monotrait Ratio (HTMT) composite reliability were used [90]. Cronbach's Alpha coefficients ranged from 0.838 to 1.000. All scores were greater than the minimum score of 0.7. The Rho A also exceeded that value. The composite reliability was over 0.7 and passed a minimum level of adequacy. This has shown that there is consistency within the data. Results of average variance extracted (AVEs) were greater than the suggested minimum of 0.5 (see Table 4) [11,74–77,79–81].

Table 4. Construct validity and reliability.

	Cronbach's Alpha	rho_A	Composite Reliability	Average Variance Extracted (AVE)
POL	0.838	0.885	0.881	0.597
RES	0.881	0.928	0.904	0.581
SYS	0.887	0.888	0.923	0.751

Source: Results from SmartPLS software 3.3.3.

We also examined the discriminatory validity of the constructs using the Heterotrait-Monotrait Ratio (HTMT). The values were below 0.85, which shows adequate discriminatory validity [90,91] (see Table 5).

Table 5. Discriminant validity-HTMT.

	IC	POL	RES
POL	0.386		
RES	0.467	0.739	
SYS	0.601	0.691	0.398

Source: Results from SmartPLS software 3.3.3.

Complementary information about the measurement model is shown in Table A1: Indicators descriptive statistics, Table A2: Mean, STDEV (Standard Deviation), T-Values, *p*-Values, confidence intervals, Table A3: Outer Loadings-Mean, STDEV, T-Values, *p*-Values, Confidence Intervals, and Table A4: Outer VIF Values.

3.5. Assessment of the Structural Model

For the structural model, inner VIF (Variance Inflation Factor) values are examined. The results are below the recommended threshold of 3.3 [92,93]. Additionally, path coefficients are statistically significant at 95%.

Regarding the predictive accuracy, coefficient of determination (R2), the exogenous constructs (POL, RES, SYS) explain 41% of the endogenous construct (IC), which is considered a moderated effect [94,95]. Q2 statistics are used to measure the PLS path model's quality. This criterion recommends that the conceptual model predicts the endogenous latent constructs. In our model, the value for IC is 0.404. The values greater than zero for a particular endogenous latent construct are considered relevant [75]. Assessing the effect sizes (f2) shows that the effect size of POL (0.019) is small, RES (0.144) is moderate, and SYS (0.284), as shown in Table 5, is substantial [75,96].

4. Discussion of Findings

To evaluate the paths' importance, the validity of the measures was assessed based on the path coefficients and the significance of the path coefficients, and the significance level. The resulting *p*-values were obtained using SmartPLS by using a bootstrapping process and calculating the *p*-value of different paths. Path coefficients and significance levels have been determined by randomly sampling 10.000 instances into the model. The results are shown in Table 6 and are supported by Figure 3.

Table 6. Hypothesis results.

Hypothesis		Coefficient	Standard Deviation	T Statistics	*p* Values	VIF	f Square	CI 2.5% Lower	CI 97.5% Upper
H1	POL-> IC	−0.158	0.052	3.061	0.002	2.260	0.019	−0.257	−0.054
H2	RES-> IC	0.369	0.044	8.316	0.000	1.611	0.144	0.275	0.459
H3	SYS-> IC	−0.526	0.040	13.052	0.000	1.655	0.284	−0.600	−0.448

Source: Results from SmartPLS software 3.3.3.

Figure 3 shows the results of the outer model in factor loadings and *p*-values, and the inner model in path coefficients and *p*-values. The size of the arrows represents the absolute value of each path. As mentioned before, indicators are significant for each construct. In Table 5, we summarize the results for each proposed hypothesis.

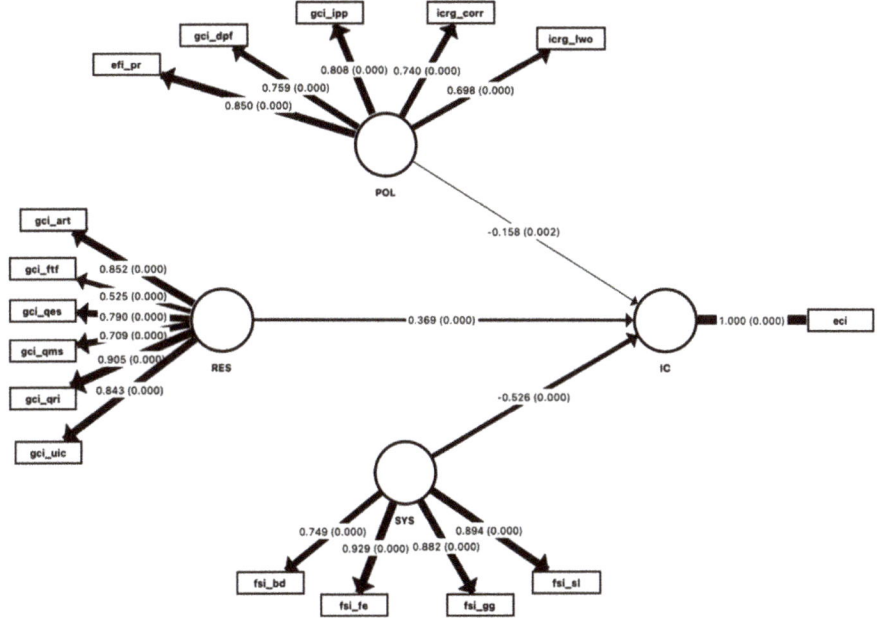

Figure 3. Model results. Source: Results from SmartPLS software 3.3.3.

Our findings are in line with the evidence from the literature that suggests that political conditions (POL) may harm the way countries compete in the international arena [19,97–103]. In the case of the analyzed emerging economies, property rights, diversion of public funds, intellectual property protection, corruption, and law and order negatively affect international competitiveness. All the indicators measured are relevant, but the higher loads are in those related to property rights and corruption. Our analysis also shows that an adequate scientific and technological framework (RES) enhances the emerging economies' international competitiveness [104–107]. The endowment of research and training services, FDI, and technology transfer, quality of the education system, quality of STEM education, quality of research and scientific institutions, and university-industry collaboration are essential factors to compete internationally. In this case, the more relevant indicators are the quality of research and scientific institutions, availability of research and training services, and university-industry collaboration.

Systemic conditions deter international competitiveness. Structural extractive frameworks impede the development of conditions required for an adequate global competition insertion [108–111]. Emerging economies are constrained by brain drain, groups of grievance, factionalized elites, and state legitimacy, as shown in this study's results. The loadings in this construct show the relevance of factionalized elites and state legitimacy in the structural systemic conditions to compete.

5. Conclusions

Research in this field is challenging because the frequent changes in the research context and the significant shifts in formal and informal institutional environments in emerging economies require alternative analysis methods. PLS-SEM exploratory modeling can handle complex models and relaxes the demands on data and relationships' specification, making it very useful for this study.

The proposed model using SEM-PLS to estimate and evaluate the correlation between selected indicators and the proposed constructs to measure institutional quality shows that the independent latent variables explain a significant proportion of the dependent construct's variability.

The analysis shows that political conditions could harm emerging economies' ability to compete with complex products in the international market. As shown in Table A1, the median value of the proposed indicators is slightly inclined to low performance, which allows us to infer that a lower quality of political conditions harms the ability to compete internationally with complex products. The indicator that has the most negative effect is property rights, which is coherent. If the firms' knowledge is not protected, innovation and productive transformation are not encouraged. In the same path, the indicator with a less adverse effect is the diversion of public funds because it affects the competitive environment.

It is also evident that the STEM resources have slightly good performance, which confirms that an adequate infrastructure for science, technology, engineering, and mathematics fosters the countries' ability to develop more complex products. In this construct, the quality of the education system somewhat contributes to the economic complexity. The quality of research institutions is the most critical indicator of the economic complexity to compete internationally.

Finally, extractive systemic conditions, which means the state's capture by elites and delegitimization of the state, are critical impediments to compete for global markets. In this construct, state legitimacy has the worst impact. If the market cannot believe in the state, it will not be possible to transform the productive structure. Although the group of grievance indicator has a lesser negative effect, it is also a condition that harms the effective transformation required for more economic complexity.

The model results, analyzed employing the PLS-SEM method, confirm the literature findings regarding the institutional framework's role, measured by political, resources, and systemic conditions. This paper demonstrates the importance of institutions in fostering the competitive economic strength of emerging economies.

A way of action could be the strengthening of regulations to increase the property rights protection and control of the investment of public funds. This could lead to a better perception of the state's legitimacy, which would promote the research and development through the participation of different stakeholders, including academia, civil society, and research institutions.

6. Contributions and Limitations of This Study

This study contributes in various ways to the existing literature. First, it sheds light on the importance of analyzing the political conditions in emerging economies to compete in the global markets. Second, it highlights the negative effect of extractive systemic conditions on international competitiveness. Third, it confirms the importance of STEM resources to generate complex products to compete internationally. Finally, it shows the deployment of an alternative method to evaluate the intricate relationships between institutional quality and international competitiveness. PLS-SEM allowed us to explore emerging economies' conditions even under the limitations described below.

A limitation of the current study is the small number of observations (528) divided into five distinct regions. Another limitation of the research is that it only focused on a few selected indicators according to the literature reviewed. This research's limitations could be overlooked in the future by adding more constructs, variables, and observations. The paper can be enriched by adding intra-regional and inter-regional approaches to control by the occurrence of particular circumstances (i.e., informal institutions or economic development).

Author Contributions: Conceptualization, R.E.B.R. Methodology, R.E.B.R. Software, R.E.B.R. Validation M.I.B.C. and F.C.V. Formal analysis, R.E.B.R. and M.I.B.C. Writing—original draft preparation, R.E.B.R. Writing—review and editing, R.E.B.R., M.I.B.C., and F.C.V. Supervision, M.I.B.C. and F.C.V. All authors have read and agreed to the published version of the manuscript.

Funding: This research received no external funding.

Institutional Review Board Statement: Not applicable.

Informed Consent Statement: Not applicable.

Data Availability Statement: Restrictions apply to the availability of these data.

Conflicts of Interest: The authors declare no conflict of interest.

Appendix A

Table A1. Indicators descriptive statistics.

	Mean	Median	Min	Max	Standard Deviation	Excess Kurtosis	Skewness	Number of Observations
eci	0.113	0.124	−2.764	1.695	0.750	0.312	−0.298	528
efi_pr	44.068	40.000	5.000	90.000	17.468	0.135	0.388	528
fsi_bd	5.378	5.200	2.100	8.500	1.541	−0.942	0.126	528
fsi_fe	6.108	6.500	1.100	10.000	2.169	−0.846	−0.286	528
fsi_gg	6.361	6.300	3.000	10.000	1.785	−0.959	0.111	528
fsi_sl	6.302	6.500	1.600	9.500	1.742	−0.737	−0.399	528
gci_art	4.151	4.149	2.340	6.084	0.583	0.705	−0.150	528
gci_dpf	3.299	3.157	1.219	6.603	0.938	1.239	1.000	528
gci_ftf	4.698	4.754	2.477	6.092	0.587	0.203	−0.393	528
gci_ipp	3.631	3.600	1.629	6.160	0.833	0.256	0.503	528
gci_qes	3.609	3.554	2.092	5.881	0.725	0.267	0.509	528
gci_qms	3.986	4.125	1.876	6.082	0.891	−0.798	−0.252	528
gci_qri	3.901	3.877	2.178	5.934	0.647	−0.007	0.310	528
gci_uic	3.560	3.479	2.072	5.472	0.614	0.261	0.462	528
icrg_corr	2.437	2.500	0.500	4.500	0.668	1.052	0.469	528
icrg_lwo	3.535	4.000	1.000	5.000	1.039	−0.857	−0.382	528

Source: Results from SmartPLS software 3.3.3.

Table A2. Mean, STDEV, T-Values, p-Values, confidence intervals.

	Original Sample (O)	Sample Mean (M)	Standard Deviation (STDEV)	T Statistics (|O/STDEV|)	p Values	CI 2.5% Lower	CI 97.5% Upper
POL -> IC	−0.158	−0.154	0.052	3.035	0.002	−0.257	−0.054
RES -> IC	0.369	0.368	0.047	7.858	0.000	0.275	0.459
SYS -> IC	−0.526	−0.524	0.039	13.622	0.000	−0.600	−0.448

Source: Results from SmartPLS software 3.3.3.

Table A3. Outer Loadings: Mean, STDEV, T-Values, p-Values, confidence intervals.

	Original Sample (O)	Sample Mean (M)	Standard Deviation	T Statistics	p Values	CI 2.5% Lower	CI 97.5% Upper
eci <- IC	1	1	0		0	1	1
efi_pr <- POL	0.85	0.85	0.014	60.122	0	0.821	0.877
fsi_bd <- SYS	0.749	0.749	0.019	38.825	0	0.71	0.784
fsi_fe <- SYS	0.929	0.929	0.008	121.245	0	0.912	0.942
fsi_gg <- SYS	0.882	0.882	0.011	76.927	0	0.857	0.902

Table A3. *Cont.*

	Original Sample (O)	Sample Mean (M)	Standard Deviation	T Statistics	*p* Values	CI 2.5% Lower	CI 97.5% Upper
fsi_sl < -SYS	0.894	0.894	0.008	105.738	0	0.876	0.909
gci_art < -RES	0.852	0.851	0.013	63.767	0	0.823	0.875
gci_dpf < -POL	0.759	0.755	0.036	21.29	0	0.677	0.816
gci_ftf < -RES	0.525	0.524	0.042	12.523	0	0.435	0.6
gci_ipp < -POL	0.808	0.806	0.026	30.721	0	0.747	0.849
gci_qes < -RES	0.79	0.787	0.025	32.217	0	0.735	0.831
gci_qms < -RES	0.709	0.708	0.028	24.924	0	0.649	0.761
gci_qri < -RES	0.905	0.905	0.008	119.89	0	0.889	0.918
gci_uic < -RES	0.843	0.843	0.015	56.113	0	0.811	0.87
icrg_corr < -POL	0.74	0.738	0.032	23.271	0	0.67	0.794
icrg_lwo < -POL	0.698	0.697	0.029	23.855	0	0.636	0.75

Source: Results from SmartPLS software 3.3.3.

Table A4. Outer VIF values.

	VIF
eci	1.000
efi_pr	1.995
fsi_bd	1.464
fsi_fe	7.036
fsi_gg	4.096
fsi_sl	3.700
gci_art	2.674
gci_dpf	2.690
gci_ftf	1.317
gci_ipp	2.857
gci_qes	3.226
gci_qms	2.751
gci_qri	3.460
gci_uic	3.419
icrg_corr	1.715
icrg_lwo	1.475

Source: Results from SmartPLS software 3.3.3.

References

1. Mihailova, I.; Panibratov, A.; Latukha, M. Dismantling institutional complexity behind international competitiveness of emerging market firms. *Thunderbird Int. Bus. Rev.* **2020**, *62*, 77–92. [CrossRef]
2. Buitrago R., R.E.; Barbosa Camargo, M.I. Institutions, institutional quality, and international competitiveness: Review and examination of future research directions. *J. Bus. Res.* **2021**, *128*, 423–435. [CrossRef]
3. Buitrago R., R.E.; Barbosa Camargo, M.I. Home country institutions and outward fdi: An exploratory analysis in emerging economies. *Sustainability* **2020**, *12*, 10010. [CrossRef]
4. Xu, K.; Hitt, M.A.; Brock, D.; Pisano, V.; Huang, L.S.R. Country institutional environments and international strategy: A review and analysis of the research. *J. Int. Manag.* **2021**, *27*, 100811. [CrossRef]
5. Zhu, H.; Ma, X.; Sauerwald, S.; Peng, M.W. Home Country Institutions behind Cross-Border Acquisition Performance. *J. Manag.* **2019**, *45*, 1315–1342. [CrossRef]

6. Sun, S.L.; Peng, M.W.; Lee, R.P.; Tan, W. Institutional open access at home and outward internationalization. *J. World Bus.* **2015**, *50*, 234–246. [CrossRef]
7. Porter, M.E.; Sachs, J.D.; Schwab, K. *Global Competitiveness Report 2001–2002*; Oxford University Press: Oxford, UK, 2002; ISBN 019521837X.
8. Moon, H.C.; Rugman, A.M.; Verbeke, A. A generalized double diamond approach to the global competitiveness of Korea and Singapore. *Int. Bus. Rev.* **1998**, *7*, 135–150. [CrossRef]
9. OECD. *Technology and the Economy: The Key Relationships*; OECD: Paris, France, 1992.
10. Lee, L.; Petter, S.; Fayard, D.; Robinson, S. On the use of partial least squares path modeling in accounting research. *Int. J. Account. Inf. Syst.* **2011**, *12*, 305–328. [CrossRef]
11. Hair, J.; Sarstedt, M.; Pieper, T.M.; Ringle, C.M. The Use of Partial Least Squares Structural Equation Modeling in Strategic Management Research: A Review of Past Practices and Recommendations for Future Applications. *Long Range Plan.* **2012**, *45*, 320–340. [CrossRef]
12. Hair, J.F.; Astrachan, C.B.; Moisescu, O.I.; Radomir, L.; Sarstedt, M.; Vaithilingam, S.; Ringle, C.M. Executing and interpreting applications of PLS-SEM: Updates for family business researchers. *J. Fam. Bus. Strateg.* **2020**. [CrossRef]
13. Peng, D.X.; Lai, F. Using partial least squares in operations management research: A practical guideline and summary of past research. *J. Oper. Manag.* **2012**, *30*, 467–480. [CrossRef]
14. Kaufmann, L.; Gaeckler, J. A structured review of partial least squares in supply chain management research. *J. Purch. Supply Manag.* **2015**, *21*, 259–272. [CrossRef]
15. North, D.C. The New Institutional Economics. *J. Inst. Theor. Econ.* **1986**, *142*, 230–237.
16. North, D.C. Institutional Change: A Framework of Analysis. In *Institutional Change: Theory and Empirical Findings*; Routledge: London, UK, 1993; pp. 35–46.
17. Acemoglu, D.; Johnson, S.; Robinson, J.A. The Colonial Origins of Comparative Development: An Empirical Investigation. *Am. Econ. Rev.* **2001**, *91*. [CrossRef]
18. Acemoglu, D.; Johnson, S.; Robinson, J.A. Institutions and economic development. *Financ. Dev.* **2003**, *3*, 2301–2370.
19. Acemoglu, D.; Robinson, J.A. *Why Nations Fail: The Origins of Power, Prosperity and Poverty*; Crown Publishers: New York, NY, USA, 2012.
20. Hnatkovska, V.; Loayza, N. Volatility and Growth. In *Managing Economic Volatility and Crises: A Practitioner's Guide*; Aizenman, J., Pinto, B., Eds.; Cambridge University Press: Cambridge, UK, 2005; pp. 65–100. ISBN 9780511510755.
21. Ramey, G.; Ramey, V.A. Cross-Country Evidence on the Link between Volatility and Growth. *Am. Econ. Rev.* **1995**, *85*, 1138–1151.
22. North, D.C. *Institutions, Institutional Change and Economic Performance*; Cambridge University Press: Cambridge, UK, 1990; ISBN 9780521397346.
23. Acemoglu, D.; Johnson, S.; Robinson, J.A. Reversal of fortune: Geography and institutions in the making of the modern world income distribution. *Q. J. Econ.* **2002**, *117*, 1231–1294. [CrossRef]
24. Acemoglu, D.; Johnson, S.; Robinson, J.A. Chapter 6 Institutions as a Fundamental Cause of Long-Run Growth. In *Handbook of Economic Growth*; Elsevier: Amsterdam, The Netherlands, 2005; Volume 1, pp. 385–472. ISBN 9780444520418.
25. Acemoglu, D.; Johnson, S. Unbundling Institutions. *J. Political Econ.* **2005**, *113*, 949–995. [CrossRef]
26. Anca, H.D.B. Literature review of the evolution of competitiveness concept. *Ann. Univ. Oradea Econ. Sci.* **2012**, *21*, 41–46.
27. Fagerberg, J. International competitiveness. *Econ. J.* **1988**, *98*, 355–374. [CrossRef]
28. Krugman, P.R. Competitiveness: A Dangerous Obsession. *Foreign Aff.* **1994**, *73*, 28. [CrossRef]
29. Balassa, B. Trade Liberalisation and "Revealed" Comparative Advantage. *Manch. Sch.* **1965**. [CrossRef]
30. Ito, K.; Pucik, V. R&D spending, domestic competition, and export performance of Japanese manufacturing firms. *Strateg. Manag. J.* **1993**, *14*, 61–75. [CrossRef]
31. Hausmann, R.; Hwang, J.; Rodrik, D. What You Export Matters. *J. Econ. Growth* **2007**, *12*, 1–25. [CrossRef]
32. Costantini, V.; Mazzanti, M. On the green and innovative side of trade competitiveness? The impact of environmental policies and innovation on EU exports. *Res. Policy* **2012**, *41*, 132–153. [CrossRef]
33. Freeman, C. Technological infrastructure and international competitiveness. *Ind. Corp. Chang.* **2004**, *13*, 541–569. [CrossRef]
34. Amable, B.; Verspagen, B. The role of technology in market shares dynamics. *Appl. Econ.* **1995**. [CrossRef]
35. Amendola, G.; Dosi, G.; Papagni, E. The dynamics of international competitiveness. *Weltwirtsch. Arch.* **1993**. [CrossRef]
36. Ervits, I.; Zmuda, M. A cross-country comparison of the effects of institutions on internationally oriented innovation. *J. Int. Entrep.* **2018**, *16*, 486–503. [CrossRef]
37. Guerrieri, P.; Meliciani, V. International Competitiveness in Producer Services. 2004. Available online: https://papers.ssrn.com/sol3/papers.cfm?abstract_id=521445 (accessed on 12 January 2021).
38. Hollingsworth, J.R. Doing institutional analysis: Implications for the study of innovations. *Rev. Int. Political Econ.* **2000**, *7*, 595–644. [CrossRef]
39. Ingram, P.; Silverman, B.S. Introduction: The new institutionalism in strategic management. In *The New Institutionalism in Strategic Management*; Emerald (MCB UP): Bingley, UK, 2002; Volume 19, pp. 1–30. ISBN 978-0-7623-0903-0.
40. Peng, M.W.; Wang, D.Y.L.L.; Jiang, Y. An institution-based view of international business strategy: A focus on emerging economies. *J. Int. Bus. Stud.* **2008**, *39*, 920–936. [CrossRef]
41. Porter, M.E. *Competitive Strategy*; Free Press: New York, NY, USA, 1990; ISBN 0029253608.

42. Porter, M.E.; Van der Linde, C. Toward a new conception of the environment-competitiveness relationship. *J. Econ. Perspect.* **1995**, *9*, 97–118. [CrossRef]
43. Wan, W.P.; Hoskisson, R.E. Home country environments, corporate diversification strategies, and firm performance. *Acad. Manag. J.* **2003**, *46*, 27–45. [CrossRef]
44. Graham, C.; Naim, M. The political economy of institutional reform in Latin America. In *Beyond Tradeoffs: Market Reforms and Equitable Growth in Latin America*; Birdsall, N., Graham, C., Sabot, R.H., Eds.; Brookings Institution Press: Washington, DC, USA, 1998; p. 376.
45. North, D.C. A transaction cost theory of politics. *J. Theor. Polit.* **1990**, *2*, 355–367. [CrossRef]
46. Peng, M.W.; Heath, P.S. The growth of the firm in planned economies in transition: Institutions, organizations, and strategic choice. *Acad. Manag. Rev.* **1996**. [CrossRef]
47. Rodrik, D. Democracies pay higher wages. *Q. J. Econ.* **1999**. [CrossRef]
48. Blume, L.; Müller, J.; Voigt, S.; Wolf, C. The economic effects of constitutions: Replicating-and extending-persson and tabellini. *Public Choice* **2009**. [CrossRef]
49. Besley, T.; Persson, T.; Sturm, D.M. Political competition, policy and growth: Theory and evidence from the US. *Rev. Econ. Stud.* **2010**. [CrossRef]
50. Cuervo-Cazurra, A.; Alfonso Dau, L. *Multinationalization in Response to Reforms*; Academy of Management: Briarcliff Manor, NY, USA, 2009.
51. Cuervo-Cazurra, A. Corruption in international business. *J. World Bus.* **2016**, *51*, 35–49. [CrossRef]
52. Gordon, R.J. *Five Puzzles in the Behavior of Productivity, Investments and Innovation*; National Bureau of Economic Research: Cambridge, MA, USA, 2004.
53. Lim, W.H.; Moon, J.T. Korea as a Knowledge Economy: Assessment and Lessons. In *Designing New Economic Framework*; Korea Development Institute (KDI): Yeongi-gun, Korea, 2004.
54. OECD. *The Source of Economic Growth in OECD Countries*; OECD: Paris, France, 2003.
55. OECD. *The New Economy: Beyond the Hype*; OECD: Paris, France, 2001.
56. Warhuus, J.P.; Basaiawmoit, R.V. Entrepreneurship education at Nordic technical higher education institutions: Comparing and contrasting program designs and content. *Int. J. Manag. Educ.* **2014**. [CrossRef]
57. Winkler, C.; Troudt, E.; Schweikert, C.; Schulman, S. Infusing Business and Entrepreneurship Education into a Computer Science Curriculum—A Case Study of the Stem Virtual Enterprise. *J. Bus. Entrep.* **2015**, *27*, 1–21.
58. Acemoglu, D.; Johnson, S. Institutions, Corporate Governance. *Corp. Gov. Cap. Flows Glob. Econ.* **2003**, *1*, 327.
59. Acemoglu, D.; Robinson, J.A. Persistence of Power, Elites, and Institutions. *Am. Econ. Rev.* **2008**, *98*, 267–293. [CrossRef]
60. Li, S. *Managing International Business in Relation-Based versus Rule-Based Countries*; Business Expert Press: New York, NY, USA, 2009.
61. Kramer, R.M. Trust and distrust in organizations: Emerging perspectives, enduring questions. *Annu. Rev. Psychol.* **1999**, *50*, 569–598. [CrossRef]
62. Lin, X.; Wang, C.L. Enforcement and performance: The role of ownership, legalism and trust in international joint ventures. *J. World Bus.* **2008**. [CrossRef]
63. Muethel, M.; Bond, M.H. National context and individual employees' trust of the out-group: The role of societal trust. *J. Int. Bus. Stud.* **2013**. [CrossRef]
64. Li, J.; Moy, J.; Lam, K.; Chris Chu, W.L. Institutional pillars and corruption at the societal level. *J. Bus. Ethics* **2008**. [CrossRef]
65. Meyer, K.E. Institutions, Transaction Costs, and Entry Mode Choice in Eastern Europe. *J. Int. Bus. Stud.* **2001**, *32*, 357–367. [CrossRef]
66. The Fund for Peace Fragile States Index. Available online: https://fragilestatesindex.org/ (accessed on 12 January 2021).
67. World Economic Forum Global Competitiveness Index. Available online: https://www.weforum.org/reports/ (accessed on 12 January 2021).
68. The PRS Group International Country Risk Guide (ICRG). Available online: https://www.prsgroup.com/explore-our-products/international-country-risk-guide (accessed on 12 January 2021).
69. The Heritage Foundation Index of Economic Freedom. Available online: https://www.heritage.org/index/ (accessed on 12 January 2021).
70. Hidalgo, C.A.; Hausmann, R. The building blocks of economic complexity. *Proc. Natl. Acad. Sci. USA* **2009**, *106*, 10570–10575. [CrossRef]
71. Sheng, L.; Yang, D.T. Expanding export variety: The role of institutional reforms in developing countries. *J. Dev. Econ.* **2016**, *118*, 45–58. [CrossRef]
72. Zhu, S.; Fu, X. Drivers of Export Upgrading. *World Dev.* **2013**, *51*, 221–233. [CrossRef]
73. Hausmann, R.; Hidalgo, C.A. *The Atlas of Economic Complexity*; The MIT Press: Cambridge, MA, USA, 2011; ISBN 9780615546629.
74. Hair, J.; Risher, J.J.; Sarstedt, M.; Ringle, C.M. When to use and how to report the results of PLS-SEM. *Eur. Bus. Rev.* **2019**, *31*, 2–24. [CrossRef]
75. Hair, J.; Hult, G.T.; Ringle, C.; Sarstedt, M. *A Primer on Partial Least Squares Structural Equation Modeling (PLS-SEM)*; Sage Publications: Thousand Oaks, CA, USA, 2016; ISBN 9781483377445.
76. Kock, N. Factor-based structural equation modeling with WarpPLS. *Australas. Mark. J.* **2019**. [CrossRef]
77. Monecke, A.; Leisch, F. SemPLS: Structural equation modeling using partial least squares. *J. Stat. Softw.* **2012**, *48*. [CrossRef]

78. Kock, N. Non-normality propagation among latent variables and indicators in PLS-SEM simulations. *J. Mod. Appl. Stat. Methods* **2016**, *15*, 299–315. [CrossRef]
79. Busu, C.; Busu, M. Research on the factors of competition in the green procurement processes: A case study for the conditions of romania using pls-sem methodology. *Mathematics* **2021**, *9*, 16. [CrossRef]
80. Chung, K.C.; Liang, S.W.J. Understanding factors affecting innovation resistance of mobile payments in Taiwan: An integrative perspective. *Mathematics* **2020**, *8*, 1841. [CrossRef]
81. Palos-Sanchez, P.; Saura, J.R.; Ayestaran, R. An Exploratory Approach to the Adoption Process of Bitcoin by Business Executives. *Mathematics* **2021**, *9*, 355. [CrossRef]
82. Dunning, J.H. A new Zeitgeist for international business activity and scholarship. *Eur. J. Int. Manag.* **2007**, *1*, 278–301. [CrossRef]
83. Dunning, J.H. New directions in international-business research: A personal viewpoint. *Res. Glob. Strateg. Manag.* **2008**, *14*, 247–257.
84. Aharoni, Y.; Brock, D.M. International business research: Looking back and looking forward. *J. Int. Manag.* **2010**, *16*, 5–15. [CrossRef]
85. Hair, J.F.; Sarstedt, M.; Ringle, C.M.; Mena, J.A. An assessment of the use of partial least squares structural equation modeling in marketing research. *J. Acad. Mark. Sci.* **2012**, *40*, 414–433. [CrossRef]
86. Richter, N.F.; Sinkovics, R.R.; Ringle, C.M.; Schlägel, C. A critical look at the use of SEM in international business research. *Int. Mark. Rev.* **2016**, *33*, 376–404. [CrossRef]
87. Ringle, C.M.; Wende, S.; Becker, J.-M. *SmartPLS 3*; SmartPLS GmbH: Bönningstedt, Germany, 2015.
88. Jarvis, C.B.; Mackenzie, S.B.; Podsakoff, P.M.; Giliatt, N.; Mee, J.F. A Critical Review of Construct Indicators and Measurement Model Misspecification in Marketing and Consumer Research. *J. Consum. Res.* **2003**, *30*, 199–218. [CrossRef]
89. Coltman, T.; Devinney, T.M.; Midgley, D.F.; Venaik, S. Formative versus reflective measurement models: Two applications of formative measurement. *J. Bus. Res.* **2008**, *61*, 1250–1262. [CrossRef]
90. Henseler, J.; Ringle, C.M.; Sarstedt, M. A new criterion for assessing discriminant validity in variance-based structural equation modeling. *J. Acad. Mark. Sci.* **2015**, *43*, 115–135. [CrossRef]
91. Kline, R.B. *Principles and Practice of Structural Equation Modeling*; The Guilford Press: New York, NY, USA, 2011; Volume 20, ISBN 9781606238776.
92. Hair, J.F.; Sarstedt, M.; Ringle, C.M.; Gudergan, S.P. *Advanced Issues in Partial Least Squares Structural Equation Modeling*; Sage Publications, Inc.: Thousand Oaks, CA, USA, 2018; ISBN 9781626239777.
93. Hair, J.F.; Howard, M.C.; Nitzl, C. Assessing measurement model quality in PLS-SEM using confirmatory composite analysis. *J. Bus. Res.* **2020**, *109*, 101–110. [CrossRef]
94. Chin, W.W. The partial least squares approach to structural equation modeling. In *Modern Methods for Business Research*; Marcoulides, G.A., Ed.; Lawrence Erlbaum Associates: Mahwah, NJ, USA, 1998; pp. 295–336.
95. Henseler, J.; Sarstedt, M. Goodness-of-fit indices for partial least squares path modeling. *Comput. Stat.* **2013**, *28*, 565–580. [CrossRef]
96. Chin, W.W.; Dibbern, J. An Introduction to a Permutation Based Procedure for Multi-Group PLS Analysis: Results of Tests of Differences on Simulated Data and a Cross Cultural Analysis of the Sourcing of Information System Services Between Germany and the USA. In *Handbook of Partial Least Squares*; Springer: Berlin/Heidelberg, Germany, 2010; pp. 171–193.
97. Downing, J.A. Dimensions of Competitive Advantage. *J. New Bus. Ideas Trends* **2018**, *16*, 1–8.
98. Gedefaw Birhanu, A.; Wezel, F.C. The competitive advantage of affiliation with business groups in the political environment: Evidence from the Arab Spring. *Strateg. Organ.* **2020**, 1–23. [CrossRef]
99. Yasar, M.; Paul, C.J.M.; Ward, M.R. Property Rights Institutions and Firm Performance: A Cross-Country Analysis. *World Dev.* **2011**, *39*, 648–661. [CrossRef]
100. Miles, M.P.; Darroch, J. Competitive Advantage. In *Wiley Encyclopedia of Management*; John Wiley & Sons, Ltd.: Chichester, UK, 2015; pp. 1–7.
101. Useche, A.J.; Reyes, G.E. Corruption, competitiveness and economic growth: Evidence from Latin American and Caribbean countries 2004–2017. *J. Glob. Compet. Gov.* **2019**, *14*, 95–115. [CrossRef]
102. Ulman, S.-R. The Impact of the National Competitiveness on the Perception of Corruption. *Procedia Econ. Financ.* **2014**, *15*, 1002–1009. [CrossRef]
103. Hegemann, P.; Berumen, S.A. A neoschumpeterian review of the impact of corruption on competitiveness and foreign direct investmentUna revisión neoschumpeteriana del impacto de la corrupción sobre la competitividad y la inversión extranjera directa. *Papeles Eur.* **2011**, *22*, 39–60. [CrossRef]
104. Takala, J.; Koskinen, J.; Liu, Y.; Tas, M.S.; Muhos, M. Validating knowledge and technology effects to operative sustainable competitive advantage. *Manag. Prod. Eng. Rev.* **2013**, *4*, 45–54. [CrossRef]
105. Argote, L.; Ingram, P. Knowledge transfer: A basis for competitive advantage in firms. *Organ. Behav. Hum. Decis. Process.* **2000**, *82*, 150–169. [CrossRef]
106. Bilgihan, A.; Wang, Y. Technology induced competitive advantage: A case of US lodging industry. *J. Hosp. Tour. Technol.* **2016**, *7*, 37–59. [CrossRef]
107. Bhatt, G.D.; Grover, V. Types of information technology capabilities and their role in competitive advantage: An empirical study. *J. Manag. Inf. Syst.* **2005**, *22*, 253–277. [CrossRef]

108. Tadei, F. Measuring Extractive Institutions: Colonial Trade and Price Gaps in French Africa. *Eur. Rev. Econ. Hist.* **2017**. [CrossRef]
109. Mizuno, N.; Naito, K.; Okazawa, R. Inequality, extractive institutions, and growth in nondemocratic regimes. *Public Choice* **2016**, *170*, 115–142. [CrossRef]
110. Vanino, E.; Lee, S. Extractive institutions in non-tradeable industries. *Econ. Lett.* **2018**, *170*, 10–13. [CrossRef]
111. Bowness, J. Foreign Direct Investment and Extractive Institutions Lessons from Latin America. *Potentia J. Int. Aff.* **2019**, *10*, 36–49. [CrossRef]

Article

Effect of Quality Lean Manufacturing Tools on Commercial Benefits Gained by Mexican Maquiladoras

Jorge Luis García Alcaraz [1,*], Flor Adriana Martínez Hernández [1], Jesús Everardo Olguín Tiznado [2], Arturo Realyvásquez Vargas [3], Emilio Jiménez Macías [4] and Carlos Javierre Lardies [5]

[1] Department of Industrial Engineering and Manufacturing, Autonomous University of Ciudad Juárez, Juarez 32310, Mexico; al194607@alumnos.uacj.mx
[2] Faculty of Engineering, Architecture and Design, Autonomous University of Baja California, Ensenada 22860, Mexico; jeol79@uabc.edu.mx
[3] Department of Industrial Engineering, Instituto Tecnológico de Tijuana, Tijuana 22414, Mexico; arturo.realyvazquez@tectijuana.edu.mx
[4] Department of Electric Engineering, University of La Rioja, 26006 La Rioja, Spain; emilio.jimenez@unirioja.es
[5] Department of Mechanical Engineering, University of Zaragoza, 50009 Zaragoza, Spain; sabicjl@unizar.es
* Correspondence: jorge.garcia@uacj.mx

Citation: García Alcaraz, J.L.; Martínez Hernández, F.A.; Olguín Tiznado, J.E.; Realyvásquez Vargas, A.; Jiménez Macías, E.; Javierre Lardies, C. Effect of Quality Lean Manufacturing Tools on Commercial Benefits Gained by Mexican Maquiladoras. *Mathematics* 2021, *9*, 971. https://doi.org/10.3390/math9090971

Academic Editor: María del Carmen Valls Martínez

Received: 25 March 2021
Accepted: 23 April 2021
Published: 26 April 2021

Publisher's Note: MDPI stays neutral with regard to jurisdictional claims in published maps and institutional affiliations.

Copyright: © 2021 by the authors. Licensee MDPI, Basel, Switzerland. This article is an open access article distributed under the terms and conditions of the Creative Commons Attribution (CC BY) license (https://creativecommons.org/licenses/by/4.0/).

Abstract: Companies implement lean manufacturing (LM) tools in their production processes to reduce waste; however, it is difficult to quantify the effect on benefits gained after their implementation. This article proposes a structural equations model (SEM) that relates three LM tools associated with quality as total quality management (*TQM*), waste, and right first time (*RFT*) as independent variables associated with commercial benefits gained as a dependent variable. Those four variables were related by six hypotheses that were validated with information from 169 responses to a survey applied to the Mexican maquiladora industry. Partial least squared was used to validate the hypotheses as direct effects. The sum of indirect and total effects was also estimated, and a sensitivity analysis was developed for relationships between variables. Findings indicate that *TQM* directly affects waste reduction, drives doing *RFT*, and directly and indirectly affects the commercial benefits gained.

Keywords: lean manufacturing; quality management; commercial performance; wastes; DIRFT

1. Introduction

The industry in a nation plays an essential role in its development, and therefore its administrative and operational practices are often studied. Specifically, programs have focused on improving national industries in Mexico such as the maquiladora industry. Foreign companies establish subsidiaries in Mexican territory, where manufacturing activities require a high level of workforce. These maquiladoras are closer to the world's largest market, the United States of America [1], and take advantage of various tariff benefits when exporting their products due to the free trade agreements that Mexico has with the United States of America and Canada, which are low or preferential. These tariff rates are different if these companies export their products from the headquarters in their country of origin.

Currently, there are 5153 maquiladora companies established in Mexican territory, 495 of which are in the State of Chihuahua, representing 9.6% of the national total and specifically in Ciudad Juárez, there are 332, representing 6.44% of the national total and 67.1% of the state total. Maquiladora companies directly provide 2,689,209 direct jobs nationwide, 477,480 corresponding to the State of Chihuahua and 361,619 to Ciudad Juárez; hence the importance of studying this sector [2].

Along with maquiladoras being established in Mexico, some philosophies, techniques, and tools applied to production systems are arriving. One of the most important is lean manufacturing (LM), which consists of tools that help to identify and eliminate operations that do not add value to the product, service, or process. However, as Alfieri, et al. [3]

mentioned, the concept is still unclear after decades of application. Nevertheless, there is consensus that LM is focused on *Wastes* elimination, inventories, and space reduction, generating robust production systems, and providing agility and flexibility.

LM tools are classified according to their focus on the production process; some are focused on the production system's operational stability, others have focused on the flow of materials and on quality [3]. Together, all are focused on obtaining a more significant reduction of waste and guaranteeing product quality at a lower cost, having a customer focus with small production batches, and providing high safety and morale in workers [4].

Companies that apply LM tools hope to obtain some operational, economic, commercial, social, or environmental benefit. Many studies have been carried out in this regard over the last two decades. For example, Melton [5] in 2005 had already considered what lean thinking offered to industries, while Rajenthirakumar and Shankar [6] analyzed the benefits that LM tools provide to a company that manufactures durable products. Recently, Islam [4] examined whether the manufacturer or suppliers obtained more benefits by applying LM to each other, while Palange and Dhatrak [3] concluded that LM was vital to achieving the productivity rates of the companies. However, Hao, et al. [7] stated that it is often challenging to award to LM or some of its tools the benefits obtained since several are implemented simultaneously in the production lines.

Other studies have focused on identifying the benefits that companies gain from a specific LM tool. For example, García-Alcaraz et al. [8] reported the benefits obtained by just-in-time (JIT) in operational performance; Singh et al. [9] analyze the impact of Total Quality Management (*TQM*) on organizational performance, and Sahoo and Yadav [10] report the benefits of Total Productive Maintenance (TPM) and *TQM* on operational performance.

Specifically, the commercial and economic effects that LM has on companies has also been reported. For example, as early as the past decade, Meade, et al. [11] simulated the financial effects obtained from LM implementation to determine its viability; Fullerton, et al. [12] related LM tools to operational performance and how to generate financial performance, and recently, Shashi et al. [13] identified the economic benefits obtained from LM in small and medium enterprises in India.

However, there are also barriers in LM implementation that avoid receiving the financial and *Commercial benefits*. For example, Elkhairi et al. [14] mentioned that the most common barriers were a lack of planning, experience, managerial commitment, misunderstanding of LM, lack of resources, and resistance to change; while Abu et al. [15] indicated that companies without the technical knowledge to implement LM sometimes subcontract external consultants who do not know the real needs of the company. Then, there are barriers such as lack of training and education, strong unions that do not accept improvement changes, and long and complicated hierarchical structures in which communication is slow with information lost.

In the maquiladora industry, one of the most used LM tools is *TQM*, which focuses on generating awareness at the organizational level about the importance of developing products and services that meet customer expectations [16]. Although some authors declare that *TQM* is a LM tool such as Lynn [17] and Sisternas [18], others such as Salleh et al. [19] consider that *TQM* and LM are different techniques. In this research, the authors decided to integrate *TQM* as a LM tool because that is the industrial practice in the maquiladora sector.

Although the relationship between LM tools and company performance makes common sense, very few authors have studied it. For example, York and Miree [20] reported the relationship between *TQM* and financial performance; Singh, Kumar, and Singh [9] analyzed *TQM* and its impact on the general performance of companies; Moitra [21] examined the role of human resources in the success of *TQM*; Green et al. [22] reported that *TQM* impacted environmental sustainability; and García-Alcaraz et al. [23] analyzed the *TQM* structure and its effect on customer satisfaction.

Although *TQM* is a complete LM tool, it is sometimes studied separately and requires support from other tools that focus on eliminating *Wastes*, and *RFT*, which guarantees obtaining *Commercial benefits* that mean the best economic performance. The relationship of these variables makes common sense, since, if there are no *Wastes* in resources and the products are produced *Right first time*, it means that there are no rework or losses in the raw material, and, therefore, greater benefits and reputation are obtained for the company. However, the relationships between these variables have not been quantified, so this article presents a structural equation model (SEM) with three independent variables associated with quality such as *TQM*, *Wastes*, and *RFT* related to each other and to the *Commercial benefits* that companies obtain.

Quantifying the relationships between *TQM*, *Wastes*, *RFT*, and *Commercial benefits*, our findings will support the managers' decision-making process in maquiladoras to identify those essential activities for achieving the commercial objectives. This will allow them to focus on human, material, and economic resource management.

2. Hypothesis and Literature Review

2.1. TQM

TQM's programs are a set of techniques, procedures, and methods that serve as the basis for ensuring quality in products and services, and proof of this is that many standards govern them such as ISO-9001 for management systems, ISO 19011 for audits of these systems, and ISO16946 for quality in the automotive sector, among others. However, standards alone do not guarantee quality as they are just guidelines, and their success depends on the people who implement them [21], indicating that several critical success factors (CSFs) of *TQM* are required. Sreedharan et al. [24] noted that the main ones were management engagement, communication, training, customer focus, and organizational culture.

Therefore, *TQM* requires dissemination throughout the organizational structure, and from receiving the raw material up until it becomes a finished product, always focused on the customer. *TQM* practices gain knowledge that companies must manage to grow and solve future problems [25] as this tool acts as a facilitator [26].

2.2. Waste

A production process that generates *Wastes* has poor quality and is not standardized [27], concluding that some LM-based tools have not been applied correctly. Waste is any element within the production process that does not add value to the final product, but adds cost [28]. There are seven types of *Waste* reported in the literature: overproduction, waiting, transportation, inadequate processes, unnecessary inventory, excessive movement of elements, and excess defects [29], which represent economic losses for the company, and value stream mapping is a tool that can help to discover them [30].

Given that eliminating all *Wastes* and increasing quality cost money, these two variables are related because a product generated with *Wastes* in the production process and defects is usually cheaper than those that do not have them. Therefore, there must be a balance among them (quality and *Wastes*) [31], and that is not a new problem. Womack and Jones [32] indicated that they considered *Waste* reduction to be the basis for generating *Economic benefits* for the company. Then, the following hypothesis is proposed.

Hypothesis 1 (H1). *TQM implementation in the maquiladora industry has a direct impact on waste reduction.*

2.3. Doing It RFT

As the name implies, *RFT* emphasizes that production processes are correct. Its implementation requires analysis of the available quality indices, focusing on identifying problems associated with human error to propose improvement strategies systematically. This technique has its origin with Philip B. Crosby, who presented the motto of "zero defects" and *RFT*. However, today, many applications of *RTF* are seen in the medical and health areas, where errors can be fatal causes.

Industrial applications of *RFT* have been observed in reports by Moshiri et al. [33] in additive manufacturing processes in Industry 4.0, and Eldessouky et al. [34] proposed a contextual framework for *RFT* when the process generates *waste* and the raw material is expensive. Therefore, there is no doubt that *RFT* is part of *TQM*'s programs. The number of defects in a product is a critical metric [35] as well as the number of accidents that occur [36], indicating that the company has an efficient system for solving problems in its production process through its leaders [37] and *TQM*, so the following hypothesis was proposed.

Hypothesis 2 (H2). *TQM implementation in the maquiladora industry has a direct and positive impact on RFT.*

Another critical aspect of *RFT* is that *Wastes* are not allowed from materials when they are expensive. If companies focus on eliminating *Wastse* in their production process, then the products may have the quality according to customer specifications [29,38]. In other words, if the number of products requiring rework is minimized due to programs focused on improving the process, then there is a greater likelihood of getting it right the first time. Additionally, if production methods have reduced transport times and operators have been trained, then the likelihood of getting it right the first time increases [39]. Besides, other quality-focused tools that support *TQM* such as six sigma help eliminate *Wastes* [40]. Therefore, this research proposes the following hypothesis:

Hypothesis 3 (H3). *Wastse implementation in the maquiladora industry has a direct and positive impact on RFT reduction in the maquiladora industry.*

2.4. Commercial Benefits

Commercial benefits are the means of generating economic income in maquiladora companies. This is why managers strive to achieve them, which is reflected in reducing the costs of acquiring materials, using electricity, the rate paid for treatment and dumping of *Wastes*, and penalties for environmental fails. These problems affect the return on sales and investment and economic benefits, among others.

García-Alcaraz, Montalvo, Sánchez-Ramírez, Avelar-Sosa, Saucedo, and Alor-Hernández [23] declared that the organizational structure for *TQM* was the basis for achieving customer satisfaction and achieving a better market coverage. That indicates that the quality of a company's products is the first thing that customers evaluate. Magdy and Tamer [41] stated that *TQM* is directly linked to the performance of companies, as it is how value is added to services. However, as indicated by Sila [35], the performance of companies in implementing *TQM* may be affected by the sector and the country; that is, by religious and cultural aspects. Then, the following hypothesis was proposed.

Hypothesis 4 (H4). *TQM implementation in maquiladora industries has a direct and positive impact on Commercial benefits obtained.*

The benefits obtained from quality programs are not based on luck; they are the product of a broad cultural and philosophical deployment in this concept. For example, Kappelman and Prybutok [42] and Kassicieh and Yourstone [43] indicated that staff training is vital in avoiding mistakes, which was recently confirmed by García-Alcaraz et al. [44]. Trained operators will make fewer mistakes in their work, and therefore there will be low rework rates with savings in energy and labor, among others. In conclusion, there are protocols to follow to reduce *Wastes* that comply with the standards in the production process, thus allowing for obtaining social and commercial acceptance to take place before the customers. Then, the following hypothesis was proposed.

Hypothesis 5 (H5). *RFT implementation in the maquiladora industry has a direct and positive impact on the Commercial benefits they obtain.*

Quality could not exist if there is *Wastes* in the production process. For example, if there is a significant transport in the materials, then the cost is increased, or if people make a lot of movement, they will be unproductive [45] and unsustainable [44]. Similarly, proper inventory management favors a circulation of the company's material and economic goods, lowering the storage costs and final price. Sreedharan, Sunder, and Raju [24] saw the elimination of *Wastes* as a key to LM's success, while Sahoo and Yadav [10] mentioned that the elimination of leisure time from machines in the production system facilitated deliveries on time and in the quantity and quality required. As such, the following hypothesis was proposed.

Hypothesis 6 (H6). *Wastes implementation in the maquiladora industry has a direct and positive impact on the Commercial benefits they obtain.*

Figure 1 graphically represents all of the proposed hypotheses.

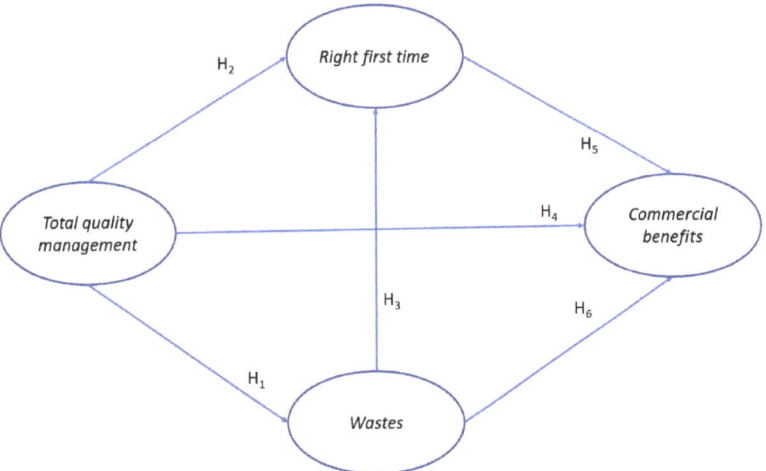

Figure 1. Proposed model.

3. Methodology

3.1. Questionnaire Design

To validate the hypotheses statistically in Figure 1, a questionnaire was designed to obtain information from the maquiladora companies. It started with a literature review in scientific databases to learn about the activities required to implement the LM tools and their benefits. This literature review corresponded to rational validation.

These activities allowed for the validation of the level of implementation of the LM tools analyzed. The first draft of the questionnaire was generated and presented to seven managers of regional maquiladora companies and four academics to obtain validation by experts. After two rounds with the experts, a final questionnaire containing *TQM*, *RFT*, *Wastes*, and *Commercial benefits* with five, seven, six, and seven items, respectively. Such a questionnaire had to be answered on a six-point Likert scale, where one means that the activity is not performed or the profit is never obtained, while six indicates that the activity is always performed or that the benefit is always obtained. The final questionnaire with all LM tools is available in Martínez Hernández and García Alcaraz [46].

The questionnaire consists of three sections: Section 1 seeks to identify the demographic data of the respondent such as gender, years in the position, and number of employees in the company, among others. Section 2 contains the three tools and their activities, and finally, Section 3 evaluates the *Commercial benefits* obtained and thanks the respondent for their participation.

3.2. Application of the Questionnaire

With support from IMMEX (Manufacturing Industry, Maquiladora and Export Services), only one manager or engineer from each regional maquiladora was invited to answer the questionnaire. The questionnaire was designed on an online platform, and the electronic link was shared with potential respondents during the June–August 2020 period.

As inclusion criteria, respondents had to have at least one year of experience in their job position, were active, had participated in at least two quality projects completed to ensure that they understood the benefits gained, and only one person per company could respond to the survey. Thus, sampling was initially stratified, focusing only on a particular sector of managers or engineers within the company. Then, the snowball technique was used as the respondent was asked about other colleagues in different companies who could answer the questionnaire [47].

An email was sent with the electronic link to prospective respondents inviting them to participate in the investigation, and 15 days were given to receive a response. If an answer was not received, a second email was sent reminding them of the invitation, and if the answer was not received in another 15 days, that case was discarded.

3.3. Obtaining Information and Debugging It

In early September 2020, a database of the platform on which the questionnaire is broadcast is downloaded, read in SPSS software v.25®. Database debugging consists of the following activities [48]:

- Identification of uncommitted respondents. The standard deviation was obtained from each case and those where it was less than 0.5 were omitted from the analysis.
- Identification of missing values. If the percentage was lower than 10%, they were replaced by the median, but if the rate was higher, then that case was removed from the analysis.

Extreme values identification. The median replaced the standardized values greater than four in absolute value.

3.4. Descriptive Analysis of the Sample and Items

Crosstables were built to describe the sample with information from the demographic section. The median was reported as a central trend measure for describing the items since the information was on a Likert scale and the interquartile range as a deflection measure. High values in the median indicate that the activity is always performed or that a profit is always obtained. In contrast, low values in the median suggest that the activity is not executed or that the benefit is not obtained.

3.5. Validation of Latent Variables

With a debugged database, each latent variable in the model was validated, and the following indexes were used [49]:

- Cronbach alpha index and composite reliability index to measure the reliability and internal consistency of variables, Values greater than 0.7 were accepted, being iteratively obtained.
- R-squared and adjusted R-squared to measure the parametric predictive validity. Values greater than 0.02 were accepted and significantly associated with p-value.
- Q-square to measure non-parametric predictive validity. This should be similar to the R-square value.
- Average extracted variance (AVE) to measure the discriminant validity of each latent variable, which must be greater than 0.5.
- Variance inflation indexes (VIFs) to measure collinearity in each construct, which should be lower than 3.3.

This paper contains Supplementary Materials reporting the T ratios for the path coefficients and their confidence intervals; factor loadings, their T ratios and confidence intervals; and PLSc reliabilities (Dijkstra's rho), PLSc loadings, *p*-values for loadings (one-tailed), *p*-values for loadings (two-tailed), among others.

3.6. Structural Equation Modeling

The structural equation modeling methodology was chosen to validate the hypotheses in Figure 1 and the partial least squares technique integrated into WarpPLS v7.0® software was used. Latent variables already validated were integrated into the model, and the following indexes with a 95% confidence level were reviewed before being analyzed [48]:

- Average path coefficient (APC) to measure the dependency between latent variables. *p*-values must be less than 0.05.
- Average R-squared (ARS) and average adjusted R-squared (AARS) to measure predictive validity, and associated *p*-values lower than 0.05.
- Average block VIF (AVIF) and average full collinearity VIF (AFVIF) to measure collinearity between variables. Values lower than 3.3 were accepted.
- Tenenhaus GoF (GoF) to measure the fit of the data to the model, which must be greater than 0.36.

To validate the scenarios raised, the direct effects between the variables analyzed at a confidence level of 95% were calculated, so a standardized value was obtained, β, as a dependency measure. The null hypothesis H_0 was tested, $\beta = 0$, against the alternative hypothesis H_1 that $\beta \neq 0$. If it was statistically proven that $\beta = 0$, then it was concluded that there is no relationship between the analyzed variables. Otherwise, if $\beta \neq 0$, then it was supposed that there is a relationship between variables.

However, indirect effects were also calculated, which occur through a mediating variable and can have two or more segments, which help to understand the effects that do not directly occur. Finally, the total effects was estimated, which was the sum of direct and indirect effects. Indirect and total effects were estimated under the same assumptions of direct effects. Finally, each of the estimated effects was associated with the effect size (ES) as a measure of the variance explained in the dependent variable, which helps determine the level of importance they have.

3.7. Sensitivity Analysis

WarpPLS analysis uses standardized values of latent variables, so it is possible to obtain the probabilities of scenarios and calculate the conditional probabilities. Specifically, the sensitivity analysis evaluates the probability of simultaneously finding the two related variables in their low scenario, high scenario, or their combinations (i.e., $P(Z_i > 1)$ and $P(Z_d > 1)$, $P(Z_i > 1)$ and $P(Z_d < -1)$, $P(Z_i < -1)$ and $P(Z_d > 1)$ y $P(Z_i < -1)$ and $P(Z_d < -1)$. Similarly, conditional probabilities for $P(Z_d > 1)/P(Z_i > 1)$, $P(Z_d > 1)/P(Z_i < -1)$, $P(Z_d < -1)/P(Z_i > 1)$, and $P(Z_d < -1)/P(Z_i < -1)$) were calculated, where Z_i represents an independent variable, Z_d represents a dependent variable, -1 represents a low scenario, and 1 represents a high scenario.

4. Results

4.1. Descriptive Analysis of the Sample

A total of 169 complete surveys were collected at the end of August 2020, of which 57 respondents were women and 112 men; 74 held the position of managers and 95 engineers. The dataset collected are available in Martínez Hernández and García Alcaraz [50]. Table 1 illustrates the industrial sector and the years of experience for respondents in that job position. Observe that everyone had at least two years of experience, so it was concluded that the respondents were a reliable source of information.

Table 1. Number of respondents by industry sector and years of experience.

Industrial Sector	Job Experience in Years			Total
	2 and <5	5 and <10	>10	
Automotive	26	38	10	74
Medical	19	16	5	40
Machining	14	12	7	33
Electronic	2	8	0	10
Logistic	5	1	1	7
Electric	3	1	1	5
Total	69	76	24	169

It was also noted that the automotive and medical sectors had the most participation. In this research, the electric industrial sector focuses on producing wires, harnesses, and electric motor assemblies. In contrast, the electronics industrial sector focuses on programmable products, which integrate microcircuits or control motherboards.

4.2. Descriptive Analysis of Items

Table 2 illustrates the descriptive analysis of the items. The median and inter quarterly range (IR) are shown, as the information came from assessments. The values are sorted in descending order in each variable according to the median value, and it was observed that all activities were implemented, or the expected benefits were obtained since all medians were always greater than 4. Furthermore, it was noted that the activities associated with *TQM* had the highest medians and that *Wastes* had the lowest values.

Table 2. Descriptive analysis of the items.

Total Quality Management	Median	IR
TQM5. The organization focuses on meeting the needs of customers, involving employees	5.23	1.42
TQM3. The concept of total quality from raw material collection to after-sales customer service is promoted	5.18	1.58
TQM4. Decision-making for improvement is justified by facts and data	4.98	1.57
TQM2. Participatory management is promoted aimed at continuous improvement in all operations	4.87	1.61
Right first time		
RFT5. Training and awareness is carried out in relation to the quality and need to do well the activities	4.92	1.67
RFT3. Compliance with quality standards is verified with a zero-defect approach	4.83	1.55
RFT4. There is a standardized protocol for sampling when you want to do an analysis	4.81	1.74
RFT2. Ensures proper process operation to prevent defects	4.77	1.53
Wastes		
W8. Waste is identified in the production process and supply chain	4.81	1.55
W5. Improvements are encouraged to reduce Waste	4.70	1.54
W4. Product rework is reduced to the acceptable minimum	4.52	1.89
W6. Seeks to minimize the transport of material	4.50	1.80
Commercial benefits		
BCR1. There is a reduction in the cost of acquiring materials	4.73	1.74
BCR6. Average profit growth has been had in the last two years	4.68	1.77
BCR5. There has been an average return on sales and investment in the last two years	4.65	1.77
BCR2. There is a reduction in the cost of using energy	4.62	1.91

4.3. Variables Validation

Table 3 illustrates the validation of latent variables, which was obtained iteratively. The row corresponding to the number of items was divided into two columns for each variable, where the first value indicates the initial number of items in the questionnaire. In contrast, the second value indicates the number of items left after the validation process. According to the information in Table 2, it can be concluded that all variables had sufficient predictive, content, and convergent validity and that they had no problems of collinearity. Based on the above, the variables were integrated into the model.

Table 3. Validation of latent variables.

Index	Best Value If	TQM		RFT		Wastes		Commercial Benefits	
Number of items		6	3	7	4	8	4	7	3
R-squared	>0.02			0.612		0.259		0.459	
Adjusted R-squared	>0.02			0.607		0.255		0.449	
Composite reliability	>0.7	0.888		0.917		0.911		0.907	
Cronbach's alpha	>0.7	0.832		0.879		0.869		0.845	
Average variance extracted	>0.5	0.666		0.734		0.719		0.765	
Full collinearity VIF	<3.3	1.804		2.704		2.152		1.806	
Q-squared	>0.02			0.613		0.260		0.463	

4.4. SEM

The model was generated with the final variables and items after the validation process. The efficiency indices on the evaluated model are illustrated below. It was observed that they met the high- and low-established cutoff values, concluding that it had sufficient predictive validity, a good data fit to the model, and no collinearity problems. The evaluated model is illustrated in Figure 2.

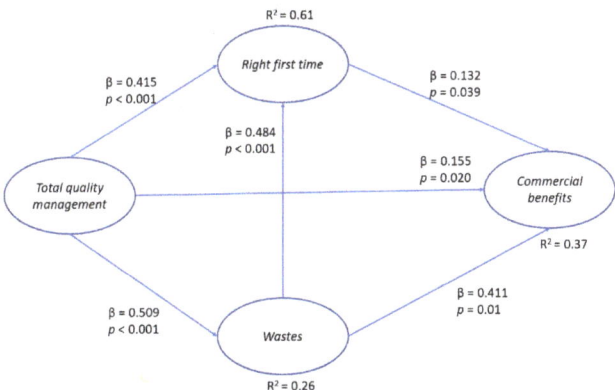

Figure 2. Evaluated model.

- Average path coefficient (APC) = 0.363, $p < 0.001$
- Average R-squared (ARS) = 0.443, $p < 0.001$
- Average adjusted R-squared (AARS) = 0.437, $p < 0.001$
- Average block VIF (AVIF) = 1.754, ideally <=3.3
- Average full collinearity VIF (AFVIF) = 2.116, ideally ≤ 3.3
- Tenenhaus GoF (GoF) = 0.565, large ≥ 0.36

4.4.1. Direct Effects

According to Figure 2, all relationships were statistically significant, so based on the direct effects, it can be concluded that the relationships established as hypotheses were valid with 95% confidence. Table 4 summarizes the information, the size effect size (ES) as a measure of variance explained, and conclusions of the hypotheses.

Table 4. Direct effect and hypotheses conclusions.

Independent Variable	Dependent Variable	β (p-Value)	Effect Size	Conclusion
TQM	Wastes	0.509 ($p < 0.001$)	0.259	Accept
TQM	RFT	0.415 ($p < 0.001$)	0.275	Accept
Wastes	RFT	0.484 ($p < 0.001$)	0.337	Accept
TQM	Commercial benefits	0.120 ($p = 0.046$)	0.046	Accept
RFT	Commercial benefits	0.316 ($p < 0.001$)	0.196	Accept
Wastes	Commercial benefits	0.334 ($p < 0.001$)	0.204	Accept

4.4.2. The Sum of Indirect Effects and Total Effects

Table 5 illustrates the sum of indirect effects, the total effects, their p-value, and associated ES. It should be noted that all the effects were statistically significant to a 95% confidence level. Here, it is important to emphasize that the relationship between TQM and Commercial benefits had a p-value of 0.046 (almost not significant) in the direct effect, but in the indirect and total effect, this effect was large and statistically significant.

Table 5. Sum of indirect and total effects.

	Sum of Indirect Effects		
	TQM	RFT	Wastes
RFT	0.246 ($p < 0.001$) ES = 0.163		
Commercial benefits	0.379 ($p <0.001$) ES = 0.187		0.153 ($p = 0.002$) ES = 0.093
	Total Effects		
RFT	0.662 ($p < 0.001$) ES = 0.438		0.484 ($p < 0.001$) ES = 0.337
Wastes	0.509 ($p < 0.001$) ES = 0.259		
Commercial benefits	0.499 ($p < 0.001$) ES = 0.246	0.316 ($p < 0.001$) ES = 0.196	0.487 ($p < 0.001$) ES = 0.297

4.4.3. Sensitivity Analysis

Table 6 illustrates the sensitivity analysis of the probability that independent and dependent variables will occur in high ($p > 1$) and low ($p < -1$) scenarios. By column, are the independent variables and, per row, the dependent variables. In this analysis, the probability that independent and dependent variables appear simultaneously in a scenario is denoted by the symbol "&", while the conditional probability of the dependent variable occurring since the independent variable has occurred is represented by "if".

Table 6. Sensitivity analysis.

		Sign/Value		TQM		Wastes		RFT	
				+ 0.166	− 0.136	+ 0.148	− 0.160	+ 0.178	− 0.154
Wastes	+	0.142		& = 0.083 If = 0.500	& = 0.000 If = 0.000				
	−	0.160		& = 0.018 If = 0.107	& = 0.065 If = 0.478				
RFT	+	0.148		& = 0.059 If = 0.357	& = 0.000 If = 0.000	& = 0.089 If = 0.625	& = 0.006 If = 0.037		
	−	0.142		& = 0.000 If = 0.000	& = 0.077 If = 0.565	& = 0.000 If = 0.000	& = 0.095 If = 0.593		
Commercial benefits	+	0.178		& = 0.077 If = 0.464	& = 0.000 If = 0.000	& = 0.077 If = 0.542	& = 0.006 If = 0.037	& = 0.065 If = 0.440	& = 0.000 If = 0.000
	−	0.154		& = 0.006 If = 0.036	& = 0.065 If = 0.478	& = 0.000 If = 0.0000	& = 0.071 If = 0.444	& = 0.006 If = 0.040	& = 0.077 If = 0.542

For example, the probability of simultaneously finding *TQM* and *Wastes* at high levels was 0.083 [P(*TQM*+∩*Wastes*+)]. Still, the probability of *Wastes* occurring at a high level, since *TQM* occurred at its high level, was also 0.500 [P(*Wastes*+/*TQM*+)], and hence the importance of *TQM* programs, as they guarantee the success of other LM tools such as *waste* reduction. However, suppose that *TQM* occurs at a low level due to quality deployment problems. In that case, it is possible to obtain *Wastes* in a low level with a probability of 0.178, which is a risk for managers. The additional probabilities were similarly interpreted.

5. Discussion of Results and Conclusions

The implementation of LM tools such as *TQM* must be based on the actual data of the company's quality situation and support many other tools. This study linked three LM tools associated with quality to *Commercial benefits*.

5.1. Conclusions from Descriptive Analysis

The descriptive analysis shows that *TQM* is the tool with the highest average values, indicating that respondents considered that activities were always carried out to achieve them. *TQM* had the lowest inter-quarterly range in item *TQM5*, meaning that there was consensus that the companies were focused on meeting the needs of customers. With regard to the activity that had the lowest median, this referred to item W6 (*Wastes*) with a value of only 4.50, focused on minimizing the transport of material; moreover, BCR2 (*Commercial benefits*) was the item with an enormous IR value, indicating that there was a lower consensus regarding its actual average value.

Concluding from an univariate point of view, to achieve success in *TQM* implementation, maquiladora companies need to focus on meeting customer needs to promote quality throughout the company. To achieve *RFT*, companies need to focus on education and training, verifying compliance with standards. To reduce *Wastes*, companies need to focus on identifying the *waste* origin and implementing improvements. Finally, the most important *Commercial benefits* gained from applying LM tools are reducing material costs, increasing sales, and return on investment.

5.2. Conclusions from SEM and Sensitivity Analysis

The relationship between *TQM* and *Wastes* in H_1 was statistically significant, meaning that focus on quality customers and promoting its concept within the company allowed a reduction in *Wastes* in the production process, delays in delivery time, rework due to human errors, and transport of material. Additionally, the sensitivity analysis showed that *TQM*+ encourages activities aimed at eliminating *Wastes*+ with a probability of 0.500, but generates very little probability of *Wastes*−; also, *TQM*− never helps eliminate *Wastes*+ since the probability is null, and there exists the risk of having *Waste*− with a probability of 0.478. In this sense, this work coincides with Johri and Kumar [51], who established that compliance with international quality standards allowed for improvements in the operational performance of companies such as certified recognition for their products and management.

The relationship between *TQM* and *RFT* in H_2 proves that focusing on customers, spreading the concept of quality throughout the company, and a decision-making process focused on continuous improvement with training support leads companies to generate protocols for quality assurance and comply with the product standards. The sensitivity analysis showed that when *TQM*+ occurs, managers have a 0.357 chance of having *RFT*+; however, *TQM*− is never associated with *RFT*+, since the probability is null. Additionally, *TQM*− can generate *RFT*− with a probability of 0.565. The results match the Moitra [18] report, indicating that errors and defects can be prevented from the root, training operators to ensure product quality.

Wastes have also been shown in this study to positively affect *RFT*, as activities aimed at identifying and eliminating *Wastes* lead to *RFT* products and services being generated. From the sensitivity analysis, we observed that *Wastes*+ favored *RFT*+ with a probability of 0.625 and that it never generated $RFT-$, since the probability was null. Additionally, it can be noted that *Wastes*− only had a probability of generating $RFT+$ of 0.037, but 0.593 of $RFT-$ happens. In this sense, the results coincide with Singh and Hussain [29] and Purushothaman et al. [52], who mentioned that focusing on dismissing *Wastes* leads to the generation of quality products the first time and with minimum human error.

One of the most exciting relationships was $TQM \rightarrow$ *Commercial benefits*, since the direct effect was only 0.120 and had a *p*-value of 0.046, very close to 0.05, set as a maximum cutoff, indicating no direct relationship. However, the indirect effects showed that the relationship between those variables was high and is given through the *RFT* and *Wastes* tools. This means that focusing on the customer and promoting quality throughout the company favors reducing costs associated with materials and energy consumption in the production process and logically increases sales. However, it is first necessary to develop employee training, standardize processes, and eliminate *Wastes* that does not add value. From the sensitivity analysis, it can be concluded that $TQM+$ favors *Commercial benefits*+, since the conditional probability was 0.464, and, in addition, the probability of generating *Commercial benefits*− was 0.036; also, $TQM-$ values did not generate *Commercial benefits*+ because the probability was null, but it could generate *Commercial benefits*−, since their conditional probability was 0.478. Our results coincide in this regard with Singh, Kumar and Singh [9] and Kaouthar and Lassaad [53], who investigated TQM's effect on the performance of manufacturing companies in India and Tunisia, respectively.

RFT has also been proven to affect *Commercial benefits* directly; that is, employee training, adherence to standards, and decreasing defects allow a reduction in the cost of used material and energy consumption, which means an increase in sales. The sensitivity analysis concluded that $RFT+$ generated *Commercial benefits*+, as the probability was 0.440, and was unlikely to generate *Commercial benefits*−, since the probability was 0.040. However, $RFT-$ never caused *Commercial benefits*+, since the probability was null, and there exists the risk of having *Commercial benefits*−, since the probability was 0.542.

Finally, it is statistically and empirically demonstrated that *Wastes* had a direct and positive effect on *Commercial benefits*. The elimination of *Wastes* generated in the production process and improvements focused on minimizing the transport of material and human resources generated a reduction in the cost of materials by avoiding accidents and energy loss in conveyor equipment. The sensitivity analysis concluded that *Wastes*+ generated *Commercial benefits*+ with a conditional probability of 0.542, but never created *Commercial benefits*−, since the probability was null. Additionally, *Wastes*− could not cause *Commercial benefits*+, and there was a risk of 0.444 to obtain *Commercial benefits*−. This work coincides with Eldessouky, Flynn, and Newman [34], who stated that getting things right required great coordination of many machines and high levels of precision, as everyday standards are higher, production batches are smaller, and products require more components.

Finally, it is important to mention that this work has the limitation of analyzing only three LM tools associated with quality in maquiladora companies and many others are being implemented. Proof of this is that the value of R-square in the dependent latent variables was not the unit, indicating that they were not fully explained, so in future research, other tools will be integrated, seeking to have greater explanatory power.

5.3. Practical Implications

Quality has been a concern of production managers since it is the basis on which the product is accepted or rejected by customers, and is directly related to the economic income of the company. In this research, it has been empirically and statistically demonstrated that *TQM* is a fundamental basis for the financial sustainability of maquiladora companies, both directly and indirectly.

Since *TQM* implementation requires economic and human resources, this study has shown that in the opinion of managers and engineers, the investments made to achieve quality in production processes will always reduce costs when acquiring materials, energy used, and penalization for environmental accidents. Still, above all, there will be an increase in sales in a competitive market since the products meet the customer expectations.

However, *TQM* does not act alone in a production process to generate quality products, but tools such as *RFT* and *Wastes* reduction are vital to support.

Proof of the above is that the direct effect between *TQM* and *Commercial benefits* was only 0.120 and the indirect effects through *RFT* and *Wastes* were 0.379; that is, the indirect effect was greater than the direct one by more than 300%, which indicates that support tools such as *RFT* and *Wastes* strengthen the impact of *TQM* on the economic benefits obtained from its implementation.

Therefore, when applying *TQM* as part of a program focused on improving the quality of their products and production processes, it should be supported with other tools such as *RFT* and *Wastes*, as they allow for the identification of activities that do not add value, but a lot of costs that may represent a loss in customer preferences.

Supplementary Materials: The following are available online at https://www.mdpi.com/article/10.3390/math9090971/s1, Full validation index for latent variables appears at García-Alcaraz, J.L. Data validation for paper: Effect of Quality Lean Manufacturing Tools on Commercial Benefits Gained by Mexican Maquiladoras, University of Ciudad Juárez: Ciudad Juárez, Mexico, 2021, Mendeley Data, V1, https://doi.org/10.17632/7tkj88dmf6.1.

Author Contributions: Conceptualization, J.L.G.A. and F.A.M.H.; Methodology, J.L.G.A. and F.A.M.H.; Software, A.R.V. and J.E.O.T.; Validation, E.J.M. and C.J.L.; Formal analysis, J.L.G.A.; Investigation, F.A.M.H.; Resources, E.J.M.; Data curation, F.A.M.H.; Writing—original draft preparation, J.L.G.A. and F.A.M.H.; Writing—review and editing, E.J.M.; Visualization, C.J.L. and J.E.O.T.; Supervision, J.L.G.A.; Project administration, J.L.G.A. All authors have read and agreed to the published version of the manuscript.

Funding: This research received no external funding.

Institutional Review Board Statement: Not applicable.

Informed Consent Statement: Not applicable.

Data Availability Statement: The full survey applied in this research is available at a repository located at https://doi.org/10.6084/m9.figshare.14301275 (accessed on 18 March 2021). Dataset in SAV version for SPSS software is available at https://doi.org/10.6084/m9.figshare.14308736.v1 (accessed on 18 March 2021).

Acknowledgments: The authors thank all of the managers and engineers who dedicated their time to answering the questionnaire applied in this research. We hope that the results are helpful to them.

Conflicts of Interest: The authors declare no conflict of interest.

References

1. Diaz-Reza, J.R.; García-Alcaraz, J.L.; Mendoza-Fong, J.R.; Maldonado-Macías, A.A.; Sánchez-Ramírez, C. The role of information sharing in the supply chain from maquiladoras in northern mexico. In *Handbook of Research on Industrial Applications for Improved Supply Chain Performance*; Luis, G.-A.J., Leal, J.G., Liliana, A.-S., Briones, P.A.J., Eds.; IGI Global: Hershey, PA, USA, 2020; pp. 175–199.
2. IMMEX. *Index Juárez—Monthly Statistic Information (January 22, 2021)*; Asociación de Maquiladoras AC: Ciudad Juárez, Mexico, 2021; pp. 1–3.
3. Palange, A.; Dhatrak, P. Lean manufacturing a vital tool to enhance productivity in manufacturing. *Mater. Today Proc.* **2021**, in press. [CrossRef]
4. Islam, A.S.M.T. End of the day, who is benefited by lean manufacturing? A dilemma of communication and pricing in buyer-supplier relationship. *Manuf. Lett.* **2019**, *21*, 17–19. [CrossRef]
5. Melton, T. The benefits of lean manufacturing: What lean thinking has to offer the process industries. *Chem. Eng. Res. Des.* **2005**, *83*, 662–673. [CrossRef]
6. Rajenthirakumar, D.; Shankar, R. Analyzing the benefits of lean tools: A consumer durables manufacturing company case study. *Ann. Fac. Eng. Hunedoara* **2011**, *9*, 335–339.
7. Hao, Z.; Liu, C.; Goh, M. Determining the effects of lean production and servitization of manufacturing on sustainable performance. *Sustain. Prod. Consum.* **2021**, *25*, 374–389. [CrossRef]

8. García-Alcaraz, J.L.; Macías, A.A.M.; Luevano, D.J.P.; Fernández, J.B.; López, A.J.G.; Macías, E.J. Main benefits obtained from a successful JIT implementation. *Int. J. Adv. Manuf. Technol.* **2016**, *86*, 2711–2722. [CrossRef]
9. Singh, V.; Kumar, A.; Singh, T. Impact of TQM on organisational performance: The case of indian manufacturing and service industry. *Oper. Res. Perspect.* **2018**, *5*, 199–217. [CrossRef]
10. Sahoo, S.; Yadav, S. Influences of tpm and TQM practices on performance of engineering product and component manufacturers. *Procedia Manuf.* **2020**, *43*, 728–735. [CrossRef]
11. Meade, D.J.; Kumar, S.; Houshyar, A. Financial analysis of a theoretical lean manufacturing implementation using hybrid simulation modeling. *J. Manuf. Syst.* **2006**, *25*, 137–152. [CrossRef]
12. Fullerton, R.R.; Kennedy, F.A.; Widener, S.K. Lean manufacturing and firm performance: The incremental contribution of lean management accounting practices. *J. Oper. Manag.* **2014**, *32*, 414–428. [CrossRef]
13. Shashi, P.C.; Cerchione, R.; Singh, R. The impact of leanness and innovativeness on environmental and financial performance: Insights from indian smes. *Int. J. Prod. Econ.* **2019**, *212*, 111–124. [CrossRef]
14. Elkhairi, A.; Fedouaki, F.; Alami, S.E. Barriers and critical success factors for implementing lean manufacturing in smes. *IFAC PapersOnLine* **2019**, *52*, 565–570. [CrossRef]
15. Abu, F.; Gholami, H.; Saman, M.Z.M.; Zakuan, N.; Streimikiene, D. The implementation of lean manufacturing in the furniture industry: A review and analysis on the motives, barriers, challenges, and the applications. *J. Clean. Prod.* **2019**, *234*, 660–680. [CrossRef]
16. Jun, M.; Cai, S.; Shin, H. TQM practice in maquiladora: Antecedents of employee satisfaction and loyalty. *J. Oper. Manag.* **2006**, *24*, 791–812. [CrossRef]
17. Lynn, R. Useful Lean Manufacturing Tools. Available online: https://www.planview.com/resources/guide/what-is-lean-manufacturing/lean-manufacturing-tools/ (accessed on 17 April 2021).
18. Sisternas, P. The Main Lean Manufacturing Tools for Your Company [in Spanish]. Available online: https://www.emprendepyme.net/las-principales-herramientas-de-lean-manufacturing-para-tu-empresa.html (accessed on 16 April 2021).
19. Salleh, N.A.M.; Kasolang, S.; Jaffar, A. Simulation of integrated total quality management (TQM) with lean manufacturing (LM) practices in forming process using delmia quest. *Procedia Eng.* **2012**, *41*, 1702–1707. [CrossRef]
20. York, K.M.; Miree, C.E. Causation or covariation: An empirical re-examination of the link between TQM and financial performance. *J. Oper. Manag.* **2004**, *22*, 291–311. [CrossRef]
21. Moitra, T. From employees to customers: Impact of HRM on TQM. *HCM Sales Mark. Alliance Excell. Essent.* **2019**, *18*, 18–21.
22. Green, K.W.; Inman, R.A.; Sower, V.E.; Zelbst, P.J. Impact of jit, TQM and green supply chain practices on environmental sustainability. *J. Manuf. Technol. Manag.* **2019**, *26*. [CrossRef]
23. García-Alcaraz, J.L.; Montalvo, F.J.F.; Sánchez-Ramírez, C.; Avelar-Sosa, L.; Saucedo, J.A.M.; Alor-Hernández, G. Importance of organizational structure for TQM success and customer satisfaction. *Wirel. Netw.* **2019**, *27*, 1601–1614. [CrossRef]
24. Sreedharan, R.V.; Sunder, V.M.; Raju, R. Critical success factors of TQM, six sigma, lean and lean six sigma: A literature review and key findings. *Benchmarking* **2018**, *25*, 3479–3504. [CrossRef]
25. Qasrawi, B.T.; Almahamid, S.M.; Qasrawi, S.T. The impact of TQM practices and km processes on organisational performance: An empirical investigation. *Int. J. Qual. Reliab. Manag.* **2017**, *34*, 1034–1055. [CrossRef]
26. Ooi, K.-B. TQM: A facilitator to enhance knowledge management? A structural analysis. *Expert Syst. Appl.* **2014**, *41*, 5167–5179. [CrossRef]
27. Katare, V.D.; Madurwar, M.V. Process standardization of sugarcane bagasse ash to develop durable high-volume ash concrete. *J. Build. Eng.* **2021**, *39*, 102151. [CrossRef]
28. Sutharsan, S.M.; Prasad, M.M.; Vijay, S. Productivity enhancement and waste management through lean philosophy in indian manufacturing industry. *Mater. Today Proc.* **2020**, *33*, 2981–2985. [CrossRef]
29. Singh, S.; Hussain, C.M. Chapter three—Zero waste manufacturing. In *Concepts of Advanced Zero Waste Tools*; Hussain, C.M., Ed.; Elsevier: Amsterdam, The Netherlands, 2021; pp. 45–67.
30. Amjad, M.S.; Rafique, M.Z.; Khan, M.A. Leveraging optimized and cleaner production through industry 4.0. *Sustain. Prod. Consum.* **2021**, *26*, 859–871. [CrossRef]
31. Wang, Y.; Huang, A.; Quigley, C.A.; Li, L.; Sutherland, J.W. Tolerance allocation: Balancing quality, cost, and waste through production rate optimization. *J. Clean. Prod.* **2021**, *285*, 124837. [CrossRef]
32. Womack, J.; Jones, D. *Lean Thinking: Banish Waste and Create Wealth in Your Corporation*; Free Press: New York, NY, USA, 2003; p. 400.
33. Moshiri, M.; Charles, A.; Elkaseer, A.; Scholz, S.; Mohanty, S.; Tosello, G. An industry 4.0 framework for tooling production using metal additive manufacturing-based first-time-right smart manufacturing system. *Procedia CIRP* **2020**, *93*, 32–37. [CrossRef]
34. Eldessouky, H.M.; Flynn, J.M.; Newman, S.T. On-machine error compensation for right first time manufacture. *Procedia Manuf.* **2019**, *38*, 1362–1371. [CrossRef]
35. Sila, I. Country and sector effects on the relationships among TQM practices and key performance measures. *Int. J. Product. Perform. Manag.* **2018**, *67*, 1371. [CrossRef]
36. Álvarez-Santos, J.; Miguel-Dávila, J.-Á.; Herrera, L.; Nieto, M. Safety management system in TQM environments. *Saf. Sci.* **2018**, *101*, 135–143. [CrossRef]

37. Kumar, V.; Sharma, R.R.K. Relating management problem-solving styles of leaders to TQM focus: An empirical study. *TQM J.* **2017**, *29*, 218–239. [CrossRef]
38. Singh, S.; Ramakrishna, S.; Gupta, M.K. Towards zero waste manufacturing: A multidisciplinary review. *J. Clean. Prod.* **2017**, *168*, 1230–1243. [CrossRef]
39. Um, N.; Park, S.-O.; Yoon, C.-W.; Jeon, T.-W. A pretreatment method for effective utilization of copper product manufacturing waste. *J. Environ. Chem. Eng.* **2021**, 105509. [CrossRef]
40. Bañuelas, R.; Antony, J.; Brace, M. An application of six sigma to reduce waste. *Qual. Reliab. Eng. Int.* **2005**, *21*, 553–570. [CrossRef]
41. Magdy, K.; Tamer, M.S. The moderating effect of structural barriers on TQM-performance relationship in egyptian service organizations. *Int. J. Qual. Serv. Sci.* **2018**, *349*. [CrossRef]
42. Kappelman, L.; Prybutok, V. Empowerment, motivation, training, and TQM program implementation success. *Ind. Manag.* **1995**, *37*, 12–15.
43. Kassicieh, S.K.; Yourstone, S.A. Training, performance evaluation, rewards, and TQM implementation success. *J. Qual. Manag.* **1998**, *3*, 25–38. [CrossRef]
44. García-Alcaraz, J.L.; Flor-Montalvo, F.J.; Avelar-Sosa, L.; Sánchez-Ramírez, C.; Jiménez-Macías, E. Human resource abilities and skills in TQM for sustainable enterprises. *Sustainability* **2019**, *11*, 6488. [CrossRef]
45. Bari, M.W.; Fanchen, M.; Baloch, M.A. TQM soft practices and job satisfaction; mediating role of relational psychological contract. *Procedia Soc. Behav. Sci.* **2016**, *235*, 453–462. [CrossRef]
46. Hernández, F.A.M.; Alcaraz, J.L.G. *Survey to Determine the Level of Implementation of Lean Manufacturing Practices*; Autonomous University of Ciudad Juárez: Ciudad Juárez, Mexico, 2021.
47. Parsazadeh, N.; Ali, R.; Rezaei, M.; Tehrani, S.Z. The construction and validation of a usability evaluation survey for mobile learning environments. *Stud. Educ. Eval.* **2018**, *58*, 97–111. [CrossRef]
48. Kock, N. *Warppls 6.0 User Manual*; ScriptWarp Systems: Laredo, TX, USA, 2018.
49. Kock, N. Factor-based structural equation modeling with warppls. *Australas. Mark. J.* **2019**, *27*, 57–63. [CrossRef]
50. Hernández, F.A.M.; Alcaraz, J.L.G. *Dataset with Lean Manufacturing Tools and Its Commercial Benefits*; Autonomous University of Ciudad Juárez: Ciudad Juárez, Mexico, 2021.
51. Johri, S.; Kumar, D. Evaluation of effect of iso 9001:2008 standard implementation on TQM parameters in manufacturing & production processes performance in small enterprises. *Mater. Today Proc.* **2021**, in press. [CrossRef]
52. Purushothaman, M.B.; Seadon, J.; Moore, D. Waste reduction using lean tools in a multicultural environment. *J. Clean. Prod.* **2020**, *265*, 121681. [CrossRef]
53. Kaouthar, L.; Lassaad, L. Impact of TQM/six sigma practices on company's performance: Tunisian context. *Int. J. Qual. Reliab. Manag.* **2018**, *35*, 1881–1906. [CrossRef]

Article

Exploring the Effect of Status Quo, Innovativeness, and Involvement Tendencies on Luxury Fashion Innovations: The Mediation Role of Status Consumption

Andreea-Ionela Puiu [1,*], Anca Monica Ardeleanu [2], Camelia Cojocaru [2] and Anca Bratu [2]

[1] Economics I Doctoral School, The Bucharest University of Economic Studies, 010374 Bucharest, Romania
[2] Faculty of Business and Administration, University of Bucharest, 030018 Bucharest, Romania; monica.ardeleanu@faa.unibuc.ro (A.M.A.); camelia.cojocaru@faa.unibuc.ro (C.C.); anca.bratu@faa.unibuc.ro (A.B.)
* Correspondence: puiuionela18@stud.ase.ro

Citation: Puiu, A.-I.; Ardeleanu, A.M.; Cojocaru, C.; Bratu, A. Exploring the Effect of Status Quo, Innovativeness, and Involvement Tendencies on Luxury Fashion Innovations: The Mediation Role of Status Consumption. *Mathematics* **2021**, *9*, 1051. https://doi.org/10.3390/math9091051

Academic Editor: María del Carmen Valls Martínez

Received: 6 April 2021
Accepted: 30 April 2021
Published: 7 May 2021

Publisher's Note: MDPI stays neutral with regard to jurisdictional claims in published maps and institutional affiliations.

Copyright: © 2021 by the authors. Licensee MDPI, Basel, Switzerland. This article is an open access article distributed under the terms and conditions of the Creative Commons Attribution (CC BY) license (https://creativecommons.org/licenses/by/4.0/).

Abstract: The article explores the mechanisms that affect consumers' interest in luxury clothing innovations. The actual research aims to investigate the effect of status quo and clothing involvement on consumer brand loyalty. More, it was intended to quantify the influence of the level of engagement concerning clothing acquisition and the status quo tendency on the consumers' level of interest toward innovative luxury fashion products. The models were analyzed through the partial-least square-path modeling method. The results revealed that status quo bias and consumers' involvement in fashion influence their loyalty to brands and level of innovativeness. The novelty of the present research comes from the analysis of the impact of the status quo manifest variables on consumers' innovative tendencies. Moreover, it was found that status consumption fully mediates the relationships among the investigated predictors and considered outcome variables. The mediator manifests the highest effect size of all investigated predictors. The actual paper advances research in a direction that was not sufficiently addressed in the past, introducing the status quo construct as the main predictor of peoples' inclination to be loyal to a brand or to manifest a tendency toward innovativeness. Moreover, the article emphasizes the essential role manifested by social status in foreseeing a behavioral response.

Keywords: luxury fashion goods; status consumption; status quo; clothing innovativeness; clothing involvement; PLS-PM

1. Introduction

Consumption appoints a process of spending for getting utility or for satisfying specific needs. Consumers buy products for a variety of reasons. Their behavior manifests an essential element in the survival of an industry in the market.

More than to satisfy existential needs and carry out practical functions, several goods are bought because of their potential to respond to social or psychological necessities [1]. One of the several social motives that urge consumers to buy items is represented by their desire to confirm their social status [2–4]. This necessity of social status confirmation is more visible when it comes to luxury goods. Luxury is characterized by conspicuousness, social value, prestige, and wealth, playing an essential role in social stratification [5]. Consumers consider luxury a vital element in defining themselves as members of a high social class [6].

Luxury commodities are components of a novel social protocol, where individuals' identity is appreciated considering the visible brands worn in their daily lives [3]. Fashion goods signal a consumer's social standing, transmitting at the same time symbolic value [2]. Individuals consume luxury goods to achieve higher social status and to confirm their wealth [5]. In this respect, clothing brands try to offer consumers status-seeking reasons to create a loyalty relationship concerning their brand [7].

Of interest in studying social status consumption are the motives that drive consumers to show loyalty to a brand or to search for novel and innovative substitutes available on the market. The attention was directed to the status quo dimension and the consumer interest in fashion. Status quo manifest constructs refer to people's satisfaction with the available fashion products and their satisfaction concerning the extent of innovations when referring to luxury fashion items [8]. On the other hand, interest in fashion appoints the degree to which a specific buying decision is indispensable for an individual [9].

Multiple existing studies revealed that status consumption is a strong motivator of consumer behavior [5,10]. In this study, it was proposed to investigate the role of status consumption in the change of the consumers' clothing brand loyalty and their level of innovativeness when dealing with luxury clothes.

A high level of status consumption serves as a justification for brand loyalty. Regarding the influence of the status quo on status consumption, it was supposed a negative relationship because a desire for change is inconsistent with holding the existing habits without assuming a behavioral change.

The focus of our study is on the luxury fashion brands, which is a product category where status consumption manifests an essential role. Luxury articles of clothing are connected to an individual social and personal identity [6,7]. Moreover, the research was conducted on students, youths being an essential market because of two reasons: for their current spending behavior and their future spending behavior as adults. Even if European youths spend less on clothes, they are still a relevant market [8]. This aspect explains why several companies design luxury fashion products especially for teens. Existing studies revealed that social motives enhance the acquisition of luxury brands of late adolescents and young adults [11]. Social constraints do not determine the purchase of luxury brands of middle-aged adults [12]. As a result of the emergence of the teenager's market, it is considered essential to understand their position in relation to luxury fashion innovations [13]. Moreover, the study was conducted on Romania, a country that is part of the European Union, an extensive and highly internationalized market. Studying the behavior response of consumers about the adoption of innovative luxury fashion products manifests a wide interest for international fashion clusters.

The obtained results of the present research are in line with existing studies. We found a positive relationship among status consumption and consumers' level of involvement, their innovativeness tendency, and loyalty to brands. In addition, the research was advanced in a direction that earlier studies do not considered sufficiently, analyzing the mediation role of status consumption when discussing mechanisms that affect consumers' interest in luxury clothing innovations. Moreover, we introduced the status quo constructs as main predictors of peoples' inclination to be loyal to a brand or exhibit a tendency toward innovativeness. Practical implications derive from the fact that innovative individuals manifest high brand loyalty, being in the interest of the luxury fashion providers to support a high level of novelty of their products and to align their marketing strategies to the segment of customers who are interested in status confirmation.

Section 2, Literature Review, discusses the theoretical background and the hypotheses development. Section 3, Materials and Methods, discusses the characteristics of the gathered sample and the method used to construct the latent variables. Section 4, Results, presents the outcomes of the partial-least-squares-path model (PLS-PM). Section 5, Discussions, shows the interpretation of the findings and extensive discussions of our results. Section 6, Conclusions, briefly summarizes the paper and discusses limits and future perspectives of research.

2. Literature Review

Luxury could be a term difficult to define because it has subjective meanings to different consumers. In addition, almost all luxury brands have products that start at low price points, whether it is about keyrings, pairs of socks, or sports cars, so that consumers position themselves differently to the category of luxury goods. Luxury designates limited supply goods at higher prices that contribute to consumer satisfaction by adding pleasure to their life, without being available for everyone [14].

Luxury appoints a concept of refinement, elegance, opulence, and wealth that is desirable but not necessary in peoples' existence [15]. It is claimed that luxury derives from the scarcity of resources used in the production process or from their high price [16]. However, luxury appoints more than materials, skills in producing those goods, or other attributes. Luxury holds social meanings [16].

Luxury and affluence goods were mostly related to a higher rank in the social hierarchy, giving rise to the concept of consumption for social status confirmation [17]. This concept describes any consumption activity realized to show off wealth to other people when using the good in public, even if it is about wearing a brand dress or driving an expensive car [18]. The status consumption concept is widely available in the existing literature. Some authors [19] described status consumption as being a mechanism through which someone owns the power of imposing respect, consideration, and envy from others.

Several authors that studied the human condition figured out that the intrinsic desire for social position or higher status in the social hierarchy is an effective motivator of behavior [20]. A study found the desire for status as a principal motivator for brand choice [21], while another one showed that the desire for status determines a purchasing behavior [22]. In direct connection to this study, clothing serves as a social symbol of status or social position [23]. Earlier studies reported that status consumption is positively related to the consumer level of involvement and their innovativeness tendency [24,25].

Related to our study, innovative luxury clothing refers to those items carried out using innovative materials and techniques combined in the production process to deliver superior products that enable a novel fashion experience when consumers wear or interact with them.

Consumers involved in fashion will experience earlier innovative luxury articles of clothing compared to consumers that manifest a low involvement in fashion, who will procrastinate regarding this decision. In this respect, consumers' involvement in fashion or their interest concerning this area will affect their acquisition of luxury fashion goods. Consumers' luxury preferences are related to their level of involvement [26].

Moreover, the connection set up with a brand describes a psychological preference toward famous brand goods [18]. Brand loyalty influences in a significant manner the purchase decision. Loyal consumers to a brand expect that brand to satisfy their necessities concerning the social standing and prestige confirmation [1].

The individuals' willingness to accept novel ideas or products is strongly related to their innovativeness [27]. Fashion innovators try to differentiate themselves from the rest of the consumers by looking for novel and innovative styles to conserve their status as innovators [28]. In our study, we paid attention to the concept of fashion innovativeness in the context of luxury fashion items. Fashion innovators are more mindful of brand names when realizing a purchase decision, manifesting a higher need for uniqueness, value, and status confirmation [29]. In other words, innovators are interested in their look, behavior, style, and the extent to which these are following their social status [3]. This tendency toward innovativeness is contradictory to consumers' bias toward the status quo, which appoints their preference for products or services already tested.

All these previously discussed aspects conduct to a brand or product choice that will assure an improvement of the consumers' social standing through social prestige.

Hypothesis Development

Considering the theoretical aspects already presented in the literature review section, the paper will continue with the hypothesis development.

Involvement is a broadly investigated construct linked to consumer behavior. Involvement defines a mental predisposition of being interested in or enthusiastic about a given product category [30]. Consumers that look for status are aware of clothing [24]; therefore, they are conscious about which clothing brands offer status and develop a preference in this respect. In addition, the involvement concept may be a relevant variable in the interpretation of dress codes specific to a certain social class [31].

Considering those aspects, we formulated the first hypothesis:

Hypothesis 1 (H1). *Clothing involvement manifests a positive influence on status consumption.*

On the other side, innovativeness tendency appoints the interest in new products, figuring out consumers to be among the first buyers of the new commodity. Those early buyers are characterized by involvement, knowledge, and seeking out information. They are opinion formers for the product category [32,33]. Being one of the first buyers of a newly appeared product was associated with social status [34–37]. Thus, it is argued that status consumption may mediate the relationship between the consumers' innovativeness and their level of involvement in the acquisition of luxury clothing.

Hypothesis 2 (H2). *Status consumption mediates the relationship between the level of innovativeness and clothing involvement.*

Hypothesis 2a (H2a). *Clothing involvement manifests a positive influence on the level of innovativeness.*

Hypothesis 2b (H2b). *Status consumption manifests a positive influence on the level of innovativeness.*

Over time, consumers may develop connections with the brands that they grow accustomed to buying [38,39]. A few researchers investigated the brand loyalty dimension; nevertheless, the investigation of the relationship with variables included in the present model was not previously studied—in particular, the relationship between brand loyalty and the status quo features. Consumers that are interested in status confirmation will search for those brands that confirm their social position. They will set up a sort of loyalty with that brand as time as the status confirmation endures [25]. This aspect fits in the case of clothing acquisition, garments being a visible symbol of social status. We hypothesized the following aspects:

Hypothesis 3 (H3). *Status consumption mediates the relationship between clothing involvement and clothing brand loyalty.*

Hypothesis 3a (H3a). *Status consumption positively influences clothing brand loyalty.*

Hypothesis 3b (H3b). *There is a positive relationship between clothing involvement and clothing brand loyalty.*

Moreover, high social status is associated with a conservative attitude, consumers being prone to status quo bias and resistant to innovations. In this respect, our proper contribution comes from investigating the status quo influence on status consumption and its indirect effect on luxury fashion innovativeness and clothing brand loyalty.

Status quo appoints an essential reference point in the decision process realized by consumers. People are inclined to favor already tried and familiar products, rather than new or innovative alternatives, even if the last would provide them with more utility or satisfaction. This situation occurs because of two reasons: firstly, because people experience emotional attachment to the product that they hold [40], and secondly, because a novel experience comes with a level of prospect and uncertainty. Risk is difficult to set up before testing the product. Human beings are inclined to minimize the risk and uncertainty, as buyers will strive for coherence and prefer the default option [41].

From another perspective, the existence of multiple available alternatives leads to a complex process of selecting information and realization of fast and correct comparisons among available possibilities. Since consumers may feel oversaturated by the existing available options, their receptiveness to innovations diminishes. In this case, the preference for status quo is increasing.

The earlier study finds two features that compose the status quo satisfaction dimension, namely satisfaction with existing products and satisfaction with the extent of innovations [8,42]. It is proposed to investigate if this aspect is also applicable for luxury fashion products. It is expected to find a negative relationship between satisfaction with existing luxury fashion products and status consumption. As previously discussed, consumers tend to show low interest in innovation when they are in their comfort zone.

Hypothesis 4 (H4). *Status consumption mediates the relationship between status quo manifest constructs and consumer level of innovativeness.*

Hypothesis 4a (H4a). *Status quo constructs negatively influence consumers' level of innovativeness.*

Hypothesis 5 (H5). *Status consumption mediates the relationship between status quo manifest constructs and consumers' brand loyalty.*

Hypothesis 5a (H5a). *Status quo manifest constructs positively influence consumers' loyalty to brands.*

To specify and clarify the social motives that underlie loyalty to a brand or tendency toward innovativeness, the focus of our study is represented by clothing, which is an area suitable to teenagers. Since this period of adolescence and young adulthood is a period of forming attitudes concerning the identity development mechanisms, clothing plays an important role in their connection to their ideal social groups [43].

3. Materials and Methods

The data were collected through a questionnaire that was distributed in the period November–December 2020. It used a combination of purposive and snowball sampling methods, respondents' participation in this study being voluntary. The size of the sample is large enough to sustain the structural analysis because of the following reasons. Firstly, the WarpPLS software supplies a function that allows determining the sample size such that to be representative. In our case, considering a significance level of 0.050, a power level of 0.950, and an inverse square root method [44], the software recommends a minimum required sample size of 279 respondents. Secondly, the partial-least square-path modeling analysis that was conducted is based on variances. This method allows the usage of smaller sample sizes compared with covariance-based structural equation modeling that requires larger sample sizes [45,46].

The data that were used in this study include 383 students enrolled in universities with economic and business profiles from Romania. Regarding the demographic features of the sample, the participants were 73.62% women and 26.37% men (Table 1). Most of them were aged between 18 and 27 years old (57.18%) and were from an urban area (79.11%).

The demographical characteristics of the sample were somehow expected. Earlier studies [47] revealed that women are more involved than men in buying luxury fashion brands. People from urban areas are more likely [48] to invest in luxury items than people from rural areas. Regarding the revenues, 13.83% of the respondents earn a net monthly income lower than 1000 USD, while 12.80% of the respondents reported a net monthly income higher than 2000 USD (Table 1). Respondents that reported monthly incomes higher than 2000 USD were preponderantly women aged between 28–37 years from urban areas. The respondents that reported between 1000 USD and 2000 USD were mostly women aged between 18 and 37 years from urban areas. The majority of our respondents (73.36%) earned a net monthly income in the interval 1000–2000 USD, their revenues being in accordance with the mean income, reported at the national level in the 3rd trimester, by the Romanian National Institute of Statistics [49].

Table 1. Summarizing the demographics of the sample.

Item	Class	N [1] = 383	
		No.	Percentage (%)
Gender	Women	281	73.62
	Men	101	26.37
Age	18–27 years	219	57.18
	28–37 years	102	26.63
	38–47 years	62	16.18
Residence	Urban	303	79.11
	Rural	80	20.89
Income	Lower than 1000 USD	53	13.83
	1000–2000 USD	281	73.36
	More than 2000 USD	49	12.80

[1] N = sample size.

We employed three models of the same variance-based, structural equation model. The difference between the models was represented by the outcome variable. The first model used clothing brand loyalty as the outcome, while the second and the third ones used two different dimensions of luxury clothing innovativeness. All three models have three predictors and one mediator variable.

The items that were used to measure consumers' brand loyalty were adapted from research that aimed to conceptualize the brand commitment and habit influence toward an attitude–behavior adoption [50]. The exploratory analysis allowed us to keep all items of the original scale. Regarding the measurement of luxury clothing innovativeness dimension, we adapted the Goldsmith and Hofacker scale [32], the difference being that our exploratory analysis split the original dimension into two distinct concepts and investigated each of them as an independent outcome variable. The items that correspond to the direct variables are presented in Table 2.

Table 2. Outcome variables: clothing brand loyalty and luxury clothing innovativeness.

Dimension	Item	
Clothing Brand Loyalty (CBL) [50]	I consider myself to be loyal to a specific luxury clothing brand.	CB1
	If my preferred brand of an item of clothing was not available at the store, I would shop at other stores until I found my brand.	CB2
Innovativeness Interest—Luxury Clothing Innovativeness (CIN_a) [51]	In general, I am among the last in my group of friends to buy a new outfit or fashion.	CN1
	Compared to my friends, I do little shopping for new fashions.	CN2
	In general, I am the last in my group of friends to know the names of the latest designers and fashion trends.	CN3
Innovativeness Awareness—Luxury Clothing Innovativeness (CIN_b) [51]	I know more about new fashions before other people do.	CN4
	If I heard that an innovative luxury outfit was available through a local clothing or department store, I would be interested enough to buy it.	CN5
	I will consider buying an innovative luxury fashion item, even if I have not heard of it yet.	CN6

The clothing involvement scale developed by Mittal and Lee [51] offers us the first predictor. Following an exploratory analysis, were kept only two items of three. The second and the third predictors were adapted from an existing status quo scale [8,42]. We emphasized two distinct latent variables that measure consumers' tendency toward status quo: namely, satisfaction with existing luxury fashion products and their satisfaction with the extent of fashion innovations. To measure those dimensions, we used the 7-point Likert scale, where 1 means "total disagreement" and 7 corresponds to "total agreement" (Table 3). Regarding the level of involvement influence on the luxury clothing innovativeness, we expected a moderate positive effect [25]. Concerning the impact of innovativeness on clothing brand loyalty, we anticipated a positive but modest relationship [25]. There is no former forecasting about the inherent existing relationship between status quo components and the considered outcome variables.

Table 3. The components of latent variable predictors: clothing involvement and status quo.

Latent Variable	Manifest Variable	Item
Clothing Involvement [51]	I have a strong interest in clothing.	CINV1
	Clothing is important to me.	CINV2
Satisfaction with existing luxury fashion products [8,42]	In the past, I was very satisfied with available luxury fashion products.	SQSP4
	In my opinion, past luxury fashion products were completely satisfactory so far.	SQSP5
	Past luxury fashion products fully met my requirements.	SQSP6
Satisfaction with the extent of fashion luxury innovations [8,42]	My personal need for innovations in the field of luxury fashion products has been by far not covered in the past.	SQSI1
	I consider the number of innovations in the field of luxury fashion products as being too low.	SQSI2
	I consider the pace of innovations in luxury fashion products as being too low.	SQSI3

Status quo is the principal predictor: a dimension consisting of two manifest variables, each one formed of three items. The difference between the satisfaction with the extent of fashion luxury innovations and satisfaction with existing luxury fashion products relates to the fact that the second manifest variable relates to situation-specific factors [8].

Status consumption is a latent predictor that was used as a mediator. Existing studies revealed that a desire for social position appoints a strong motivator to adopt a behavioral change [52]. Suitable for this study is the fact that clothing is a universal emblem of status [23], while social rank connects to brand loyalty and people's devotion to fashion pieces [53].

Social consumption is delimited as a motivational agent that encourages consumers to improve their social positions through the acquisition and consumption of products that symbolize status [54,55]. The status consumption dimension was measured through the scale developed by Eastman, Goldsmith, and Flynn [54]. The original scale contained five items, but the exploratory factor analysis dropped two items to keep the factor loadings above 0.7 (Table 4).

Table 4. The mediator variable "status consumption".

Latent Variable	Manifest Variable	Items
Status Consumption [54]	I would pay more for a product if it had status.	SC1
	I am interested in new products with status.	SC2
	A product is more valuable to me if it has some snob appeal.	SC3

Previous studies revealed a positive influence of status consumption on clothing brand loyalty and the level of clothing innovativeness [25]. Regarding the impact of status quo latent variables, we expected a negative influence on status consumption because people who are inclined to status quo manifest a tendency to strive for consistency, while high scores on the status consumption scale are associated with complex personalities [24]. Regarding the clothing involvement, we expected a positive influence on status consumption [25].

Gender and residence were used as control variables. Earlier studies revealed that women are more willing to buy luxury fashion clothes than men [47]. Regarding residence, existing studies revealed that people from urban areas are more prone to invest in luxury items than people from rural areas [48].

Table 5 creates a summary of the latent variable predictors, which were formed as a weighted average of their related manifest variables [56].

Table 5. An outline of latent predictors, with abbreviations and descriptions.

Latent Structure	Observed Variables
SC	Status consumption. Refers to people interested in confirming their social position; SC1, SC2, SC3.
SQSP	Satisfaction with existing luxury fashion products. Part of the status quo dimension; SQSP4, SQSP5, SQSP6.
SQSI	Satisfaction with the extent of fashion luxury innovations. Part of the status quo dimension; SQSI1, SQSI2, SQSI3
CINV	Clothing involvement. Describes a subjective disposition of being interested and excited about a particular product category; CINV1, CINV2.
CBL	Clothing brand loyalty. Refers to consumers' relationship with the brand that they buy; CBL1, CBL2.
CIN_a	Innovativeness interest. Part of the luxury clothing innovativeness scale. Captures consumers interest toward exiting innovation in luxury fashion clothing; CIN1, CIN2, CIN3.
CIN_b	Innovativeness Awareness. Part of the luxury clothing innovativeness scale. Captures consumers awareness toward exiting innovation in luxury fashion clothing; CIN4, CIN5, CIN6.

The research model is presented in Figure 1.

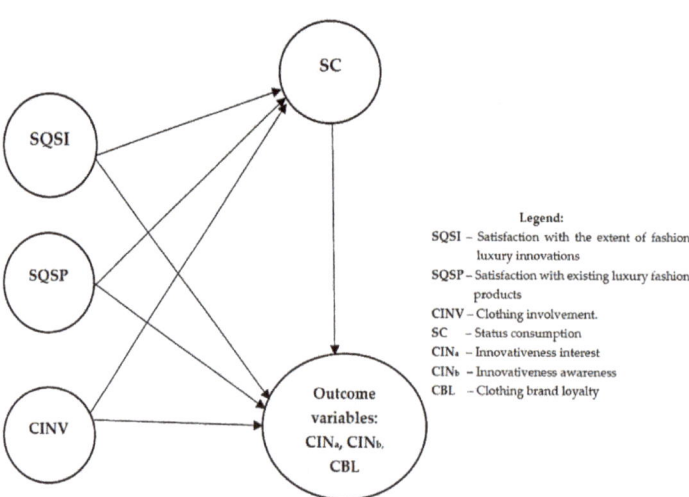

Figure 1. The research model.

4. Results

The PLS-PM analysis proposes maximizing the explained variance of the dependent latent variables [57]. Our outcome variables are innovativeness interest (CIN_a), innovativeness awareness (CIN_b), and clothing brand loyalty (CBL). We used status consumption

(SC) as a mediator in the relationship between the predictors and outcome variables. Status quo with its two manifest variables is the main predictor.

Briefly speaking, the estimation technique describes an iterative algorithm that is based on the ordinary least squares' approach. A PLS-PM analysis considers a measurement model and a structural model. The measurement model estimates the relationship between the latent variables and their corresponding manifest variables evaluated considering the composite criteria. The structural model investigates the existing relationships between latent variables.

In the incipient stage of our analysis, the model was estimated using the R software, version 3.6.3 [58] with the "lavaan" package [59]. After that, the obtained results were checked using WarpPLS software, version 7.0 [60].

The reliability results of the measurement model were exposed in Table 6. In terms of the composite reliability, the scores ranged from 0.791 to 0.965, those registered values being above the score of 0.7 recommended by the theoretical background [61]. Concerning the Cronbach alpha indicator, all manifest variables registered scores above the 0.70 threshold [62]. In the case of the average variance extracted (AVE), we considered relevant a score higher than 0.50 [63], which was a criterion met for all latent variables.

Table 6. Evaluation of the measurement model.

Variable	Abbreviation	Composite Reliability	Cronbach's Alpha	Average Variance Extracted (AVE)
Status consumption	SC	0.897	0.791	0.711
Clothing involvement	CINV	0.965	0.927	0.932
Innovativeness interest	CIN_a	0.913	0.856	0.777
Innovativeness awareness	CIN_b	0.866	0.767	0.683
Clothing brand loyalty	CBL	0.894	0.763	0.808
Satisfaction with the extent of fashion luxury innovations	SQSI	0.845	0.725	0.646
Satisfaction with existing luxury fashion product	SQSP	0.932	0.890	0.821

Table 7 presents diagonal components of the inter-correlation matrix. It is desirable to register higher values of the square roots of all the averages variance extracted values than their off-diagonal components in their corresponding lines and columns. Moreover, all off-diagonal coefficients of correlation were below the 0.8 recommended cut-off [64].

Table 7. Inter-correlation of variables constructs.

Variable	SC	CINV	CIN_a	CIN_b	CBL	SQSI	SQSP
SC	0.843	0.344	−0.228	0.572	0.515	0.369	0.357
CINV		0.965	−0.410	0.467	0.321	0.221	0.312
CIN_a			0.882	−0.389	−0.170	−0.008	−0.224
CIN_b				0.826	0.620	0.387	0.441
CBL					0.899	0.355	0.402
SQSI						0.804	0.186
SQSP							0.906

Table 8 shows the loadings and cross-loadings of the manifested variables considered in the present research. All loadings registered values higher than 0.7, with values ranging from 0.740 to 0.981. The fact that items that correspond to the same dimension register values higher than cross-constructs loadings show the convergent validity of those indicators and infer that they cluster in distinct latent dimensions.

Table 8. Inter-item correlations of variable constructs.

Variable	SQSP	SQSI	CINV	CIN$_a$	CIN$_b$	SC	CBL
SQSP6	0.932	0.152	0.264	−0.198	0.417	0.350	0.370
SQSP5	0.929	0.147	0.252	−0.183	0.355	0.295	0.352
SQSP4	0.856	0.210	0.338	−0.232	0.430	0.325	0.372
SQSI1	0.107	0.770	0.141	0.102	0.276	0.281	0.283
SQSI2	0.118	0.830	0.135	−0.010	0.273	0.252	0.268
SQSI3	0.223	0.809	0.258	−0.106	0.384	0.357	0.306
CINV1	0.329	0.245	0.965	−0.402	0.463	0.324	0.310
CINV2	0.273	0.182	0.965	−0.389	0.439	0.339	0.310
CIN2	−0.202	0.020	−0.363	0.911	−0.344	−0.215	−0.181
CIN3	−0.216	−0.053	−0.407	0.873	−0.464	−0.253	−0.182
CIN1	−0.175	0.011	−0.312	0.860	−0.219	−0.135	−0.085
CIN4	0.326	0.267	0.444	−0.415	0.797	0.395	0.462
CIN5	0.456	0.387	0.403	−0.362	0.872	0.538	0.575
CIN6	0.304	0.299	0.312	−0.187	0.807	0.479	0.497
SC1	0.335	0.298	0.248	−0.204	0.510	0.893	0.507
SC3	0.342	0.406	0.275	−0.191	0.530	0.904	0.483
SC4	0.212	0.212	0.365	−0.184	0.396	0.719	0.291
CBL1	0.391	0.332	0.291	−0.147	0.573	0.500	0.899
CBL2	0.331	0.307	0.286	−0.159	0.543	0.427	0.899

Table 9 presents the results of the first structural model. This model uses clothing brand loyalty as the outcome variable. Table 10 presents the results of the second structural model, where innovativeness interest is the outcome variable. The third structural model, which uses innovativeness awareness as a direct variable, is presented in Table 11. The models' figures are placed in the Appendix A. Figure A1 corresponds to the first structural equation model, Figure A2 corresponds to the second structural model, and Figure A3 refers to the third model.

Table 9. The first structural equation model.

	Direct Effects		Indirect Effects	Direct Effect Sizes		Total Effects (Direct Effect + Indirect Effect via Status Consumption)
	SC	CBL	CBL	SC	CBL	CBL
SC	-	0.339 *** [0.049] (<0.001)		-	0.176 [0.049]	0.339 *** [0.049] (<0.001)
SQSI	0.289 *** [0.049] (<0.001)	0.164 *** [0.049] (<0.001)	0.098 ** [0.036] (0.003)	0.110 [0.049]	0.093 [0.049]	0.262 *** [0.049] (<0.001)
SQSP	0.239 *** [0.049] (<0.001)	0.243 *** [0.049] (<0.001)	0.081 * [0.036] (0.012)	0.087 [0.049]	0.135 [0.049]	0.324 *** [0.049] (<0.001)
CINV	0.218 *** [0.050] (<0.001)	0.117 * [0.050] (0.010)	0.074 * [0.036] (0.020)	0.076 [0.050]	0.063 [0.050]	0.191 *** [0.050] (<0.001)
Gender	-	−0.044 [0.051] (0.196)	-	-	0.002 [0.051]	−0.044 [0.051] (0.196)
Residence	-	0.05 [0.051] (0.457)	-	-	0.000 [0.051]	0.05 [0.051] (0.457)
R^2/Adj R^2	0.273/0.267	0.373/0.363	-	-	-	-

Note: *** p value < 0.001; ** p value < 0.01; * p value < 0.05; · p value < 0.10; [ ]—standard error.

Table 10. The second structural equation model.

	Direct Effects		Indirect Effects	Direct Effect Sizes		Total Effects (Direct Effect + Indirect Effect via Status Consumption
	SC	CIN$_a$	CIN$_a$	SC	CIN$_a$	CIN$_a$
SC	-	0.389 ** [0.050] (0.002)		-	0.233 [0.050]	0.389 *** [0.050] (<0.001)
SQSI	0.239 *** [0.049] (<0.001)	0.165 * [0.050] (0.006)	0.113 *** [0.036] (<0.001)	0.110 [0.049]	0.110 [0.050]	0.278 *** [0.050] (<0.001)
SQSP	0.239 *** [0.049] (<0.001)	0.187 ** [0.050] (0.004)	0.093 ** [0.036] (0.005)	0.087 [0.049]	0.123 [0.050]	0.280 *** [0.050] (<0.001)
CINV	0.218 *** [0.050] (<0.001)	0.242 *** [0.049] (<0.001)	0.085 * [0.036] (0.009)	0.076 [0.050]	0.153 [0.049]	0.327 *** [0.049] (<0.001)
Gender	-	0.022 [0.051] (0.336)	-	-	0.002 [0.051]	0.022 [0.051] (0.336)
Residence	-	−0.052 [0.051] (0.154)	-	-	0.001 [0.051]	−0.052 [0.051] (0.154)
R^2/Adj R^2	0.273/0.267	0.192/0.179	-	-	-	-

Note: *** p value < 0.001; ** p value < 0.01; * p value < 0.05; · p value < 0.10; [ ]—standard error.

Table 11. The third structural equation model.

	Direct Effects		Indirect Effects	Direct Effect Sizes		Total Effects (Direct Effect + Indirect Effect via Status Consumption)
	SC	CIN$_b$	CIN$_b$	SC	CIN$_b$	CIN$_b$
SC	-	0.339 *** [0.048] (<0.001)		-	0.176 [0.048]	0.339 *** [0.048] (<0.001)
SQSI	0.289 *** [0.049] (<0.001)	0.164 *** [0.050] (<0.001)	0.098 ** [0.036] (0.003)	0.110 [0.049]	0.093 [0.049]	0.262 *** [0.049] (<0.001)
SQSP	0.239 *** [0.049] (<0.001)	0.243 *** [0.050] (<0.001)	0.081 * [0.036] (0.012)	0.087 [0.049]	0.135 [0.049]	0.324 *** [0.049] (<0.001)
CINV	0.218 *** [0.050] (0.001)	0.117 * [0.050] (0.010)	0.074 * [0.036] (0.020)	0.076 [0.050]	0.063 [0.049]	0.191 *** [0.049] (<0.001)
Gender	-	−0.044 [0.051] (0.196)	-	-	0.002 [0.051]	−0.044 [0.051] (0.196)
Residence	-	0.05 [0.051] (0.457)	-	-	0.000 [0.051]	0.05 [0.051] (0.457)
R^2/Adj R^2	0.273/0.267	0.495/0.487	-	-	-	-

Note: *** p value < 0.001; ** p value < 0.01; * p value < 0.05; · p value < 0.10; [ ]—standard error.

All three models previously presented estimated the direct, indirect, and total effects and their level of statistical significance. It was also reported the effect size of each direct path. Tables 9–11 indicated very food explanatory power, with R-squared values of 37.3 (clothing brand loyalty); 27.3 (status consumption); 19.2 (innovativeness interest); and 49.5 (innovativeness awareness).

Table 12 realizes a summary of the already exposed results in Tables 9–11 by revealing the effect and the level of significance of each relationship in a simpler format, having the three versions of the model in the same table.

Table 12. Simplified structural equation model (with the mediator).

Predictor	Status Consumption	Direct Effects			Indirect Effects			Total Effects		
		CBL	CIN_a	CIN_b	CBL	CIN_a	CIN_b	CBL	CIN_a	CIN_b
SQSI	+ ***	+ ***	+ *	+ ***	+ **	+ ***	+ **	+ ***	+ ***	+ ***
SQSP	+ ***	+ ***	+ **	+ ***	+ *	+ **	+ *	+ ***	+ ***	+ ***
CINV	+ ***	+ *	+ ***	+ *	+ *	+ *	+ *	+ ***	+ ***	+ ***
Residence	None	+	-	+	None	None	None	+	-	+
Gender	None	-	+	-	None	None	None	-	+	-

Note: *** p value < 0.001; ** p value < 0.01; * p value < 0.05; · p value < 0.10; + positive coefficient; - negative coefficient.

SQSI is a positive and significant predictor of the status consumption and the mediator variable, and it manifests a statistically significant and positive total effect on all outcome variables. It manifests statistically significant total effects on clothing brand loyalty, innovativeness interest, and innovativeness awareness. In addition, the SQSI predictor manifests a direct and indirect effect on all outcome variables. It was considered that status consumption fully mediates the relationship between satisfaction with the extent of fashion luxury innovations and all outcome variables. Satisfaction with the extent of fashion luxury innovations increases consumers' level of clothing brand loyalty, as well as their level of innovativeness. As expected, the relationship is mediated by status consumption. Even if there was no former expectation regarding the inherent potential relationship between the status quo manifest variables and the considered outcome variables, the positive relationship between status quo and status consumption was unexpected. Status consumption designates people's motivator to adopt a behavioral change to improve their social position. This positive relationship between status quo and status consumption reveals that people are satisfied with their acquisitions. The prestige and social status that they acquire while purchasing the existing luxury fashion innovations is satisfactory.

Satisfaction with existing luxury fashion products manifests a positive influence on status consumption (the mediator) and has a statistical total effect on all outcome variables. There is a direct and positive effect on all outcome variables. The indirect effect on the clothing brand loyalty and innovativeness awareness (via the mediator, status consumption) has a slight statistical significance. In this case, status consumption fully mediates the relationship between satisfaction with existing luxury fashion products and the considered outcomes.

Regarding the clothing involvement, we expected a positive influence on status consumption [25]. That hypothesis was confirmed by our results. It was observed that clothing involvement exhibits positive total effects on all outcome variables. When decomposed, the indirect effect is marginally significant in the case of all outcome variables. The direct effect is highly significant in the case of innovativeness interest outcome variables and marginally significant for the rest of the outcome variables. As expected, the relationship between all three outcome variables and clothing involvement is mediated by status consumption.

Gender manifests negative total effects on clothing brand loyalty and innovativeness awareness, but the influence is not statistically significant. The relationship with the innovativeness interest is positive, but again, it is not statistically significant. There are no indirect effects, only direct effects.

A similar situation emerged regarding the residence—total effects are not significant, and there are only direct effects on the outcome variables. The influence on clothing brand loyalty and innovativeness awareness is negative, while the impact on innovativeness involvement is positive. In those models, residence and gender are not good predictors of the outcome variables.

5. Discussion

The considered models showed that status quo manifest constructs present a strong statistical significance in their relationship with consumers' loyalty to brands. The satisfaction experienced by consumers determines them to create a sort of emotional connection with the brand. This relation is available over time, as the excitement that is offered by that brand remains constant.

On the other hand, status quo influence on consumers' innovativeness awareness and their innovativeness interest was positive, contrary to our expectations. This was interpreted in the light that the existing innovations in the luxury fashion markets are not such disruptive to disorder the consumers' equilibrium such that to determine their resistance to the available novelties. It was interesting to note the positive relationship between status quo and status consumption, suggesting that consumers manifest satisfaction with the acquisition that they gather, their desire for social confirmation and prestige being satisfied by the already exiting fashion innovations.

The positive influence of the clothing involvement on the investigated predictors was expected [25] and confirmed by our research. Clothing involvement exhibited a positive effect on the mediator: status consumption.

Previous studies hypothesized the positive influence of the innovativeness in fashion on consumers' brand loyalty [17,55]. Moreover, we hypothesized the positive impact of the innovativeness on status consumption when discussing luxury fashion products [17,56], and the influence that was also confirmed by our research. An innovative consumer results in high brand loyalty, being in the interest of the luxury fashion providers to maintain a high level of fashion design. Of course, this aspect may not prove applicable if referring to other types of products than luxuries.

Existing studies find a positive relationship between consumers' status consumption and their loyalty to brands [57]. Consumers that manifest interest in status-seeking discover which brands provide and confirm this status and develop a connection with them as time as this status confirmation persists. This aspect is more visible when referring to clothes because those act as a dominant symbol of social status [27]. Our research also confirmed this positive influence among status confirmation and brand loyalty.

The hypothesis testing results suggest that there is a need to pay attention to the status consumption. Its direct effects on clothing brand loyalty and consumer level of innovativeness are both significant. Moreover, it fully mediates all the tested relationships among predictors and the outcome variables, indicating that the consumers' connection with a brand or its interest concerning innovations in the area of luxury fashion innovations is dictated by social status confirmation.

As the luxury market advances, consumers exhibit interest in luxury products because of their symbolic value, status consumption manifesting an essential role in attracting consumers that are eager to show their place in society. This aspect could be in the interest of the marketers, to align their marketing strategies to the segment of customers who are interested in status confirmation.

All three models have good explanatory power. The predictors explained 50% of the variation of the innovativeness awareness. Furthermore, 27.3% of the mediator variation is explained by the predictors included in the analysis. The control variables do not manifest a statistically significant influence.

The contribution of each predictor to the overall explanatory power of the model represents an essential feature. To be considered as being an effective contribution, the effect size of that predictor should be above 0.02 [65]. In all models, it seems that the status consumption manifests the greatest effect size of all predictors. In addition, in the first and the third models, satisfaction with existing luxury fashion products manifests the second most powerful effect size.

6. Conclusions

The existing literature revealed that status consumption represents an effective motivator to adopt a behavioral change. Status consumption was identified as a good predictor of consumers' brand loyalty and their innovativeness capacities. This paper proposed to take the available literature one step further. It investigated the relationship between consumers' default tendencies, their level of involvement in clothing, and their loyalty to brands or innovativeness tendencies, using the status consumption as a mediator. The focus of the present research was designated by luxury fashion items, which is a product category where status consumption manifests an essential role. The analysis was conducted on teenagers because there are only a few studies that investigate their position regarding luxury clothing innovations and because of the development of the teenagers market, it is essential to identify and understand their attitude concerning this topic. Moreover, the actual study was conducted on Romania, which is an emerging country that is part of the European Union, a highly internationalized market. In this context, the behavioral response of the youth consumers may be in the interest of the fashion providers and international clothing clusters. The gathered data of the present study consisted of a sample of 383 students enrolled in different universities in Romania, most of them aged between 18 and 27 years old. It was used as a method in the PLS-PM analysis. There were three outcome variables: clothing brand loyalty, innovativeness interest, and innovativeness awareness. The principal predictor was the status quo with its two manifest variables: satisfaction with the extent of innovations and satisfaction with the existing luxury fashion products. The control variables were gender and residence. Status consumption was used as a mediator between the considered predictors and our outcome variables.

The obtained results highlighted a positive relationship among status consumption and consumers' loyalty to brands, their level of involvement, and innovativeness tendencies. Moreover, it was revealed that status consumption is a good mediator of the relationship among proposed predictors and considered outcome variables. In addition, status consumption manifested the highest effect size of all investigated predictors.

The results suggest that there is a need to direct our attention to the status consumption variable. The fact that it fully mediates all the tested relationships among predictors and the outcome variables indicates that the consumers' connection with a brand or its interest concerning luxury fashion innovations is related to social status confirmation.

The present study is an exploratory one, and there are several limitations of this research. For the future, it is proposed to test these hypotheses on distinct groups of consumers to check the reliability of the results. Even if the results sustained the hypotheses, the study is not without limitations. The obtained results have a limited degree of generality considering the convenience sample and its relative homogeneity. It is quite probably to not gather the same results on different groups of consumers. In addition, our findings are limited to one product category. It could not be assumed that the same results would be obtained for other product categories. In the end, the limitation is figured out by the fact that the model does not hold all predictable essential predictors. It is necessary to conduct further theoretical investigations to fully understand the phenomena before testing new models.

Author Contributions: Conceptualization, A.-I.P.; Methodology, A.-I.P.; Software, A.-I.P.; Validation, A.-I.P.; Investigation, A.-I.P.; Resources, A.-I.P.; Data curation, A.-I.P.; Writing—original draft preparation, A.-I.P.; Writing—review and editing, A.M.A., C.C. and A.B. All authors have read and agreed to the published version of the manuscript.

Funding: The University of Bucharest financially supported this research.

Conflicts of Interest: The authors declare no conflict of interest.

Appendix A

Analysis results presented in figures.

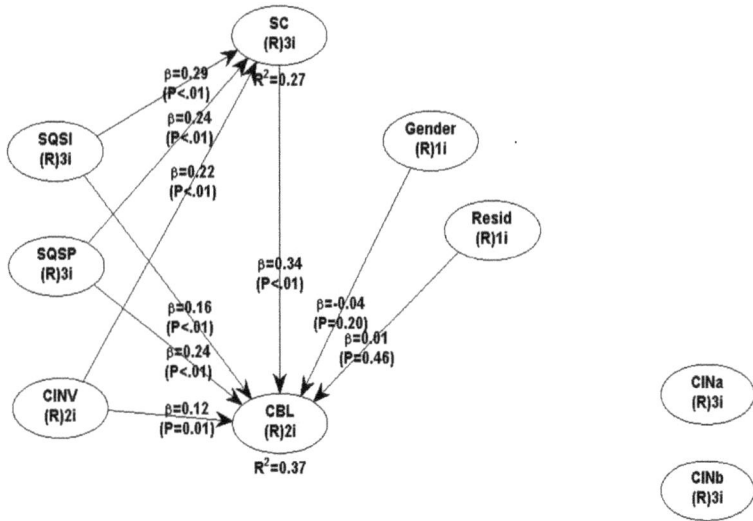

Figure A1. The first structural equation model—clothing brand loyalty (CBL) is the outcome variable.

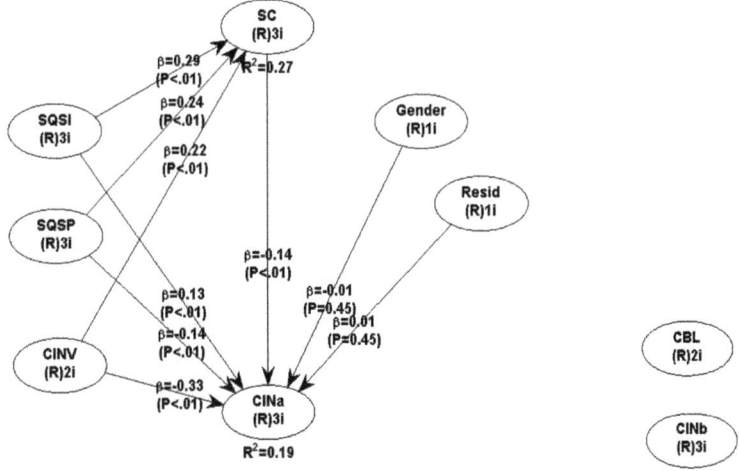

Figure A2. The second structural equation model—innovativeness interest (CIN_a) is the outcome variable.

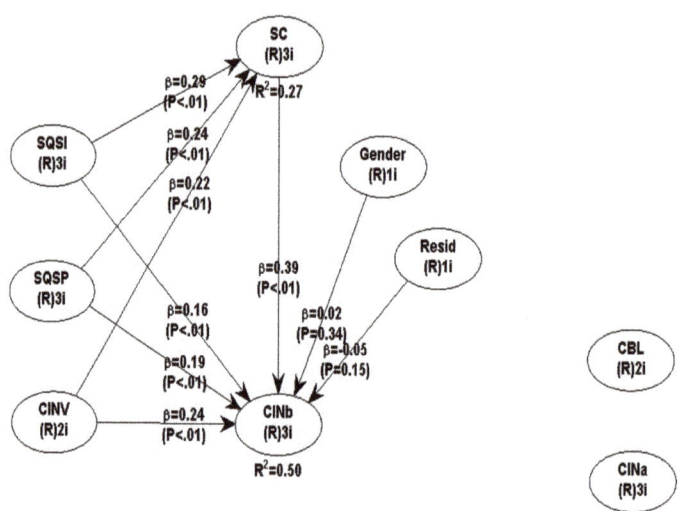

Figure A3. The third structural equation model—innovativeness awareness (CIN$_b$) is the outcome variable.

References

1. Ostrovskiy, A.; Garkavenko, V.; Rybina, L. Influence of Socio-Psychological Factors on Consumers Purchasing Behavior in Kazakhstan. *Serv. Ind. J.* **2019**, 1–26. [CrossRef]
2. Solomon, M.R. The Role of Products as Social Stimuli: A Symbolic Interactionism Perspective. *J. Consum. Res.* **1983**, *10*, 319. [CrossRef]
3. Husic, M.; Cicic, M. Luxury Consumption Factors. *J. Fash. Mark. Manag.* **2009**, *13*, 231–245. [CrossRef]
4. Nelissen, R.M.A.; Meijers, M.H.C. Social Benefits of Luxury Brands as Costly Signals of Wealth and Status. *Evol. Hum. Behav.* **2011**, *32*, 343–355. [CrossRef]
5. Veblen, T.; Mills, C.W. *The Theory of the Leisure Class*; Transaction Publishers: New Brunswick, NJ, USA, 1992; ISBN 978-1-56000-562-9.
6. Butcher, L.; Phau, I.; Teah, M. Brand Prominence in Luxury Consumption: Will Emotional Value Adjudicate Our Longing for Status? *J. Brand Manag.* **2016**, *23*, 701–715. [CrossRef]
7. Klabi, F. To What Extent Do Conspicuous Consumption and Status Consumption Reinforce the Effect of Self-Image Congruence on Emotional Brand Attachment? Evidence from the Kingdom of Saudi Arabia. *J. Mark. Anal.* **2020**, *8*, 99–117. [CrossRef]
8. Heidenreich, S.; Handrich, M. What about Passive Innovation Resistance? Investigating Adoption-Related Behavior from a Resistance Perspective. *J. Prod. Innov. Manag.* **2015**, *32*, 878–903. [CrossRef]
9. Schiffman, L.G.; Kanuk, L.L. *Consumer Behavior*, 9th ed.; Pearson Prentice Hall: Upper Saddle River, NJ, USA, 2007; ISBN 978-0-13-186960-8.
10. Shackle, G.L.S.; Duesenberry, J.S. Income, Saving, and the Theory of Consumer Behaviour. *Econ. J.* **1951**, *61*, 131. [CrossRef]
11. Puiu, A.-I. Romanian Young Adults' Attitudes Regarding Luxury Fashion Brands: A Behavioral Identity Perspective. *Int. J. Appl. Behav. Econ.* **2021**, *10*, 1–21. [CrossRef]
12. Schade, M.; Hegner, S.; Horstmann, F.; Brinkmann, N. The Impact of Attitude Functions on Luxury Brand Consumption: An Age-Based Group Comparison. *J. Bus. Res.* **2016**, *69*, 314–322. [CrossRef]
13. Nistor, L. Young Consumers' Fashion Brand Preferences. An Investigation among Students in Romania. *Acta Univ. Sapientiae Commun.* **2019**, *6*, 41–59. [CrossRef]
14. Okonkwo, U. *Luxury Fashion Branding: Trends, Tactics, Techniques*; Palgrave Macmillan: Basingstoke, UK; Hampshire, UK; New York, NY, USA, 2007; ISBN 978-0-230-52167-4.
15. Goody, J. From Misery to Luxury. *Soc. Sci. Inf.* **2006**, *45*, 341–348. [CrossRef]
16. Berthon, P.; Pitt, L.; Parent, M.; Berthon, J.-P. Aesthetics and Ephemerality: Observing and Preserving the Luxury Brand. *Calif. Manag. Rev.* **2009**, *52*, 45–66. [CrossRef]
17. Pino, G.; Amatulli, C.; Peluso, A.M.; Nataraajan, R.; Guido, G. Brand Prominence and Social Status in Luxury Consumption: A Comparison of Emerging and Mature Markets. *J. Retail. Consum. Serv.* **2019**, *46*, 163–172. [CrossRef]
18. Scheetz, T.K.; Dubin, R.; Garbarino, E.C. A Modern Investigation of Status Consumption. 2007. Available online: https://artscimedia.case.edu/wp-content/uploads/sites/57/2014/01/14235806/scheetzreport.pdf. (accessed on 12 January 2021).

19. Csikszentmihalyi, M.; Halton, E. *The Meaning of Things: Domestic Symbols and the Self*; 1. publ. 1981, reprinted; Cambridge Univ. Press: Cambridge, UK; New York, NY, USA; Melbourne, Australia, 1995; ISBN 978-0-521-28774-6.
20. Chen, Y.; Lu, F.; Zhang, J. Social Comparisons, Status and Driving Behavior. *J. Public Econ.* **2017**, *155*, 11–20. [CrossRef]
21. Martineau, P. *Motivation in Advertising: Motives That Make People Buy*; McGraw-Hill: New York, NY, USA, 1971; ISBN 978-0-07-040661-2.
22. Chao, A.; Schor, J.B. Empirical Tests of Status Consumption: Evidence from Women's Cosmetics. *J. Econ. Psychol.* **1998**, *19*, 107–131. [CrossRef]
23. Solomon, M.R.; Rabolt, N.J. *Consumer Behavior in Fashion*, 2nd ed.; Prentice Hall: Upper Saddle River, NJ, USA, 2009; ISBN 978-0-13-171474-8.
24. Goldsmith, R.E.; Flynn, L.R.; Eastman, J.K. Status Consumption and Fashion Behaviur: An Exploratory Study. *Proc. Assoc. Mark. Theory Pract.* **1996**, *5*, 309–316.
25. Goldsmith, R.E.; Flynn, L.R.; Kim, D. Status Consumption and Price Sensitivity. *J. Mark. Theory Pract.* **2010**, *18*, 323–338. [CrossRef]
26. Deeter-Schmelz, D.R.; Moore, J.N.; Goebel, D.J. Prestige Clothing Shopping by Consumers: A Confirmatory Assessment and Refinement of the *Precon* Scale with Managerial Implications. *J. Mark. Theory Pract.* **2000**, *8*, 43–58. [CrossRef]
27. Rogers, E.M. *Diffusion of Innovations*, 5th ed.; Free Press: New York, NY, USA, 2003; ISBN 978-0-7432-2209-9.
28. Beaudoin, P. Determinants of Adolescents' Brand Sensitivity to Clothing. *Fam. Consum. Sci. Res. J.* **2006**, *34*, 312–331. [CrossRef]
29. Workman, J.E.; Kidd, L.K. Use of the Need for Uniqueness Scale to Characterize Fashion Consumer Groups. *Cloth. Text. Res. J.* **2000**, *18*, 227–236. [CrossRef]
30. Bloch, P.H. The Product Enthusiast: Implications for Marketing Strategy. *J. Consum. Mark.* **1986**, *3*, 51–62. [CrossRef]
31. McCracken, G.D.; Roth, V.J. Does Clothing Have a Code? Empirical Findings and Theoretical Implications in the Study of Clothing as a Means of Communication. *Int. J. Res. Mark.* **1989**, *6*, 13–33. [CrossRef]
32. Goldsmith, R.E.; Hofacker, C.F. Measuring Consumer Innovativeness. *J. Acad. Mark. Sci.* **1991**, *19*, 209–221. [CrossRef]
33. Flynn, L.R.; Goldsmith, R.E. A Validation of the Goldsmith and Hofacker Innovativeness Scale. *Educ. Psychol. Meas.* **1993**, *53*, 1105–1116. [CrossRef]
34. O'Cass, A. An Assessment of Consumers Product, Purchase Decision, Advertising and Consumption Involvement in Fashion Clothing. *J. Econ. Psychol.* **2000**, *21*, 545–576. [CrossRef]
35. Gabriel, Y.; Lang, T. *The Unmanageable Consumer*, 2nd ed.; Sage Publications: Thousand Oaks, CA, USA, 2006; ISBN 978-1-4129-1892-3.
36. Hu, Y.; Van den Bulte, C. The Social Status of Innovators, Imitators, and Influentials in New Product Adoption: It's Not Just About High Versus Low—Marketing Science Institute. Available online: https://www.msi.org/working-papers/the-social-status-of-innovators-imitators-and-influentials-in-new-product-adoption-its-not-just-about-high-versus-low/ (accessed on 20 January 2021).
37. Jayarathne, P.G.S.A. Mediating Role of Fashion Consciousness on Cosmopolitanism and Status Consumption of Young Fashion Consumers in Sri Lanka. *Adv. Sci. Lett.* **2018**, *24*, 3456–3458. [CrossRef]
38. Fournier, S. Consumers and Their Brands: Developing Relationship Theory in Consumer Research. *J. Consum. Res.* **1998**, *24*, 343–353. [CrossRef]
39. Romaniuk, J.; Nenycz-Thiel, M. Behavioral Brand Loyalty and Consumer Brand Associations. *J. Bus. Res.* **2013**, *66*, 67–72. [CrossRef]
40. Kahneman, D.; Tversky, A. Prospect Theory: An Analysis of Decision under Risk. *Econometrica* **1979**, *47*, 263. [CrossRef]
41. Wang, Y. Observational Learning in the Product Configuration Process: An Empirical Study. In Proceedings of the 2018 IEEE International Conference on Industrial Engineering and Engineering Management (IEEM), Bangkok, Thailand, 16–19 December 2018; pp. 1211–1215.
42. Puiu, I.-A. Consumer Resistance to Innovation in the Fashion Industry. *Stud. Bus. Econ.* **2019**, *14*, 127–140. [CrossRef]
43. Muzinich, N.; Pecotich, A.; Putrevu, S. A Model of the Antecedents and Consequents of Female Fashion Innovativeness. *J. Retail. Consum. Serv.* **2003**, *10*, 297–310. [CrossRef]
44. Kock, N.; Hadaya, P. Minimum Sample Size Estimation in PLS-SEM: The Inverse Square Root and Gamma-Exponential Methods: Sample Size in PLS-Based SEM. *Info. Syst. J.* **2018**, *28*, 227–261. [CrossRef]
45. Goodhue, D.L.; Lewis, W.; Thompson, R. Does PLS Have Advantages for Small Sample Size or Non-Normal Data? *MIS Q.* **2012**, *36*, 981. [CrossRef]
46. Hair, J.F., Jr.; Matthews, L.M.; Matthews, R.L.; Sarstedt, M. PLS-SEM or CB-SEM: Updated Guidelines on Which Method to Use. *Int. J. Multivar. Data Anal.* **2017**, *1*, 107. [CrossRef]
47. Stokburger-Sauer, N.E.; Teichmann, K. Is Luxury Just a Female Thing? The Role of Gender in Luxury Brand Consumption. *J. Bus. Res.* **2013**, *66*, 889–896. [CrossRef]
48. Dobbs, R.; Remes, J.; Manyika, J.; Roxburgh, C.; Smith, S.; Schaer, F. *Urban World: Cities and the Rise of the Consuming Class*; McKinsey Global Institute, 2012. Available online: https://www.mckinsey.com/featured-insights/urbanization/urban-world-cities-and-the-rise-of-the-consuming-class# (accessed on 12 January 2021).
49. NIS-Romania Comunicat de Presă. Domeniul: Nivel de Trai 2021. Available online: https://insse.ro/cms/sites/default/files/com_presa/com_pdf/cs11r20.pdf (accessed on 12 January 2021).
50. Beatty, S.E.; Kahle, L.R. Alternative Hierarchies of the Attitude-Behavior Relationship: The Impact of Brand Commitment and Habit. *J. Acad. Mark. Sci.* **1988**, *16*, 1–10. [CrossRef]
51. Mittal, B.; Lee, M.-S. A Causal Model of Consumer Involvement. *J. Econ. Psychol.* **1989**, *10*, 363–389. [CrossRef]

52. Phau, I.; Siew Leng, Y. Attitudes toward Domestic and Foreign Luxury Brand Apparel: A Comparison between Status and Non Status Seeking Teenagers. *J. Fash. Mark. Manag.* **2008**, *12*, 68–89. [CrossRef]
53. O'Cass, A.; Frost, H. Status Brands: Examining the Effects of Non-Product-Related Brand Associations on Status and Conspicuous Consumption. *J. Prod. Brand Manag.* **2002**, *11*, 67–88. [CrossRef]
54. Eastman, J.K.; Goldsmith, R.E.; Flynn, L.R. Status Consumption in Consumer Behavior: Scale Development and Validation. *J. Mark. Theory Pract.* **1999**, *7*, 41–52. [CrossRef]
55. Nabi, N.; O'Cass, A.; Siahtiri, V. Status Consumption in Newly Emerging Countries: The Influence of Personality Traits and the Mediating Role of Motivation to Consume Conspicuously. *J. Retail. Consum. Serv.* **2019**, *46*, 173–178. [CrossRef]
56. Fornell, C.; Bookstein, F.L. Two Structural Equation Models: LISREL and PLS Applied to Consumer Exit-Voice Theory. *J. Mark. Res.* **1982**, *19*, 440. [CrossRef]
57. Homburg, C.; Klarmann, M.; Vomberg, A. *Handbook of Market. Research*; SAGE Publications Inc.: New York, NY, USA, 2017; ISBN 978-3-319-05542-8.
58. R Core Team. *R: A Language and Environment for Statistical Computing*; R Foundation for Statistical Computing: Vienna, Austria, 2020.
59. Rosseel, Y. Lavaan: An R Package for Structural Equation Modeling. *J. Stat. Soft.* **2012**, *48*. [CrossRef]
60. Kock, N. *WarpPLS User Manual (Latest Version: 7.0, 2020)*; ScriptWarp Systems: Laredo, TX, USA, 2020.
61. Nunnally, J.C. Psychometric Theory—25 Years Ago and Now. *Educ. Res.* **1975**, *4*, 7–21. [CrossRef]
62. Cronbach, L.J. Coefficient Alpha and the Internal Structure of Tests. *Psychometrika* **1951**, *16*, 297–334. [CrossRef]
63. Fornell, C.; Larcker, D.F. Evaluating Structural Equation Models with Unobservable Variables and Measurement Error. *J. Mark. Res.* **1981**, *18*, 39. [CrossRef]
64. Kennedy, P. *A Guide to Econometrics*, 6th ed.; Blackwell Pub.: Malden, MA, USA, 2008; ISBN 978-1-4051-8258-4.
65. Cohen, J. *Statistical Power Analysis for the Behavioral Sciences*, Rev. ed.; Academic Press: New York, NY, USA, 1977; ISBN 978-0-12-179060-8.

Article

How Governance Paradigms and Other Drivers Affect Public Managers' Use of Innovation Practices. A PLS-SEM Analysis and Model

Alberto Peralta and Luis Rubalcaba *

Department of Business and Economics, Faculty of Business, Economics, and Tourism, University of Alcala (UAH), 28802 Alcalá de Henares, Spain; alberto.peralta@edu.uah.es
* Correspondence: luis.rubalcaba@uah.es

Abstract: Using the Unified Theory of Acceptance and Use of Technology for Innovations in the Public Sector (UTAUT-IPS) model, this study examined the influences on using a specific innovation practice on public managers. We based our analysis on an end-of-2019 sample of 227 Spanish public managers, aiming to answer the question "Are public innovation and project managers driven only by a governance paradigm, influencing their intention and usage of an innovation practice?" Using the Partial Least Squares Structural Equation Modelling (PLS-SEM) algorithm, we singled out the effects of the governance paradigm, performance expectancy, and motivation, among seven other behavioral composite variables. The PLS-Prediction-Oriented Segmentation routine was used to segment our sample into three distinct groups of innovation managers: (i) those driven by nearly all influences; (ii) those driven by results and the governance paradigm; and (iii) those driven by governance and habits. The three groups highlight the different practical approaches to public innovation and co-creation initiatives, which clearly reflect the complex process of deciding which tool (or tools) should be used to implement these. Our UTAUT-IPS model helps visualize this complex decision-making process.

Keywords: public service logic; new public management; innovation; co-creation; co-production; PLS-SEM; Spain

Citation: Peralta, A.; Rubalcaba, L. How Governance Paradigms and Other Drivers Affect Public Managers' Use of Innovation Practices. A PLS-SEM Analysis and Model. *Mathematics* **2021**, *9*, 1055. https://doi.org/10.3390/math9091055

Academic Editors: David Carfì and María del Carmen Valls Martínez

Received: 18 March 2021
Accepted: 5 May 2021
Published: 7 May 2021

Publisher's Note: MDPI stays neutral with regard to jurisdictional claims in published maps and institutional affiliations.

Copyright: © 2021 by the authors. Licensee MDPI, Basel, Switzerland. This article is an open access article distributed under the terms and conditions of the Creative Commons Attribution (CC BY) license (https://creativecommons.org/licenses/by/4.0/).

1. Introduction

Public decision makers increasingly see innovation as a source of organizational change, adaptation to uncertainty, and trust from the citizenry. In scenarios that favor the adoption of innovations in the public sector, Damanpour and Schneider [1] and others [2–7] presented the conditions for their usage, namely, "macro constructs" that facilitate (or inhibit) the use of innovations by organizations, e.g., their cost, complexity, or impact. These studies also encompassed the moderating effects of a public manager's age, tenure, education, gender, pro-innovation attitude, and political orientation.

However, academics and practitioners continue to struggle with the translation of these conditions into the acceptance and adoption of innovation practices for the effective design and deployment of public innovation; that is, although the relevance of innovations for public decision makers may be understood, there is a lack of understanding about the factors that drive the selection of a particular practice—e.g., agile methods, participation or experimentation, tenders—used to implement these innovations in their organizations and environments.

Among the few, but prominent, voices attempting a theoretical answer to this question, the proponents of "service innovation studies" [8–15] propose that the selection of innovation practices is based on the governance paradigms dominating a certain public entity. Here, we adhere to Osborne's neat definition of these governance paradigms, or regimes, as the "different modes of design and implementation of public policy and delivery of public services" [16]. Therefore, Traditional Public Administration (TPA), New Public Management (NPM), and New Public Governance (NPG) or Public Service Logic (PSL) are then presented as "policy and implementation regimes" [16].

Associated with these paradigms, some authors ([17–19] and the CoVAL project) have theoretically connected linear innovation practices and models with the TPA and NPM paradigms, and connected interactive, circular, or networked practices with NPG or PSL. They even recognize that networked practices create distinctive public policy and services, integrating "multilevel, cross-border settings, in which former demarcations of policy fields become blurred." [17]. In short, they perceive the significant impact of the governance paradigms on the selection of an innovation practice and the type of policy or services developed by a public entity.

However, with the exception of these governance paradigms, we were unable to find any other references to predictors of the use of innovation practices in the public sector, although we did find such studies applied to the IS sector and others [20–22]. These predictors of the use of technologies and practices are important because the design, deployment, and success of any public innovation are intimately related to the practice(s) used to deploy them. Furthermore, our earlier research and experience indicated that multiple drivers of the acceptance and use of a practice exist: predictors, decision-maker characteristics, and other "macro" contextual drivers, such as the "macro-constructs" of Damanpour and Schneider.

In this context, we ask if governance paradigms—i.e., TPA, NPM, and NPG or PSL—are the only determinants of the use of innovation practices in public administration. Equally importantly, which are the observable and meaningful characteristics of the governance paradigms and other criteria—i.e., Habit, Performance Expectancy, Effort Expectancy, Cost, Social Influence, Hedonic Motivation, and Facilitating Conditions—that influence the use of public innovation practices? Finally, are gender, age, tenure, or experience relevant moderators of the acceptance and use of innovation tools by public innovators?

Our first expected contribution is an exploration of the measurements of the criteria using Partial Least Squares Structural Equation Modelling (PLS-SEM). In line with research on public governance structures [17,23,24], the value of our present work lies in the identification of the scales—from 114 indicators—and measurement modes of each of our model's 10 composite drivers.

In addition, we investigated the inner connections of the criteria using a structural PLS-SEM model. This model visualizes how public managers decide to adopt an NPM-, NPG- or other paradigm-related practice to achieve public innovation goals. We constructed the Unified Theory of Acceptance and Use of Technology for Innovations in the Public Sector (UTAUT-IPS) model to visualize how its criteria work and which dynamics (i.e., indirect effects) exist. Finally, this model helps understand how different groups of managers (e.g., women, men, older, younger) behave toward innovation tools.

We describe each of the composite drivers in the next section, drawing on the references that helped us develop a sound set of observable elements of the behavior of public innovators and decision makers. This previous analysis allowed us to build the scales and composites of the UTAUT-IPS model, and set out our research hypotheses. The third section describes our method, and we present our results in the fourth section. Our discussion and conclusions come in the fifth and sixth sections, respectively.

2. Theoretical Framework and Hypotheses

Due to the innovation gap, which refers to the researchers' marginalization of non-technological innovations—including social and public-sector innovations—Gallouj and Weinstein [25], Windrum and Koch [26], Djellal et al. [10], Osborne and Brown [27], Desmarchelier et al. [28], and others, have integrated public services innovation into wider "innovation studies" and, specifically, into "service innovation studies." They note the trend towards the advancement, adaptation, and re-definition of public services. This approach complements other innovation initiatives that reach most levels of public entities:

improvements (visible and invisible) in organizations and systems, changes in modes of formation of these improvements (e.g., spontaneous vs. planned), and disruptions in form and content (e.g., bottom-up vs. top-down) [17,26,29].

Beyond the description of these innovations and the drivers of their adoption [1], we know little about what inspires or influences public decision makers and managers to accept and use a specific practice to design and implement innovations. The current practices in developing public innovations include agile development, design thinking, world-cafe meetings, social hackathons, living lab workshops, and conventional tenders and bids. Here, we gather governance and other social and environmental research to identify the drivers or criteria of the acceptance and use of a practice.

2.1. Understanding the Drivers of the Use of Public Innovation Practices

Based on the above premises, the selection and analysis of the observable elements that affect the acceptance of public innovation practices are particularly relevant, and have multilevel effects and multiple origins. To the best of our knowledge, this is a novel line of research, contributing to "services innovation studies" the observable elements, grouped into (latent) drivers or criteria, of the innovators' decision to use a technology, tool, or practice to innovate in a public service.

A useful option in which to frame these elements (items) and drivers is the Unified Theory of Acceptance and Use of Technology (UTAUT [21]). This is a summary model that outperforms each of its underpinning eight psychological and sociological theories that describe the adoption of a technology. With the UTAUT model, Venkatesh et al. successfully separated the intention to use a tool from actual behavior (the usage of the tool), and corroborated the role of the intention as a predictor of usage, in the context of information systems [30–32].

Using the UTAUT model, we believe we can select and analyze the individual traits, and social and environmental elements, that comprise the drivers of the acceptance and use of a certain technology or practice in the context of public innovation. As in the case of Venkatesh et al., we also believe that intention acts as the predictor of usage of a technology in our context—we define a technology in the public innovation context as a summary of visible and invisible transformations, protocols, designs, and implementation tools, adhering to the original UTAUT definition.

The UTAUT model identifies the elements or items of "four key constructs (Performance Expectancy, Effort Expectancy, Social Influence, and Facilitating Conditions) that influence behavioral intention to use a technology and/or technology use" [33] (p. 159). UTAUT2, a later development [33], adapted the original indicators to technology consumers, and added three new constructs—Hedonic Motivation, Costs/price, and Habit [33] (p. 160).

To extend the UTAUT model, Venkatesh et al. demanded "careful theoretical consideration to the context being studied," to advance and complement existing practice with new constructs (drivers) or elements, the "scope and generalizability of UTAUT" [33] (p. 160). Following this instruction, and following other previous studies [34–36], we extended the original scope of the UTAUT model, which was restricted to IS and IT technologies, to a broader spectrum of technologies, including tools and practices commonly used in public innovation projects. (Although "technology" is, in lay terms, often regarded as software applications or hardware, it also includes practices, methods, skills, or knowledge used to accomplish any objective.)

Consequently, we concentrated on applying the UTAUT drivers to our public innovation context and developed our UTAUT for Innovation in the Public Sector (UTAUT-IPS) model. Prior to describing this model, however, we examine why "the governance paradigm" must complement the original UTAUT drivers to allow analysis of the behavior of public innovators.

2.2. Why Should the "Governance Paradigm" Variable Be Added as a New Driver of the UTAUT for Public Innovation Model?

Whether hierarchical (Traditional Public Administration—TPA), market (New Public Management—NPM) or network (New Public Governance—NPG, or Public Service Logic—PSL), governance paradigms (GPs) reflect different approaches to the nature and mode of production of public services [37–40]. They are also theoretically connected to different practices (Table 1). Conceptually, the GPs, or policy and implementation regimes, represent the "different modes of design and implementation of public policy and delivery of public services" [16]. Although they are associated with different historical moments of the public sector, we currently find hybrid forms of governance [17,41] that complement each other, including one GP within another, at the larger scale of a public organization.

Table 1. The theoretical governance paradigms (based on [16,37,42]).

Governance Paradigm	Coordination Mode	Services as a … ?	Role of Citizen	Performance Evaluation	Type of Innovation
Traditional Public Administration (TPA)	Bureaucracy Hierarchy Top-down governance Monopoly Monitoring based on control of processes	Standard "Good" or quasi-product	Passive user/consumer. Client, who expresses his preferences through office elections or as instructed	Objective, tangible output metrics (e.g., productivity, efficiency) Cost-benefit analysis Goods-Dominant Logic	Technological Linear (new product development) Internal to the public organization Top-down driven Politicians are the decision makers and process owners
New Public Management (NPM)	The market decides Competition Privatization Public-private partnerships and outsourcing Hierarchy Top-down governance Departments, areas (silos)	"Marketable good" or a market quasi-product	User and customer, who can freely choose the service and make public services compete Tester	Market and mostly economic and political metrics (e.g., outputs, costs, returns, loyalty, reputation) Goods-Dominant Logic and Market-Dominant Logic	Technological Linear (new product development) Internal to the public organization Top-down driven Managers are process owners Politicians are the decision makers
New Public Governance (NPG) Public Service Logic (PSL)	Networks (formal and informal) Multi-agent partnerships Horizontal governance	Service (immaterial, co-produced, un-stockable) Product/service bundle	Co-designer Co-producer	Multicriteria evaluation: objective and subjective Evolution/time Complex leading and lagging indicators Service-Dominant Logic	Technological and non-technological, including social innovation Linear and non-linear (new product development and new service development) Multi-agent, internal and external The network owns the process Bottom-up and top-down ways Most decisions have political implications, but their scope is usually narrower

Our research cases confirmed this coexistence at a broader scale (e.g., a municipality or region might be top-down governed across its areas, and they may have occasional social hackathons and volunteer groups in living labs, enacting bottom-up government initiatives, such as in the cases of Madrid or Valencia (Spain), or the EU Commission). Due to the potential risk associated with innovations abiding by any of the GPs of a public organization, the behavioral influence of the GPs is stochastic or, potentially, chaotic.

To simplify this apparent randomness, we decided to investigate how public administrations innovate services, organizations, or processes. Desmarchelier et al. [28] conceptually discussed the three stated paradigms of public administration and their influence on the use of innovation tools and practices in innovation projects: a concession or the conventional tender process, for example, might be associated with TPA; an adapted bidding process, following goods-dominant and market-dominant logic [43], is associated with NPM; and more interactive alternatives, such as a citizens jury for a city or the collaborative development of a library, which involve an integrative perspective and a service-dominant logic, are clearly closer to NPG or PSL [44].

We found that only one type of GP is dominant at the project level. If a top-down practice—e.g., tender, concession—is used as the innovation tool in an innovation project, then a top-down, traditional TPA paradigm is in effect. If a bottom-up practice—e.g., citizens' jury or a co-designed square in the middle of a town—is selected, it results from the bottom-up (NPG or PSL) governance approach of the project manager. At this project level, the other paradigms are apparently excluded, at least when managers choose the innovation practice used to start their project. We can conclude that the paradigm governing each specific innovation project is a clear driver of the public decision maker's behavior. Consequently, in this research, we situated our analysis at the project level, investigating the meaningful influence of the GP in the use of practices, and theorize as follows:

Hypothesis 1a (H1a). *At a given moment, the GP in effect in a public innovation project will influence the use (US) of its innovation practice (we explain US in the following section).*

To describe the indicators of GP, we followed Desmarchelier et al. [28], who identified a collection of intrinsic indicators, or predictors, of the type of GP in effect in a given service or project: the coordination mode, the nature of the product, the mode of production organization, and the mode of performance evaluation. Each predictor refines and translates the influence of the citizenry and other stakeholders on the decision makers. Then, the GP driver indicators must capture the intrinsic predictors (refer to the Appendix A for the complete list of our GP indicators).

In innovation scenarios that involve absence or scarcity of references, isolation of teams, or multiple paths or requirements derived from networking with citizens, public managers often combine the different modes of governance—e.g., TPA, NPM, or PSL [1,19,45]. As experience increases with any of these, the managers become used to the written and unwritten norms and rules of each governance paradigm, and tend to limit their choices to earlier successes. Age works similarly because older individuals might experience difficulty in applying tools or practices associated with a different form of governance. Gender might also play a role because inclusion, equity, or collaboration with minorities are linked to NPG or PSL. Thus, we hypothesize:

Hypothesis 1b (H1b). *Gender, age, and experience will moderate the relationship of the GP and the intention to use a tool or practice (BI) (we explain BI in the following section), and the effect will be stronger for older men with high levels of experience with a tool or practice.*

2.3. Other Behavioral Drivers Complementing the Governance Paradigms in the UTAUT-IPS Model

Here, we review the original drivers and indicators of UTAUT, reworking them and their original relationships in the public innovation context.

2.3.1. Usage (US) and Behavioral Intention (BI)

Usage (US) is the key dependent variable of our research; Behavioral Intention (BI) is its main predictor. Following the original tenets of the UTAUT model, we explain the relationship between an individual's intention (acceptance) to use a practice, tool, or technology and their final behavior (usage [21] p. 427); that is, how a public innovator's intention to use a particular practice to develop a new service drives their actual use of that practice.

The relevance of the BI–US pair stems from the different approaches (e.g., [20,21,43,46]) to proactively designing interventions, which mainly consist of training programs, but also to disseminating the different modes of governance within a public organization. According to UTAUT, solely acting on the acceptance of an individual leads to their use of a certain practice. In the absence of this acceptance, as a composite predictor of the behavior (i.e., the use of new practices), the intervention over the individual must control for the compounding (or contradicting) effects of several other predictors on the behavior. Thus, we theorize:

Hypothesis 2 (H2). *The acceptance (BI) of a public innovation practice has a positive relationship with the use (US) of this practice.*

To select the practices (refer to the Appendix A for our list of innovation practices and tools) in use by public innovation project managers and decision makers, we used three dimensions. The linearity dimension [28,47,48] distinguishes linear and non-linear innovation practices and their related dynamics (see Table 2). The monitoring dimension differentiates the handling of missed expectations due to fallbacks, failures, or tests [49–52]. Finally, the collaborative dimension discriminates between stage-gate practices [53] and bottom-up practices [17,18,20] (Table 2).

Table 2. Public service innovation practice types.

Type of Innovation Practice	Description
Linear and closed [18]	Although the science-push model [54] prevailed in the public sector connected to the TPA paradigm, the market demand for innovation was considered relevant following economists such as Schmookler [55]. Today, most administrations and governments act following market demands, which then call for applied research and technical development, hence innovation; this is the antecedent of the NPM paradigm. The demand-pull model and the "government" push model have very different policy implications, but they are enacted similarly: idea, concept design, development, launch, post-service. This linear deployment is popularly referred to as a traditional, stage-gate, or incremental means of executing a service (also called new product development—NPD [28,56]).
Interactive and open or networked [19]	Neither the science-push nor the demand-pull models are sufficient to explain how design and implementation of services work if a manager applies a service logic. In 1986, Kline and Rosenberg [19] proposed the interactive chain-link model; this is a highly effective representation of the complex public service production reality, with its constant feedback loops, and potential step-backs. Subsequently, Blank's [57] customer development process, agile software experimentation [49,50], and business model design [58–60] further clarified this non-linear process. All of these references assume that the relevant stakeholders must be present and contribute to the design and deployment of an innovation from the inception of the idea, rather than merely validating, agreeing, or giving feedback, in the final development steps of any service [19,52].

2.3.2. Costs (CO)

Following the concerns expressed by Venkatesh et al. [33] (p. 161), innovators bear the responsibility of the costs (budget) and economic burden of their innovation projects. They usually begin projects by seeking funding [61,62], knowing that they are primarily accountable for costs (CO). This accountability for costs then impacts on the search for funds throughout the development stages of the new service. This is a constant that affects the intention and behavior of public managers and innovators.

In UTAUT, men's sensitivity to costs and prices differs from those of women, and sensitivity also differs between younger and older individuals. This may also be applicable when deciding to use a new practice to innovate a public service because sensitivity to value differs by gender and age among managers. Thus, we theorize:

Hypothesis 3 (H3). *Gender and age will moderate the relationship of CO and BI, and the effect will be stronger among women, particularly older women.*

2.3.3. Effort Expectancy (EE)

"Effort expectancy is defined as the degree of ease associated with the use of the system" [21] (p. 450). In our context, service innovation practices help clarify a complex and often ambiguous design-implementation gap [63] and a linear or multilinear innovation process. The EE construct captures the feeling and experience of seasoned innovators regarding practices for service design and implementation.

Of note, however, EE in the original UTAUT was significant only during training and post-training, and became insignificant with use. In the early phases, gender, age, and experience (i.e., the lack of it) played a role in the relationship between effort demanded by, and the use of, a practice. Thus, in our public context, we hypothesize:

Hypothesis 4 (H4). *Gender, age, and experience will moderate the relationship of EE and BI, and the effect will be stronger for women, particularly younger women, and in their early stages of experience with a tool or technology.*

2.3.4. Facilitating Conditions (FC)

"Facilitating conditions are defined as the degree to which an individual believes that an organizational and technical infrastructure exists to support use of the system" [21] (p. 453). FCs thus help deal with the factors that, externally and internally, foster or hinder the development and implementation of public innovations and, consequently, the use of innovation practices.

We found the morphological analysis of Desmarchelier et al. [28] appropriate in establishing a comprehensive framework to analyze the support available to an innovator or innovation project. First, from their topographical variables, the actors involved and their interactions produce part of the support, or lack thereof. Regardless of it being recognized, any public innovation practice involves multiple stakeholders. These are different in nature—public, private, individuals, collectives—and intervene in different ways—directly, indirectly, or neither—but are emotionally, ideologically, or otherwise connected to it. In addition, they occupy different places, depending on the quantity and quality of the interactions with others, and their influence may be temporary or permanent. Thus, the different stakeholders produce an evolution and dynamism, and interactions ranging from ignoring others to cooperation, collaboration, and even close partnerships.

Second, functional variables identify the sector or field in which the innovation occurs. In addition, the type of innovation—policy, service, managerial system, organizational mode—translates into different degrees of support for innovation projects. Based on this, managers might find a supportive culture and systems, a collaborative reputation, or help when facing difficulties.

Our UTAUT-IPS includes these topographical and functional variables, and measures their effects in the design and execution phases, for technological and nontechnological innovations. Age and experience moderate the effects of FC: older and seasoned project managers are able to navigate bureaucratic and departmental cultures, and find support for their projects, more easily than younger and newer managers. Thus, we hypothesize:

Hypothesis 5 (H5). *Age and experience will moderate the relationship of FC and US, and the effect will be stronger for older managers, particularly with increasing experience in the use of a tool or practice.*

2.3.5. Habit (HT)

"Habit is viewed as prior behavior; and second, habit is measured as the extent to which an individual believes the behavior to be automatic" [33] (p. 161). Prior experiences are a predictor of HT because they form beliefs and influence behavior. In our context, HT is also a perceptual construct that is intrinsic to the individual [64].

Regarding HT, our context differs significantly from that of the original UTAUT user or consumer contexts. In our case, and given the complexity of any of the tools, practices, and processes of public innovation, we agree with Venkatesh et al. [33] that, whether automatic behavior or instant intention [65], the selection of a practice requires a "stable environment." In the public innovation context, this stability may come from two sources: first, innovation in public organizations is the result of a sometimes long, reflective, and compliant process that demands stability. Second, the cue processing and association—i.e., the habit—that affects the public innovator is highly dependent on the innovator's perception of change. In environments that are perceived as being stable, HT will have a large amount of behavioral control; in situations perceived as changing, sensitivity to the changes will affect HT [66].

Additionally, experience affects the sensitivity of HT in its relationship with BI and US, because more experienced users will weaken their sensitivity, increasing the control of HT over BI and US. Similarly, age increases the control of HT. Furthermore, women are more conscious and detailed about new cues or changes, lowering the strength of HT on BI and US. Thus, we hypothesize:

Hypothesis 6a (H6a). *Gender, age, and experience will moderate the relationship of HT and BI, and the effect will be stronger for older men with high levels of experience with a tool or practice.*

Hypothesis 6b (H6b). *Gender, age, and experience will moderate the relationship of HT and US, and the effect will be stronger for older men with high levels of experience with a tool or practice.*

2.3.6. Hedonic Motivation (HM)

"Hedonic motivation is defined as the fun or pleasure derived from using a technology" [32] (p. 161). This construct relates to the "perceived enjoyment" and inspiration of the innovator, using a practice or technology to develop a new service [67]. Like HT, HM is an intrinsic construct, adding items from motivation theory: it is fun to use a tool, or its use brings recognition, or the public servant fulfills their vocation. Furthermore, HM is also influenced by experience, age, and gender, and we theorize that they control the relationship HM→BI:

Hypothesis 7 (H7). *Gender, age, and experience will moderate the relationship of HM and BI, and the effect will be stronger among younger men in the early stages of experience with a tool or practice.*

2.3.7. Performance Expectancy (PE)

"Performance expectancy is defined as the degree to which an individual believes that using the system will help him or her to attain gains in job performance." [33] (p. 447)

Examination of the step that follows the achievement of success, or the failure to achieve the expected results, is an underexplored field in innovation studies. These drive intention and behavior, but we have not seen examples of the manner in which any of the types of practices steer the governance process, or set controls, to reassert the process once it succeeds (or fails) [68]. Apparently, conventional models of governance (TPA and NPM) are not prone to recognizing failures. The response of these models to a failure is to return, if possible, to the previous state (status quo ante), or the end of the project. The NPG approach appears to more adaptive to failures, most probably due to its open, networked, iterative approach. It seeks public value through collaborative processes [69], and is prone to failure or unexpected results. Nonetheless, as White [70] noted, with unclear, even absent, measures of success or failure, public innovation practices will be "commended" to the voters to decide on their adequacy.

Moreover, gender and age moderate the relationship of PE and BI (e.g., [71,72]). Thus, we hypothesize:

Hypothesis 8 (H8). *Gender and age will moderate the relationship of PE and BI, such that the effect will be stronger for men and particularly for younger men.*

2.3.8. Social Influence (SI)

"Social influence is defined as the degree to which an individual perceives that important others believe he or she should use the new system." [33] (p. 451) Then, the one-self [73] is relevant when speaking of decision making to innovate public services. Individual intrapreneurs, public innovators [74–76], or citizens identify and then act upon the need for an innovation.

The instigators or paladins of innovation in public entities are isolated individuals, who seek agreement and strength through critical masses and collectivities [77]. They clearly share some characteristics with corporate innovators [78,79]: knowledge of their organizations, and strong motivation based on beliefs, challenges, or explicit mandates. They also have power and set the expectations for the new service, in addition to influencing the value proposals, groups to develop the service, networks, and the mind-sets and cognitive mechanisms behind the new service [74]. (Innovation in the public sector is a radically different strategy than its cousin in the private sector (see [22]). In the private sector, innovation usually follows an illusion, sometimes called vision, which is generally vaguely defined and unquantifiable. In the public sector, the new services appear to address an identifiable need or demand, whether current or future.)

Then, the innovators' individual traits must correlate with the dynamics and changes that are so typical of innovation processes. Therefore, they support the argument in favor of researching the use of public innovation tools as an individual's triggered effort [74,80] (p. 1329).

In addition to public innovators, many examples exist (library of San Fermin, Artropoloops, or the experiments in Experimenta Distrito) of individual citizens triggering collective demands for new or updated public services. Private citizens regroup into virtual networks, temporary gatherings, or conventional families or associations, and have significant mobilization power and a strong voice [75,76,80]. They not only create demands that translate into mandates for conventional TPA or NPM designs; NPG, PSL, or open government practices empower them [16,29,73], and provide means to co-design and co-create policy (e.g., the Observatory of Madrid, or Madrid en Verde). Our UTAUT-IPS framework also integrates the influences of individual citizens, reflecting their social impact.

This SI construct reflects, more intensely than the remaining variables, the dynamism and time limitations of public innovation. It captures the top-down or bottom-up styles of the managers' influencers, who would be different in voluntary and mandatory situations. In addition, gender, age, and experience may also play a role [32]. Thus, we hypothesize:

Hypothesis 9 (H9). *Gender, age, experience, and voluntariness will moderate the relationship of SI and BI, and the effect will be stronger for women, particularly older women, in mandatory settings, in their early stages of experience with a practice.*

Figure 1 presents our research model, and highlights our hypotheses.

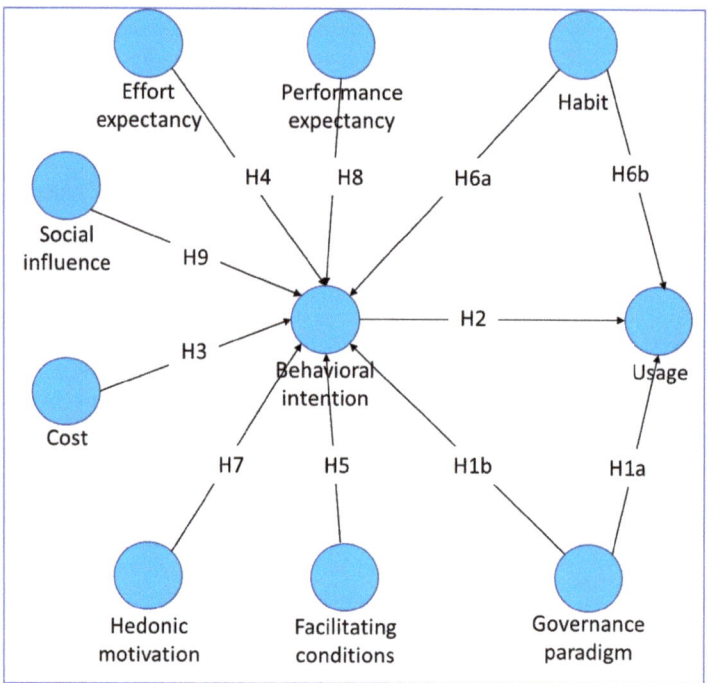

Figure 1. Research model.

3. Methodology

3.1. Survey

To select our participants, we screened social network sites where public managers presented their professional profiles and activities. We examined the public profiles of a random population of more than 4500 managers in an attempt to identify for each their responsibility for leading innovation projects in their public units. Based on this extensive analysis, we selected Spanish directors, deputy directors, managers, heads, or chiefs of innovation or projects, with 4+ years' experience in leading public projects. They had improved (rationalized), changed, complemented, or created public services, whether internal or external to their governments, public agencies, and government-owned corporations. We finally sent an anonymous survey to 1034 public officials who matched our requirements between August and December 2019.

The selection of Spain as the region for our study was motivated by the continuation of the research that we led for the CoVAL project. Based on this experience, the practices of public administrations and managers in Spain are relatively homogeneous with those of other European Union administrations [81–84]. Thus, we decided to choose these common practices and study them and their antecedents here.

3.2. Measurement

Because this is one of the first studies in the decisions field to examine the practice of innovating in public services, we decided to use partial least squares structural equations modelling (PLS-SEM [85]) due to its "ability to [create] independent latent variables directly on the basis of cross-products involving the response variable(s)" [79]. Hinseler et al. [85] recommended PLS path modeling "in an early stage of theoretical development in order to test and validate exploratory models."

As shown in Figure 2, after setting up our hypotheses and concept model, and validating it with our case studies, we designed our questionnaire. The first version was distributed in July 2019 to test the on-line tool, introductory text, and categorical scales. The prototype also helped us rework the layout and final wording of the questionnaire, limit the time spent in each section, and adjust the expected response rate. From August to December 2019, we administered the final on-line survey (Limesurvey, V. 2.73.1), aiming for a minimum sample size of 84 cases, as estimated using G*Power [86]. (F2: 0.015; alpha: 0.5; Power: 0.8; # of predictors (maximum of any latent variable): 16. G*Power, version 3.1.9.4, t-tests: linear multiple regression.)

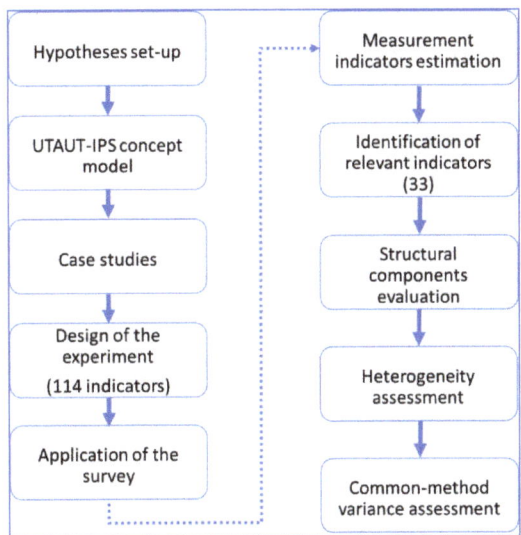

Figure 2. Steps for the validation of the UTAUT-IPS model (based on [87–89]).

To measure our constructs, following the systematic application of the nonparametric criteria of PLS-SEM [85] (p. 96), we first needed to assess our measurement indicators. Then, using the factor weighting scheme in the PLS algorithm and bootstrapping [90] included in Smart-PLS 3 software [91], we calculated the structural components and relationships of our theoretical UTAUT-IPS framework.

Our behavioral manifest items helped respondents self-report on personality traits, perceptions, and behaviors. This might lead to common method bias (CMV), which could affect the validity of our conclusions. To control for CMV, we followed the recommendations of Tehseen et al. for mixed controls [92] (p. 146). We used a Measured Latent Marker Variable (social desirability) and different formats of response (e.g., randomly presenting the items for each construct). Our questions were laid out in a general positive style, interchanged with a negative style for some items and constructs (e.g., FC7 or CO). We protected the anonymity of participants. We then used the Construct Level Correction (CLC) approach, [92] (p. 158) and Levene's test of variances [89] to assess CMV and ensure its absence in our model.

4. Results

We sufficiently met the required minimum sample size (84 at 0.8 power) with our sample of 227 cases. The sample had a similar number of cases at employee level—employee responsible for innovating processes or services, 31%; manager in support of teams creating innovative services in their agency, 31%; and senior manager, general manager of their agency, 38% (Table 3). Most interviewees were males (68%), aged between 30 and 65 (99%), with an education level of master's or higher degree (65%). Our interviewees were distributed among federal agencies (23%), regional entities (37%), and municipalities (40%) (refer to the Appendix A for an extended presentation of the sample participants).

Table 3. Distribution of observations by employee rank.

	Employee	Mid-mngr	Senior mngr	Total
Female	24	26	22	72 (32%)
ECON	10	14	8	32
Collaborative	4	6	3	13
Conventional	5	7	5	17
No practice	1	1	-	2
TBL	14	12	14	40 (56%)
Collaborative	5	4	4	13
Conventional	9	7	10	26
No practice	-	1	-	1
Male	47	44	64	155 (68%)
ECON	19	17	27	63
Collaborative	5	4	11	20
Conventional	13	13	12	38
No practice	1	-	4	5
TBL	28	27	37	92 (59%)
Collaborative	14	13	14	41
Conventional	14	13	22	49
No practice	-	1	1	2
Total	71 (31%)	70 (31%)	86 (38%)	227

Notes: Collaborative—using practices such as networks of public-private agents for social innovation, participatory practices (world cafe, open space, wise action), experimentation and iteration. Conventional—using practices such as surveys and polls, traditional prototyping, traditional public-private partnerships. No practice—using no concrete practice, but a mixed toolbox. ECON—agency concentrates only on economic goals (according to respondents). TBL—agency concentrates on economic, social, and environmental goals (according to respondents).

Next, following the guidelines of Sarsted et al. [88] (Figure 2), we present our measurement model assessment.

4.1. Measurement Model Assessment

The relevant latent construct loadings and weights of our study were 68 measurement indicators (refer to the Appendix A for the complete list) grouped into six reflective (Mode A) constructs (42) and four formative (Mode B) constructs (26).

4.1.1. Reflective (Mode A) Constructs

After assessing Jöreskog's composite reliability levels, several of our items were dropped due to low loadings, until each composite reached "satisfactory to good" marks (0.70–0.95 [93] p. 112). Cronbach's alpha values and rho-A scores of each composite were also within the same thresholds. The composites' convergent validity, measured using the Average Variance Extracted (AVE) values, showed that all constructs explained more than 50% (0.50) of the variance of their respective items. As recommended for PLS-SEM models [93], we assessed the discriminant validity of our reflective constructs, and all of them exhibited discriminant validity with HTMT scores lower than 0.8 (Table 4).

Table 4. Items forming the reflective constructs.

	Loading	Mean	SD	VIF	Cronbach's Alpha	rho_A	Composite Reliability	Average Variance Extracted (AVE)
(BI1) ← BI	0.797	0.797	0.035	2.146	0.771	0.773	0.854	0.594
(BI2) ← BI	0.774	0.773	0.040	2.079	-	-	-	-
(BI3) ← BI	0.796	0.794	0.034	2.078	-	-	-	-
(BI4) ← BI	0.712	0.711	0.046	2.128	-	-	-	-
(EE10) ← EE	0.750	0.747	0.061	1.485	0.638	0.637	0.805	0.579
(EE1) ← EE	**0.769**	**0.764**	**0.050**	**1.541**	-	-	-	-
(EE2) ← EE	0.763	0.758	0.057	1.338	-	-	-	-
(FC1) ← FC	**0.798**	**0.745**	**0.150**	**1.327**	0.710	0.742	0.814	0.525
(FC4) ← FC	0.746	0.703	0.141	1.501	-	-	-	-
(FC5) ← FC	0.622	0.588	0.226	1.328	-	-	-	-
(FC6) ← FC	0.721	0.677	0.201	1.257	-	-	-	-
(HT2) ← HT	0.804	0.802	0.041	1.626	0.757	0.781	0.859	0.671
(HT4) ← HT	0.783	0.778	0.044	1.457	-	-	-	-
(HT6) ← HT	**0.868**	**0.868**	**0.023**	**1.555**	-	-	-	-
(HM4) ← HM	0.648	0.643	0.075	1.203	0.780	0.785	0.860	0.607
(HM1) ← HM	0.813	0.809	0.047	1.757	-	-	-	-
(HM2) ← HM	0.815	0.811	0.049	1.777	-	-	-	-
(HM5) ← HM	**0.827**	**0.825**	**0.036**	**1.397**	-	-	-	-
(SI22) ← SI	1.000	1.000	0.000	1.000	1.000	1.000	1.000	1.000
HTMT Matrix (Discriminant Validity)								
	BI	EE	FC	HM	HT			
EE	0.470	-	-	-	-			
FC	0.326	0.432	-	-	-			
HM	0.313	0.693	0.317	-	-			
HT	0.558	0.204	0.369	0.132	-			
SI	0.352	0.531	0.264	0.635	0.136			

Note: Items in bold are showing the strongest path loadings/weights per composite variable.

4.1.2. Formative (Mode B) Constructs

To assess the formative mode constructs of the UTAUT-IPS model, we followed the suggested assessment procedure of Hair et al. [93] (Exhibit 5.1). We checked the formative items for collinearity by examining their VIF values, and confirmed that all of them were below the threshold value of 5. Next, we analyzed the outer indicators' weights for significance and relevance using the rules of thumb of Hair et al. [93] (Exhibit 5.7). We retained all significant weights, and only retained nonsignificant weights with significant or relatively high (≥ 0.5) loadings (Table 5) (refer to the complete lists of indicators in the Appendix A).

Table 5. Indicators forming the formative constructs.

	Weight (Loading)	Mean	SD	VIF
(CO4) → CO	0.992 (0.819)	0.669	0.635	1.098
(GP1) → GP	0.377 **	0.361	0.145	1.163
(GP7) → GP	0.259 (0.553)	0.254	0.168	1.229
(GP4) → GP	0.711 ***	0.702	0.114	1.186
(PE3) → PE	0.432 **	0.424	0.157	1.069
(PE10) → PE	0.46 **	0.457	0.160	1.006
(PE5) → PE	0.54 **	0.521	0.158	1.070
(US10) → US	0.416 **	0.384	0.151	1.730
(US1) → US	0.429 **	0.406	0.147	1.152
(US2) → US	0.217 (0.585)	0.214	0.147	1.385
(US5) → US	0.479 **	0.443	0.152	1.290

Notes: ** $p < 0.05$; *** $p < 0.001$.

4.2. Structural Model Assessment

Following Garson [87] and Hair et al. [89] (Exhibit 6.1), the second step to appropriately build the UTAUT-IPS model was the analysis of the inner relationships between its constructs, and to identify the model's capability to predict the endogenous US. We confirmed that, with the data in our sample, there were no collinearity issues in our constructs (Table 6).

Table 6. Path coefficients, effect size, and collinearity values of the inner model constructs.

Complete	Path Coefficients	f-Squared	VIF
BI → US	0.070	0.004	1.573
CO → BI	−0.168	0.043 +	1.078
CO → US	−0.001	0.000	1.122
EE → BI	0.108	0.014	1.539
EE → US	−0.022	0.001	1.557
FC → BI	−0.006	0.000	1.234
FC → US	−0.030	0.001	1.234
GP → BI	0.184 **	0.038 +	1.172
GP → US	0.390 ***	0.163 ++	1.225
HM → BI	0.025	0.002	1.844
HM → US	0.202 *	0.029 +	1.845
HT → BI	0.390 **	0.224 ++	1.095
HT → US	0.013	0.000	1.335
PE → BI	0.065	0.003	1.438
PE → US	0.181 **	0.030 +	1.444
SI → BI	0.160 **	0.016	1.534
SI → US	−0.020	0.000	1.574
$R^2 = 0.307; q^2 = 0.077$			

Notes: * $p < 0.10$; ** $p < 0.05$; *** $p < 0.001$. (+) f-squared > 0.02 (small effect); (++) f-squared > 0.15 (medium effect) [45,94].

Thus, our analysis concludes that only six path coefficients were significant and relevant, three connected with BI (see Figure 3) and three with US. The exogenous constructs GP, HT, and SI were significantly connected with BI; and GP, HM, and PE significantly connected with US. The inner or structural model showed weak coefficients of determination: 0.36 (weak [88]) for BI and 0.31 for usage (weak [88]). In addition, the model showed predicted relevance, although its effect size was small.

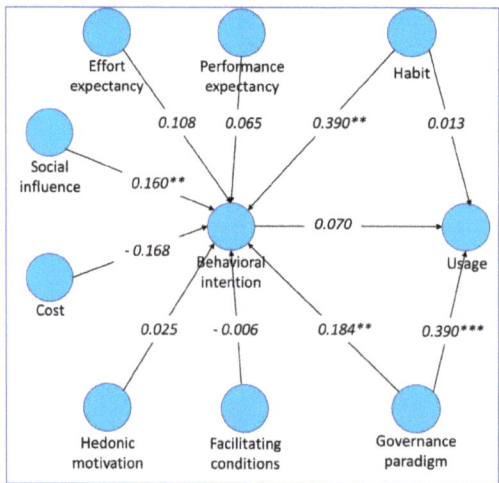

Figure 3. Structural (inner) model for the pooled sample. Note: ** $p < 0.05$; *** $p < 0.001$.

The paths of the UTAUT-IPS pooled model presented small (e.g., CO → BI, GP → BI, HM → US, PE → US) and moderate (e.g., GP → US, HT → BI) f-square effect measures [45], [94]. Although some indirect effects added strength to the GP → US and PE → US paths and some of the other relationships, none of these were significant at this stage of our analysis.

From our set of hypotheses, our sample confirmed H1a, with significant and positive effects of GP over BI and US. We were unable to confirm the relationship (H2) between BI and US, which prevents the mediation effect of BI over US.

Other unpredicted relationships were found to be significant using our model. (GP → US) was the strongest connection of UTAUT-IPS. PE → US shows the significant effect of the performance expectations on the use of innovation tools in our context, and HM → US meaningfully connects the motivation of public servants with the use of innovation tools.

4.3. Analysis by Observed Groups

The next step in our analysis was the assessment of the groups we selected using our descriptive variables (age, gender, experience, and voluntariness). We first established the measurement invariance, supported by configurational and compositional invariances. We applied the MICOM procedure [95,96] to all of the variables we controlled for—i.e., age, gender, experience, voluntariness, organizational level, geographical reach, success of last project, type of collaboration, education, and type of project objectives. Then, we applied a Multigroup Analysis (MGA [88,97]), and found that only geographical reach produced any significant differences, as confirmed by a permutation analysis [96,98]: regional representatives showed a stronger, and more positive and relevant, GP → US path than municipal interviewees.

We were unable to confirm our hypotheses—H1b (GP → BI), H3 (CO → BI), H4 (EE → BI), H5 (FC → US), H6a (HT → BI), H6b (HT → US), H7 (HM → BI), H8 (PE → BI), and H9 (SI → BI)—due to very low and, in most cases, unmeaningful path coefficients. For all of the relationships, the effect of age, gender, experience, or voluntariness—or any other moderating variable—resulted in unmeaningful differences between observable groups.

4.4. Analysis by Unobserved Groups

Complementing our observed heterogeneity analyses, we performed extensive analyses using the PLS Prediction-Oriented Segmentation (POS) routine to discover unexpected and unobserved heterogeneity in our data [37,86,93]. Our election of POS for the unobserved heterogeneity analysis was derived from our mixed formative–reflective measurement model. This is also a routine with improved explanatory power as a nonparametric method to assess nonnormal distributions. Using POS, we found three unobserved groups, which bettered our analysis, reaching substantial coefficients of determination (between 0.68 and 0.85) and predictive relevance (between 0.045 and 0.19).

4.4.1. Group #1

PLS-POS Group #1 shows stronger positive Effort Expectancy, Facilitating Conditions, Government Paradigm, Hedonic Motivation, and Social Influence paths, and strong negative Cost, Habit, Performance Expectancy paths—EE is the strongest construct (Table 7). The negative paths might be the main difference with the other groups: CO, HT, and PE inversely relate to Behavioral Intention or Usage. BI shows an insignificant relationship with US, but strengthens the paths of CO, EE, and FC, and weakens the connection of PE with US (Table 8).

Table 7. Path coefficients, effect size, and collinearity values of the inner model constructs for the PLS-POS group #1.

	Path Coefficients	f-Squared	VIF
BI → US	−0.221	0.066 +	4.068
CO → BI	−0.435 **	0.679 +++	1.131
CO → US	−0.370 **	0.394 +++	1.899
EE → BI	0.721 ***	1.367 +++	1.546
EE → US	0.361 **	0.195 ++	3.660
FC → BI	0.324 **	0.281 ++	1.525
FC → US	0.316 **	0.279 ++	1.953
GP → BI	0.123	0.055 +	1.117
GP → US	0.224 **	0.233 ++	1.178
HM → BI	0.130	0.042 +	1.624
HM → US	0.209 **	0.141 +	1.693
HT → BI	−0.021	0.001	1.296
HT → US	−0.313 **	0.414 +++	1.298
PE → BI	−0.761 ***	1.138 +++	2.073
PE → US	0.217	0.058 +	4.431
SI → BI	0.272 **	0.212 ++	1.423
SI → US	−0.046	0.007	1.724
	$R^2 = 0.817$; $q^2 = 0.190$		

Notes: ** $p < 0.05$; *** $p < 0.001$. (+) f-squared > 0.02 (small effect); (++) f-squared > 0.15 (medium effect); (+++) f-squared > 0.35 (large effect) [45,94].

Table 8. Mediation size of Behavioral Intention for the PLS-POS group #1.

	Mediation Effect Size
CO → US	26%
EE → US	44%
FC → US	23%
GP → US	12%
PE → US	78%

In Group #1, the specific predictors of BI and US (Table 7) show an extended version of the pooled sample case, which was limited to six significative paths, highlighting the complexity of the decision process and the diverse influences on the selection process of a practice to develop innovation in the public context.

4.4.2. Group #2

This group shows Government Paradigm, Hedonic Motivation, and Social Influence as the stronger constructs, of which GP is the strongest (Table 9). Due to the relevance of the GP construct, this group stands out. Performance Expectancy shows a negative, significative path that also qualifies the group. Behavioral Intention is unmeaningfully connected with Usage, and its mediating effects are weak on HM and PE (Table 10).

Table 9. Path coefficients, effect size, and collinearity values of the inner model constructs for the PLS-POS group #2.

	Path Coefficients	f-Squared	VIF
BI → US	−0.058	0.009	2.529
CO → BI	0.002	0.000	1.073
CO → US	0.088	0.049 +	1.073
EE → BI	0.067	0.009	1.225
EE → US	−0.040	0.007	1.237
FC → BI	0.086	0.016	1.166
FC → US	−0.048	0.011	1.184
GP → BI	−0.089	0.017	1.189
GP → US	0.710 ***	2.763 +++	1.209
HM → BI	−0.232 *	0.075	1.803
HM → US	0.595 ***	1.174 +++	1.938
HT → BI	0.554 ***	0.512 +++	1.517
HT → US	0.194	0.150 ++	2.295
PE → BI	0.391 **	0.237 ++	1.633
PE → US	−0.432 **	0.559 +++	2.020
SI → BI	0.089	0.012	1.698
SI → US	0.287	0.334 ++	1.718
$R^2 = 0.851$; $q^2 = 0.140$			

Notes: * $p < 0.10$; ** $p < 0.05$; *** $p < 0.001$. (+) f-squared > 0.02 (small effect); (++) f-squared > 0.15 (medium effect); (+++) f-squared > 0.35 (large effect) [45,94].

Table 10. Mediation size of Behavioral Intention for the PLS-POS group #2.

	Mediation Effect Size
GP → US	1%
HM → US	2%
PE → US	6%

In Group #2, as in the case of Group #1, the predictors of BI and US (Table 9) extend the pooled sample case list. In this group, the number of constructs is significantly lower than in Group #1, but their power to explain the behavior of its managers, measured by the coefficient of determination, is stronger.

4.4.3. Group #3

Government Paradigm and Habit are the stronger, significant constructs (Table 11) in Group #3, of which HT is the strongest construct. In this group, the negative significant loadings characterize the relationship of Cost and Effort Expectancy with Behavioral Intention. BI, as in the case of the other groups, shows an unmeaningful relationship with Usage, and only slightly mediates the relationships of GP and HT (Table 12).

Table 11. Path coefficients, effect size, and collinearity values of the inner model constructs for the PLS-POS group #3.

	Path Coefficients	f-Squared	VIF
BI → US	0.263	0.069 +	2.529
CO → BI	−0.334 *	0.328 ++	1.073
CO → US	0.065	0.052 +	1.073
EE → BI	−0.338 **	0.166 ++	1.225
EE → US	−0.150	0.005	1.237
FC → BI	−0.077	0.014	1.166
FC → US	−0.517	0.594 +++	1.184
GP → BI	0.389 ***	0.381 +++	1.189
GP → US	0.426 **	0.192 ++	1.209
HM → BI	0.469 ***	0.382 +++	1.803
HM → US	0.011	0.016	1.938
HT → BI	0.310 ***	0.251 ++	1.517
HT → US	0.501 **	0.370 +++	2.295
PE → BI	−0.116	0.039 +	1.633
PE → US	0.387	0.486 +++	2.020
SI → BI	0.214 **	0.079 +	1.698
SI → US	0.180	0.024 +	1.718
$R^2 = 0.679$; $q^2 = 0.045$			

Notes: * $p < 0.10$; ** $p < 0.05$; *** $p < 0.001$. (+) f-squared > 0.02 (small effect); (++) f-squared > 0.15 (medium effect); (+++) f-squared > 0.35 (large effect) [45,94].

Table 12. Mediation size of Behavioral Intention for the PLS-POS group #3.

	Mediation Effect Size
GP → US	32%
HT → US	19%

Group #3 predictors are different from the those of the other two groups. BI is meaningfully predicted by six predictors, some of which are negatively loaded (CO, EE), but only two constructs predict US (Table 11). Our model is evidently less powerful with this group; nonetheless, its coefficient of determination has moderate strength.

5. Discussion

In this research we used Partial Least Squares Structural Equation Modelling (PLS-SEM) [88,93,97] to identify that governance paradigms (GP) are only one, albeit highly relevant, of the determinants of the use of innovation practices in a public administration. Using this method, we constructed the Unified Theory of Acceptance and Use of Technology for Innovations in the Public Sector (UTAUT-IPS) model. This complements the GP with performance expectancy (PE) and hedonic motivation (HM).

We investigated GP as a descriptor of the policy and implementation regimes [16] or modes of production of public services [37]. Furthermore, we confirmed its importance in the use of an innovation practice (H1a) due to their strong relationship. Additionally, we featured the indicators of GP (GP1, GP4, GP7) from the theorized group of nine indicators [27,37] (refer to the complete list in the Appendix A). These reflect the involvement of users/citizens and other agents (consultants, technical staff, etc.) in the (co-)creation of public innovation.

Thus, we corroborated the few theoretical references of the influence of GP on the usage of innovation practices (e.g., [17,37,99]) in public organizations in Spain, where the traditional public administration (TPA) and new public management (NPM) paradigms dominate [100]. However, our sample innovation managers confirmed that, at the project level, their preferred implementation regime or mode of production of innovations deeply involves the citizenry—i.e., the new public governance/public service logic (NPG/PSL) regimes). From a practical perspective, innovation managers are then willing to involve citizens in the innovation of services, and the intention to involve citizenry drives the innovation practice selection.

The implementation regime that inspires managers, as identified by our empirical definition of GP, partially drives the practices and tools they use. Top-down (agile, design thinking) and bottom-up (collaboration) practices form their mixed preferences. This is a mixed toolbox, and although bottom-up practices are the most prominent, these combine with top-down practices.

Our results then expanded the theoretical prediction of Desmarchelier et al. [28]. They linked collaborative tools with the NPG/PSL paradigm. Our results blurred this connection. We demonstrated that the mixed-practices toolbox is available and used, and also dependent on other significant drivers. Practically speaking, managers appear to avoid the use of conventional tools (e.g., public–private partnerships, surveys and polls, or conventional metrics) when designing their innovation projects. They opt for a toolset combining collaborative (world cafe, open space, wise action, etc.) and novel top-down innovation tools (e.g., design-thinking, agile methodologies, or safaris).

The broader theoretical implications of the lack of a meaningful connection between Behavioral Intention (BI) and Usage (US) require further investigation. A larger sample of managers might help in an investigation into the lack of an effect of managers' age, experience, gender, and other moderators.

Nonetheless, from a methodological perspective, we can highlight three relevant implications of the effect of BI in our model. The original UTAUT model [21,33] predicts BI as a mediating variable. The dynamism between the exogenous variables and the endogenous US is reflected in the direct effects complemented with significant indirect effects. In our UTAUT-IPS model, this dynamism is clearly less evident, because no mediation effects are observable for the pooled sample. For the POS groups, in Group #1, we find BI partially mediating CO → US, EE → US, and FC → US, and fully mediating PE → US. In Group #3, BI partially mediates GP → US.

Furthermore, from a modelling perspective, these BI indirect effects might evidence the actual link between BI and US predicted by Venkatesh et al. [20,21]. BI already shows strong reliability and validity.

From the practice perspective, BI highlights four significant behavioral predictors of intention: the will of managers to continue to use a practice in subsequent projects; their scheduling of activities following the requirements of the practice; their securing of resources to use the practice in the future; and their preference for one practice over others. These predictors clearly highlight the strategies of managers when planning innovation projects.

Strengthening our UTAUT-IPS model, exogenous constructs other than BI and GP show meaningful effects on the endogenous US. These raise the model's predictive power of the behavior of three different types of innovation managers: more experienced or senior managers (Group #1), those pursuing their careers (Group #2), and novel or low-ranked managers (Group #3). In the following, we describe the influences of Costs (CO), Effort Expectancy (EE), Facilitating Conditions (FC), and Habit (HT) on managers:

- Group #1 managers are motivated by CO, EE, HT, FC, GP, and HM. Consequently, a manager in Group #1 opts for practices that are easier to use and have more resources and help. The practices should allow more collaboration with the citizenry and other stakeholders, and, to a lesser extent, allow motivational rewards to be earned from using them. In addition, the manager favors practices that lower costs and are novel or contrary to habit. The group of experienced managers remains neutral regarding stakeholders' influences; social influences are insignificant from the manager's perspective at the early project stage of innovation tool selection.
- Group #2 managers are motivated by GP, HM, and PE. For these managers, the use of a practice correlates with better chances to co-create and collaborate with citizens. In addition, this usage is linked to the motivational rewards earned by the manager from using the practice. These managers do not believe the practice they choose will

benefit their organization, increase their team's productivity, or help identify new services. Group #2 managers lack sufficient experience or do not have enough influence on outcomes because they lack sufficient rank. In addition, they are also neutral regarding social influences at the early project stage of innovation tool selection.

- Group #3 managers are motivated by HT and GP. These managers choose tools and practices that foster co-creation with citizens and other stakeholders. These tools are popular in their agency or department. Group #3 appears to be the least experienced, or lowest ranked, of the three types of managers. As with the other groups, social influences are insignificant for these managers at the early stage of innovation tool selection.

The predictive power of the UTAUT-IPS model was relatively weak for the pooled sample and is clearly weaker than the psychology and sociology models which underpin UTAUT-IPS. The observable groups (e.g., gender, age, experience, or education) were unable to improve this power with any of the underlying models, because they do not differ from each other or from the pooled sample.

However, the results found from the POS groups, and their drivers, presented a different picture. The strength of the model for these groups demonstrated the capacity of the model to capture the different drivers of the use of a practice, and explain a large portion of its variance. Our results complement the findings of Damanpour and Schneider [1] and others [2–7], and help to understand the results of this earlier research; that is, a public manager willing to produce an innovation will first need to choose a tool to design and implement it. Our contribution is the identification of the types of managers and their different behavioral drivers in using a practice to innovate in a public service.

Our research is clearly limited by its exploratory nature [101] because it deals with an under-scrutinized behavior with a novel systematic PLS-SEM approach. This study is mainly limited by its geographical reach and the characteristics of our sample participants, who were self-selected experienced public innovators and innovation managers. In addition, our model is affected by the practices we studied, the timing of measurement, and the nature of measurement. The public practices and participants we included in our research differed from those of the original UTAUT model, which only analyzed the private sector. Consequently, the two studies might complement each other, because we also used the previous recommendations for expanding the original model [33]. In an attempt to limit unexpected effects or wishful thinking, we asked our participants to relate to a recent past project, after they had chosen their tools. This may have produced common method bias, despite attempts to control for it, as explained in the Methods section. In addition, our measurement is cross-sectional, lacking dynamic or longitudinal analysis of the cyclical interactions between intention, behavior, and performance.

6. Conclusions and Future Research

This paper presents a new model, the PLS-SEM-based UTAUT-IPS model, to theorize the drivers of the use of public innovation practice. The model includes the acting government paradigms, and other personal characteristics of public managers and their context, such as performance, motivation, habit, effort, or social influence. We tested this model with a sample of 227 Spanish public managers gathered at the end of 2019. Our PLS-SEM analyses adequately assessed 124 behavioral indicators (114 items and 10 composite variables) of the managers' attitude towards collaboration, measurement of innovation, and the (non-)linearity of the innovation practices.

Our first contribution with the UTAUT-IPS model applied to our study sample demonstrates the powerful connection between the governance paradigms, or regimes, in effect in one public entity with the practices used by the innovators of that entity: agile methods, participatory practices, and collaborative prototyping. The behavioral drivers of the use of an innovation practice are linked to the type of manager: governance, cost, effort, available help, motivation, and novelty correlate with experienced managers (Group #1); governance, motivation, and expectations of performance influence less experienced managers (Group #2); and governance and habit correlate with novel managers (Group #3).

Second, with our exploratory research, we contribute by identifying the mixed toolbox of practices of innovation managers, and expose the coexistence of collaborative and conventional (top-down) practices, independent of the governance paradigm or regime, at the entity level.

The future of this research line lies in investigating the changes in the selection of innovation practices over the course of a project. We also suggest extending our scope to other geographies. Using the indicators and constructs identified here, and with an enlarged base of respondents, future investigations can confirm the connection of the governance paradigms and the other behavioral drivers. More importantly, future research should strive to validate the mediating role of the acceptance (intention) of an innovation practice, and more closely examine the characteristics of the types of public innovation manager.

Author Contributions: Conceptualization, A.P. and L.R.; methodology, A.P.; formal analysis, A.P. and L.R.; investigation, A.P.; writing—original draft preparation, A.P.; writing—review & editing, A.P. and L.R.; supervision, L.R.; funding acquisition, L.R. All authors have read and agreed to the published version of the manuscript.

Funding: This paper has received funding from the European Union's Horizon 2020 research and innovation programme under grant agreement No 770356: Co-VAL. This publication reflects the views only of the authors, and the Agency cannot be held responsible for any use, which may be made of the information contained therein. The paper has also been co-funded by the Spanish National Research Programme RTI2018-101473-B-100.

Institutional Review Board Statement: Not applicable.

Informed Consent Statement: Not applicable.

Data Availability Statement: Not applicable.

Conflicts of Interest: The authors declare no conflict of interest.

Appendix A

Table A1. Interviewee distribution by geographical scope of their government agency.

	Total
Federal	52 (23%)
Collaborative	23
Conventional	26
No practice	3
Regional	83 (37%)
Collaborative	33
Conventional	49
No practice	1
Municipality	92 (40%)
Collaborative	31
Conventional	55
No practice	6
Total	227

Note: Collaborative—using practices such as networks of public–private agents for social innovation, participatory practices (world cafe, open space, wise action), experimentation and iteration. Conventional—using practices such as surveys and polls, traditional prototyping, traditional public-private partnerships. No practice—using no concrete practice, but a mixed toolbox.

Table A2. Lists of all items per latent construct of UTAUT-IPS.

Usage (US)
USE1. Agile methods: scrum, lean, kanvan and similar
USE2. Design thinking, safaris, interviews, observation and similar
USE3. Tenders, over-arching agreements and similar
USE4. Network of public–private agents for social innovation and similar
USE5. Participatory practices: world cafe, open space, wise action and similar
USE6. Classical public–private partnerships
USE7. Classical metrics and indicators
USE8. Budget
USE9. Surveys and polls
USE10. Collaborative prototyping with citizens
USE11. Classical prototyping
USE12. Experimentation and iteration
USE13. Not really following a particular practice—just discovering

Behavioral Intention (BI)
BI1. you will continue using this practice in the next project
BI2. this practice drives your calendar of next actions for the coming 3 months
BI3. you have already secured resources for this practice in the next 3 months
BI4. you will use this practice again before any other

Performance Expectancy (PE)
PE1. meet the time requirement of your project
PE2. adapt a process/service to a new regulation
PE3. it helped our organization to reap the benefits/returns of the project
PE4. improve the adoption or use of a process/service
PE5. increase your (or team's) productivity
PE6. in my organization the use of this practice satisfies the confirmed political guidelines
PE8. better address a crisis/urgent demand
PE10. identify new un-serviced areas/users
PE11. identify duplicated or useless services/processes

Effort Expectancy (EE)
EE1. the practice sequence of steps was clear and understandable
EE2. given my progress, I expected to become a master very easily
EE7. made collaboration with users/citizens easier
EE8. made collaboration with other agents (different from users) easier
EE10. within our team, made collaboration and contribution to the dialogue easier
EE6. actionable (you can apply it rather quickly)

Social Influence (SI)
SI20. others around you were using it
SI21. people who you inquired suggested you should use this practice
SI22. you influenced the behavior of others suggesting they should use this practice
SI1 Potential users
SI2 Current users
SI03 User influencers (associations, communities)
SI4 District council
SI5 Municipality council/board
SI6 Politician
SI7 National or regional government/public institutions
SI8 Corporate influencers (businesses or business associations)
SI9 Other members of your team
SI10 Mentors/consultants
SI11 Teachers/instructors
SI12 Other fellow managers in other departments or institutions
SI13 Boss/direct managers

Table A2. *Cont.*

Facilitating Conditions (FC)
FC1. had enough budget
FC4. easily got help when you had difficulties using this practice
FC5. found your unit's culture and people supportive to use this practice
FC6. found it easy to recruit your expected number of users/citizens to use this practice
FC7. you had to work out the extra, non-budgeted costs because the practice did not help with them
FC8. found the structure and systems of your organization supportive of this practice

Hedonic Motivation (HM)
HM1. was inspiring
HM2. fun made you had a good time
HM4. brought me recognition in my organization and helped my career development
HM5. allowed me to emotionally connect with the people involved
HM6. fulfill your public service vocation
HM8. feel good because you made it against all

Costs (CO)
CO1. We had extra, non-budgeted costs due to the use of this practice
CO2. we spent more time due to the use of this practice
CO3. costs are less controlled due to this practice
CO4. we needed to revise the service after implementation because results were not satisfactory
CO5. risks of failure were higher due to this practice

Habit (HT)
HT1. engaging (I could spend a lot of time on it without much effort)
HT2. an everyday or regular activity
HT4. I am so used to this practice that it is natural for me to use it, before any other
HT6. in my unit we use this innovation practice all the time

Government Paradigm (GP)
GP1. evaluate the actual engagement of users/citizens
GP2. assess user/citizen satisfaction with the service or process, pre- and after its innovation
GP3. assess the needs of users/citizens
GP4. include users/citizens in co-creation/prototyping sessions
GP5. include users/citizens in services or processes co-production/co-implementation
GP6. include users in analysis of data on their experiences
GP7. include any other agents (consultants, technical staff or any other) in idea generation or prototyping sessions
GP8. include any non-user (consultants, technical staff or any other) in services or processes co-production/co-implementation
GP9. work with users' representatives (e.g., NGOs, associations) more than with individual end users

Lists of loadings and weights per indicator.

Table A3. Outer loadings.

Behavioral Intention (BI)	Loading	Mean	SD
[BI1] ← BI	0.797	0.797	0.035
[BI2] ← BI	0.774	0.773	0.040
[BI3] ← BI	0.796	0.794	0.034
[BI4] ← BI	0.712	0.711	0.046
Effort Expectancy (EE)	**Loading**	**Mean**	**SD**
[EE10] ← EE	0.750	0.747	0.061
[EE1] ← EE	0.769	0.764	0.050
[EE2] ← EE	0.763	0.758	0.057
[EE7] ← EE	~~0.377~~	~~0.360~~	~~0.136~~
[EE8] ← EE	~~0.608~~	~~0.597~~	~~0.098~~
[EE6] ← EE	~~0.373~~	~~0.372~~	~~0.131~~
Facilitating Conditions (FC)	**Loading**	**Mean**	**SD**
[FC1] ← FC	0.798	0.745	0.150
[FC4] ← FC	0.746	0.703	0.141
[FC5] ← FC	0.622	0.588	0.226
[FC6] ← FC	0.721	0.677	0.201
[FC7] ← FC	~~0.018~~	~~0.075~~	~~0.232~~
[FC8] ← FC	~~0.602~~	~~0.566~~	~~0.234~~
Habit (HT)	**Loading**	**Mean**	**SD**
[HT1] ← HT	~~0.364~~	~~0.360~~	~~0.136~~
[HT2] ← HT	0.804	0.802	0.041
[HT4] ← HT	0.783	0.778	0.044
[HT6] ← HT	0.868	0.868	0.023
Hedonic Motivation (HM)	**Loading**	**Mean**	**SD**
[HM1] ← HM	0.813	0.809	0.047
[HM2] ← HM	0.815	0.811	0.049
[HM6] ← HM	~~0.604~~	~~0.611~~	~~0.081~~
[HM8] ← HM	~~0.731~~	~~0.732~~	~~0.061~~
[HM4] ← HM	0.648	0.643	0.075
[HM5] ← HM	0.827	0.825	0.036
Social Influence (SI)	**Loading**	**Mean**	**SD**
[SI01] ← SI	~~0.531~~	~~0.494~~	~~0.103~~
[SI02] ← SI	~~0.528~~	~~0.500~~	~~0.087~~
[SI03] ← SI	~~0.527~~	~~0.468~~	~~0.151~~
[SI04] ← SI	~~0.367~~	~~0.291~~	~~0.200~~
[SI05] ← SI	~~0.306~~	~~0.232~~	~~0.199~~
[SI06] ← SI	~~0.399~~	~~0.342~~	~~0.146~~
[SI07] ← SI	~~0.327~~	~~0.283~~	~~0.128~~
[SI08] ← SI	~~0.396~~	~~0.358~~	~~0.120~~
[SI09] ← SI	~~0.499~~	~~0.500~~	~~0.103~~
[SI10] ← SI	~~0.440~~	~~0.430~~	~~0.094~~
[SI11] ← SI	~~0.264~~	~~0.225~~	~~0.140~~
[SI12] ← SI	~~0.551~~	~~0.521~~	~~0.091~~
[SI13] ← SI	~~0.446~~	~~0.434~~	~~0.095~~
[SI20] ← SI	~~0.241~~	~~0.257~~	~~0.131~~
[SI21] ← SI	~~0.387~~	~~0.404~~	~~0.108~~
[SI22] ← SI	1.000	1.000	0.000

Table A4. Outer weights.

Cost (CO)	Weight	Mean	SD
[CO1] → CO	−0.025	0.142	0.289
[CO2] → CO	0.117	0.161	0.297
[CO3] → CO	−0.599	−0.202	0.621
[CO4] → CO	0.992	0.669	0.635
[CO5] → CO	0.236	0.205	0.312
Government Paradigm (GP)	**Weight**	**Mean**	**SD**
[GP1] → GP	0.377	0.361	0.145
[GP7] → GP	0.259	0.254	0.168
[GP8] → GP	0.477	0.411	0.141
[GP9] → GP	0.240	0.207	0.202
[GP2] → GP	0.724	0.674	0.117
[GP3] → GP	0.514	0.466	0.137
[GP4] → GP	0.712	0.702	0.114
[GP5] → GP	0.624	0.582	0.128
[GP6] → GP	0.827	0.759	0.098
Performance Expectancy (PE)	**Weight**	**Mean**	**SD**
[PE10] → PE	0.464	0.457	0.160
[PE11] → PE	0.389	0.325	0.170
[PE1] → PE	0.263	0.279	0.187
[PE2] → PE	0.086	0.081	0.163
[PE4] → PE	0.469	0.442	0.142
[PE5] → PE	0.786	0.673	0.121
[PE6] → PE	0.298	0.302	0.218
[PE8] → PE	0.482	0.449	0.165
[PE3] → PE	0.433	0.424	0.157
Usage (US)	**Weight**	**Mean**	**SD**
[US10] → US	0.416	0.384	0.151
[US11] → US	0.529	0.460	0.133
[US12] → US	0.044	0.050	0.160
[US13] → US	0.377	0.332	0.137
[US1] → US	0.430	0.406	0.147
[US2] → US	0.217	0.214	0.147
[US3] → US	0.042	0.041	0.173
[US4] → US	−0.095	−0.101	0.176
[US5] → US	0.479	0.443	0.152
[US6] → US	0.031	0.031	0.233
[US7] → US	0.353	0.330	0.147
[US8] → US	0.317	0.280	0.163
[US9] → US	0.202	0.168	0.174

Note: Dropped items are stricken.

References

1. Damanpour, F.; Schneider, M. Characteristics of innovation and innovation adoption in public organizations: Assessing the role of managers. *J. Public Adm. Res. Theory* **2009**, *19*, 495–522. [CrossRef]
2. Zaltman, G.; Holbek, J.; Duncan, R. *Innovations and Organizations*; John Wiley & Sons: New York, NY, USA, 1973.
3. Boyne, G.A.; Gould-Williams, J.S.; Law, J.; Walker, R.M. Explaining the adoption of innovation: An empirical analysis of public management reform. *Environ. Plan. Gov. Policy* **2005**, *23*, 419–435. [CrossRef]
4. Meyer, A.D.; Goes, J.B. Organizational assimilation of innovation: A multilevel contextual analysis. *Acad. Manag. J.* **1988**, *31*, 897–923.
5. Schneider, M. Do attributes of innovative administrative practices influence their adoption? An exploratory study of US local government. *Public Perform. Manag. Rev.* **2007**, *30*, 590–614. [CrossRef]
6. Elenkov, D.S.; Manev, I.M. Top management leadership and influence on innovation: The role of socio-cultural context. *J. Manag.* **2005**, *31*, 381–402.
7. Hooijberg, R.; DiTomaso, N. Leadership in and of demographically diverse organizations. *Leadersh. Q.* **1996**, *7*, 1–19. [CrossRef]
8. Gallouj, F.; Rubalcaba, L.; Windrum, P.; Toivonen, M. Understanding social innovation in service industries. *Ind. Innov.* **2018**, *25*, 551–569. [CrossRef]

9. Rubalcaba, L.; Windrum, P.; Gallouj, F.; Di Meglio, G.; Pyka, A.; Sundbo, J.; Weber, M. *ServPPIN. The Contribution of Public and Private Services to European Growth and Welfare, and the Role of Public-Private Innovation Networks*; Servppin Final Publishable Report; European Commision: Brussels, Belgium; Available online: https://hal.archives-ouvertes.fr/hal-01111787 (accessed on 7 May 2021).
10. Djellal, F.; Gallouj, F.; Miles, I. Two decades of research on innovation in services: Which place for public services? *Struct. Chang. Econ. Dyn.* **2013**, 98–117. [CrossRef]
11. Gallouj, F.; Windrum, P. Services and services innovation. *J. Evol. Econ.* **2009**, *19*, 141–148. [CrossRef]
12. Naor, M.; Druehl, C.; Bernardes, E.S. Servitized business model innovation for sustainable transportation: Case study of failure to bridge the design-implementation gap. *J. Clean. Prod.* **2018**, *170*, 1219–1230. [CrossRef]
13. Albury, D. Creating the conditions for radical public service innovation. *Aust. J. Public Adm.* **2011**, *70*, 227–235. [CrossRef]
14. Fuglsang, L. Bricolage and invisible innovation in public service innovation. *J. Innov. Econ. Manag.* **2010**, *5*, 67–87. [CrossRef]
15. Vickers, I.; Lyon, F.; Sepulveda, L.; McMullin, C. Public service innovation and multiple institutional logics: The case of hybrid social enterprise providers of health and wellbeing. *Res. Policy* **2017**, *46*, 1755–1768. [CrossRef]
16. Osborne, S.P. *Meta-Governance and Public Management*; Routledge: Oxfordshire, UK, 2010.
17. Koppenjan, J.; Koliba, C. Transformations towards new public governance: Can the new paradigm handle complexity? *Int. Rev. Public Adm.* **2013**, *18*, 1–8. [CrossRef]
18. Martin, B.R. Twenty challenges for innovation studies. *Sci. Public Policy* **2016**, *43*, 432–450. [CrossRef]
19. Kline, S.J.; Rosenberg, N. An overview of innovation. In *The Positive Sum Strategy: Harnessing Technology for Economic Growth*; Landau, R., Rosenberg, N., Eds.; National Academy of Sciences: Washington, DC, USA, 1986; pp. 275–305.
20. Davis, F.D.; Bagozzi, R.; Warshaw, P.R. User acceptance of computer technology: A comparison of two theoretical models user acceptance of computer technology: A comparison of two. *Manag. Sci.* **1989**, *35*, 982–990. [CrossRef]
21. Venkatesh, V.; Morris, M.G.; Davis, G.B.; Davis, F.D. User acceptance of information technology: Toward a unified view. *MIS Q.* **2003**, *27*, 425–478. [CrossRef]
22. Peralta, A.; Carrillo-Hermosilla, J.; Crecente, F. Sustainable business model innovation and acceptance of its practices among Spanish entrepreneurs. Sustainable Innovation: Processes, Strategies, and Outcomes. *J. Corp. Soc. Responsab. Environ. Manag.* **2019**, *26*, 1119–1134. [CrossRef]
23. Ansell, C.; Gash, A. Collaborative platforms as a governance strategy. *J. Public Adm. Res. Theory* **2017**, *28*, 16–32. [CrossRef]
24. Mergel, I.; Desouza, K.C. Implementing open innovation in the public sector: The case of Challenge.gov. *Public Adm. Rev.* **2013**, *73*, 882–890. [CrossRef]
25. Gallouj, F.; Weinstein, O. Innovation in services. *Res. Policy* **1997**, *26*, 537–556. [CrossRef]
26. Windrum, P.; Koch, P. (Eds.) *Innovation in Public Sector Services. Entrepreneurship, Creativity and Management*; Edward Elgar Publishing Ltd.: Cheltenham, UK; Northampton, UK, 2008.
27. Osborne, S.P.; Brown, L. (Eds.) *Handbook of Innovation in Public Services*; Edward Elgar Publishing Limited: Cheltenham, UK, 2014.
28. Desmarchelier, B.; Djellal, F.; Gallouj, F. *Public Service Innovation Networks (PSINs): Collaborating for Innovation and Value Creation*; HAL: Lyon, France, 2018.
29. Gallouj, F.; Rubalcaba, L.; Windrum, P. *Public–Private Innovation Networks in Services*; Edward Elgar Publishing Limited: Cheltenham, UK, 2013.
30. Ajzen, I. The theory of planned behavior. *Organ. Behav. Hum. Decis. Process.* **1991**, *50*, 179–211. [CrossRef]
31. Shepherd, D.A.; Haynie, J.M.; McMullen, J.S. Confirmatory search as a useful heuristic? Testing the veracity of entrepreneurial conjectures. *J. Bus. Ventur.* **2012**, *27*, 637–651. [CrossRef]
32. Taylor, B. The 4 Leadership Styles, and How to Identify Yours. HBR. Available online: https://cb.hbsp.harvard.edu/cbmp/product/H031MR-PDF-ENG (accessed on 7 May 2021).
33. Venkatesh, V.; Thong, J.Y.L.; Xu, X. Consumer acceptance and use of information technology: Extending the unified theory of acceptance and use of technology. *MIS Q.* **2012**, *36*, 157–178. [CrossRef]
34. Peralta, A.; Rubalcaba, L.; Carrillo-Hermosilla, J. How governance paradigms and other drivers affect public managers acceptance of innovation practices. *Acad. Manag. Annu. Meet. 2020 (Vanc. Can.)* **2020**, *2020*, 20024. [CrossRef]
35. Peralta, A.; Carrillo-Hermosilla, J.; Crecente, F. Modelling the Entrepreneur's Decision on Which Practice Use to Create Sustainable Businesses. *Acad. Manag. Annu. Meet. 2020 (Vanc. Can.)* **2020**, *2020*, 19467. [CrossRef]
36. Lima, M.; Baudier, P. Business model canvas acceptance among French entrepreneurship students: Principles for enhancing innovation artefacts in business education. *J. Innov. Econ.* **2017**, *23*, 159. [CrossRef]
37. Desmarchelier, B.; Djellal, F.; Gallouj, F. Towards a servitization of innovation networks: A mapping. *Public Manag. Rev.* **2020**, *22*, 1368–1397. [CrossRef]
38. Ropret, M.; Aristovnik, A. Public sector reform from the post-new public management perspective: Review and bibliometric analysis. *Cent. Eur. Public Adm. Rev.* **2019**, *17*. [CrossRef]
39. Ropret, M.; Aristovnik, A.; Kovač, P. A content analysis of the rule of law within public governance models: Old vs. new EU member states. *NISPAcee J. Public Adm. Policy* **2018**, *11*, 129–152. [CrossRef]

40. Cepiku, D.; Mititelu, C. Public administration reforms in transition countries: Albania and Romania between the Weberian model and the new public management. In *Public Administration in the Balkans: From Weberian Bureaucracy to New Public Management*; Matei, L., Flogaitis, S., Eds.; ASsee Online Series, South-Eastern European administrative studies; Editura Economica: Bucharest, Romania, 2011; p. 299.
41. Peralta, A.; Gismera, L. Sustainable business model innovation and ethics: A conceptual review from the institutional theory addressing (un)sustainability. *Int. J. Innov. Sustain. Dev.* **2020**. [CrossRef]
42. Djellal, F.; Gallouj, F. Fifteen advances in service innovation studies. *Serv. Exp. Innov. Integr. Extending Res.* **2018**, 39–65. [CrossRef]
43. Page, S. What's new about the new public management? Administrative change in the human services. *Public Adm. Rev.* **2005**, *65*, 713–727. [CrossRef]
44. Osborne, S.P. From public service-dominant logic to public service logic: Are public service organizations capable of co-production and value co-creation? *Public Manag. Rev.* **2018**, *20*, 225–231. [CrossRef]
45. Gilmore, T.N.; Krantz, J. Innovation in the public sector: Dilemmas in the use of ad hoc processes. *J. Policy Anal. Manag.* **1991**, *10*, 455–468. [CrossRef]
46. Compeau, D.R.; Higgins, C.A. Computer self-efficacy: Development of a measure and initial test. *MIS Q.* **1995**, *19*, 189–211. [CrossRef]
47. Martin, B.R. Science Policy Research—Having an Impact on Policy? Sussex. 2010. Available online: http://www.ohe.org/publications/article/science-policy-research-having-an-impact-on-policy-14.cfm%5Cnpapers3://publication/uuid/D7320A30-A725-4138-AE2B-15CF5A30E257 (accessed on 7 May 2021).
48. Blank, S.; Dorf, B. *The Startup Owner's Manual, The Step-By-Step Guide for Building a Great Company*; K&S Ranch: Pescadero, CA, USA, 2012.
49. Jessop, B. Governance and meta-governance: On reflexivity, requisite variety and requisite irony. In *Governance, Governmentality and Democracy*; Bang, H., Ed.; Manchester University Press: Manchester, UK, 2002.
50. Bell, S.; Park, A. The problematic metagovernance of networks: Water reform in New South Wales. *J. Public Policy* **2006**, *26*, 63–83. [CrossRef]
51. Radin, B. *Challenging the Performance Movement: Accountability, Complexity and Democratic Values*; Georgetown University Press: Washington, DC, USA, 2006.
52. Wirtz, B.W.; Pistoia, A.; Ullrich, S.; Göttel, V. Business models: Origin, development and future research perspective. *Long Range Plan.* **2016**, *49*, 36–54. [CrossRef]
53. Cooper, R.G. What's next? After stage-gate. *Res. Manag.* **2014**, *57*, 20–31. [CrossRef]
54. Bush, V. *Science the Endless Frontier*; Office of Scientific Research and Development: Washington, DC, USA, 1945.
55. Schmookler, J. *Invention and Economic Growth*; Harvard University Press: Cambridge, MA, USA, 1966.
56. York, J.L.; Danes, J.E. Customer development, innovation, and decision-making biases in the lean startup. *J. Small Bus. Manag.* **2014**, *24*, 21–39. [CrossRef]
57. Blank, S. *The Four Steps to the Epiphany: Successful Strategies for Products that Win*, 3rd ed.; K&S Ranch: Pescadero, CA, USA, 2007.
58. Dybå, T.; Dingsøyr, T. Empirical studies of agile software development: A systematic review. *Inf. Softw. Technol.* **2008**, *50*, 833–859. [CrossRef]
59. Rounsevell, M.; Arneth, A.; Brown, D.; Noblet-Ducoudré, N.; Ellis, E.C.; Finnigan, J.; Galvin, K.; Grigg, N.; Harman, I. *Incorporating Human Behaviour and Decision Making Processes in Land Use and Climate System Models*; Global Land Project International Project Office: São José dos Campos, Brazil, 2013; Volume 7, pp. 1–25.
60. Osterwalder, A. The Business Model Ontology—A Proposition in a Design Science Approach. Ph.D. Thesis, Universite de Lausanne, Lausanne, Switzerland, 2004.
61. Page, S. Entrepreneurial strategies for managing interagency collaboration. *J. Public Adm. Res. Theory J.* **2003**, *13*, 311–339. Available online: http://www.jstor.org/stable/3525852 (accessed on 7 May 2021). [CrossRef]
62. Karadag, H. Financial management challenges in small and medium-sized enterprises: A strategic management approach. *EMAJ Emerg. Mark. J.* **2015**, *5*, 26–40. [CrossRef]
63. Geissdoerfer, M.; Savaget, P.; Evans, S. The Cambridge business model innovation process. In Proceedings of the 14th Global Conference on Sustainable Manufacturing GCSM, Stellenbosch, South Africa, 3–5 October 2016; pp. 262–269. [CrossRef]
64. Yin, R.K. Life histories of innovations: How new practices become routinized. *Public Adm. Rev.* **1981**, *41*, 21–28. [CrossRef]
65. Kim, S.S.; Malhotra, N.K. A longitudinal model of continued IS use: An integrative view of four mechanisms underlying post-adoption phenomena. *Manag. Sci.* **2005**, *51*, 741–755. [CrossRef]
66. Verplanken, B.; Wood, W. Interventions to break and create consumer habits. *J. Public Policy Mark.* **2006**, *25*, 90–103. [CrossRef]
67. Fernandez, S.; Moldogaziev, T. Using employee empowerment to encourage innovative behavior in the public sector. *J. Public Adm. Res. Theory* **2013**, *23*, 155–187. [CrossRef]
68. Sorenson, E. Metagovernance: The changing role of politicians in processes of democratic governance. *Am. Rev. Public Adm.* **2006**, *36*, 98–124. [CrossRef]
69. Yang, K. Creating public value and institutional innovations across boundaries: An integrative process of participation, legitimation, and implementation. *Public Adm. Rev.* **2016**, *76*, 873–885. [CrossRef]
70. White, L.D. *Trends in Public Administration*; McGraw-Hill: New York, NY, USA, 1933.
71. Minton, H.L.; Schneider, F.W. *Differential Psychology*; Waveland Press: Prospect Heights, IL, USA, 1980.

72. Hall, D.; Mansfield, R. Relationships of age and seniority with career variables of engineers and scientists. *J. Appl. Psychol.* **1995**, *60*, 201–210. [CrossRef]
73. Castells, M. *The Rise of the Network Society*, 2nd ed.; Wiley-Blackwell: Chichester, UK, 2010.
74. Cavalcante, S.; Kesting, P.; Ulhøi, J. Business model dynamics and innovation: (re)establishing the missing linkages. *Manag. Decis.* **2011**, *49*, 1327–1342. [CrossRef]
75. Alford, J. The multiple facets of co-production: Building on the work of Elinor Ostrom. *Public Manag. Rev.* **2014**, *16*, 299–316. [CrossRef]
76. Bovaird, T. Beyond engagement and participation: User and community coproduction of public services. *Public Adm. Rev.* **2007**, *67*, 846–860. [CrossRef]
77. Scott, T.A.; Thomas, C.W.; Magallanes, J.M. Convening for consensus: Simulating stakeholder agreement in collaborative governance processes under different network conditions. *J. Public Adm. Res. Theory* **2018**, *29*, 32–49. [CrossRef]
78. Felin, T.; Foss, N.J. Strategic organization: A field in search of micro-foundations. *Strateg. Organ.* **2005**, *3*, 441–455. [CrossRef]
79. Crossan, M.M.; Apaydin, M. A multi-dimensional framework of organizational innovation: A systematic review of the literature. *J. Manag. Stud.* **2010**, *47*, 1154–1191. [CrossRef]
80. Kearney, C.; Hisrich, R.; Roche, F. Facilitating public sector corporate entrepreneurship process: A conceptual model. *J. Enterp. Cult.* **2007**, *15*, 275–299. [CrossRef]
81. Ramió, C. Una administración pública de futuro sostenible económicamente e innovadora en el contexto de la globalización. *Cuad. Gob. Adm. Pública* **2017**, *3*, 103–122. [CrossRef]
82. Benéitez, M.B.; Engelken, M. *De la Nueva Gestión Pública al Nuevo Servicio Público: De Ciudadano a Cliente y Vuelta a Empezar*; Asociación Española de Ciencia Política y de la Administración: Madrid, Spain, 2011.
83. Graells, I.; Costa, J.; Ramilo-Araujo, M.C. Ciudadanía y administraciones en red. La administración pública ante la nueva sociedad. *Cuad. Gob. Adm. Pública* **2014**, *1*, 221–226.
84. Desmarchelier, B.; Djellal, F.; Gallouj, F. Public Service Innovation Networks (PSINs): An Instrument for Collaborative Innovation and Value Co-Creation in Public Service(s). 2018. Available online: https://halshs.archives-ouvertes.fr/halshs-01934284 (accessed on 7 May 2021).
85. Hair, J.F.; Sarstedt, M.; Hopkins, L.; Kuppelwieser, V. Partial least squares structural equation modeling (PLS-SEM): An emerging tool for business research. *Eur. Bus. Rev.* **2014**, *26*, 106–121. [CrossRef]
86. Faul, F.; Erdfelder, E.; Buchner, A.; Lang, A.-G. Statistical power analyses using G*Power 3.1: Tests for correlation and regression analyses. *Behav. Res. Methods* **2009**, *41*, 1149–1160. Available online: http://www.psychologie.hhu.de/arbeitsgruppen/allgemeine-psychologie-und-arbeitspsychologie/gpower.html (accessed on 7 May 2021). [CrossRef]
87. Garson, D.G. Partial least squares: Regression & structural equation models. In *Blue Book*; Statistical Associates Publishing: Asheboro, NC, USA, 2016.
88. Sarstedt, M.; Ringle, C.M.; Hair, J.F. *Partial Least Squares Structural Equation Modeling*; Springer: Cham, Switzerland, 2017; Volume 26.
89. Hair, J.F., Jr.; Hult, G.T.M.; Ringle, C.M.; Sarstedt, M. *A Primer on Partial Least Squares Structural Equation Modeling (Pls-Sem)*; SAGE Publications Inc.: Thousand Oaks, CA, USA, 2014.
90. Lohmöller, J.B. *Latent Variable Path Modeling with Partial Least Squares*; Physica: Heidelberg, Germany, 1989.
91. Ringle, C.M.; Wende, S.; Becker, J.-M. *SmartPLS 3*; SmartPLS GmbH: Boenningstedt, Germany, 2015.
92. Tehseen, S.; Ramayah, T.; Sajilan, S. Testing and controlling for common method variance: A review of available methods. *J. Manag. Sci.* **2017**, *4*, 142–168. [CrossRef]
93. Hair, J.F.; Hult, G.T.M.; Ringle, C.M.; Sarstedt, M. *A Primer on Partial Least Squares Structural Equation Modeling (PLS-SEM)*, 2nd ed.; SAGE Publications: Thousand Oaks, CA, USA, 2016.
94. Cohen, J. *Statistical Power Analysis for the Behavioral Sciences*; Lawrence Erlbaum: Mahwah, NJ, USA, 1988.
95. Edgington, E.; Onghena, P. *Randomization Tests*; Chapman and Hall/CRC: Boca Raton, FL, USA, 2007.
96. Hair, J.F.; Sarstedt, M.; Ringle, C.M.; Gudergan, S.P. *Advanced Issues in Partial Least Squares Structural Equation Modeling (PLS-SEM)*; SAGE Publications Inc.: Thousand Oaks, CA, USA, 2018.
97. Onghena, P. Randomization and the randomization test: Two sides of the same coin. In *Randomization, Masking and Allocation Concealment*; Chapman and Hall/CRC: Boca Raton, FL, USA, 2019; pp. 185–208. [CrossRef]
98. Becker, J.M.; Rai, A.; Ringle, C.M.; Völckner, F. Discovering unobserved heterogeneity in structural equation models to avert validity threats. *MIS Q.* **2013**, *37*, 665–694. [CrossRef]
99. Hartley, J.; Sørensen, E.; Torfing, J. Collaborative innovation: A viable alternative to market-competition and organizational entrepreneurship? *Public Adm. Rev.* **2013**, *76*, 821–830. [CrossRef]
100. Criado, I. Nuevas tendencias en la gestión pública: Innovación abierta, gobernanza inteligente y tecnologías sociales en unas administraciones públicas colaborativas. In *Monografías*; Instituto Nacional de Administración Pública: Madrid, Spain, 2016.
101. Given, L.M. Exploratory research. In *The SAGE Encyclopedia of Qualitative Research Methods*; SAGE Publications Inc.: Thousand Oaks, CA, USA, 2008; pp. 327–329.

Article

Exploring the Antecedents of Cruisers' Destination Loyalty: Cognitive Destination Image and Cruisers' Satisfaction

María Dolores Benítez-Márquez [1,*], Guillermo Bermúdez-González [2], Eva María Sánchez-Teba [3] and Elena Cruz-Ruiz [2]

1. Department of Applied Economics (Statistics and Econometrics), Faculty of Economics and Business, University of Malaga, 29070 Malaga, Spain
2. Department of Business Management, Faculty of Commerce and Management, University of Malaga, 29071 Malaga, Spain; gjbermudez@uma.es (G.B.-G.); ecruz@uma.es (E.C.-R.)
3. Department of Business Management, Faculty of Economics and Business, University of Malaga, 29070 Malaga, Spain; emsanchezteba@uma.es
* Correspondence: bemarlo@uma.es

Abstract: This study is one of the few of its kind that explores the individual impact of each of the cognitive attributes of a tourist destination's image on cruisers' destination loyalty and overall satisfaction. It also analyzes the mediating role of satisfaction between each of the attributes and loyalty. Variance-based structural equation modeling (PLS-SEM) was used for this analysis, based on a survey of 457 cruisers visiting the city of Malaga. The results confirm that three of the five attributes, the destination's environment image, the perceived value image of services and the accessibility image have a direct influence on cruisers' overall satisfaction, where environment image has the most significant impact. Moreover, the results support the mediating role of satisfaction in certain cases. There is total mediation between perceived value and loyalty, as well as between accessibility image and loyalty while there is only partial complementary mediation between environment image and loyalty. The confirmation that overall satisfaction influences loyalty enables management organizations to develop more efficient loyalty strategies for their respective destinations.

Keywords: cognitive destination image; cruise; satisfaction; loyalty; behavioral intention; structural equation modeling; PLS-SEM

1. Introduction

The relationship between satisfaction and the success of organizations has historically been considered to be a business reality, a fact that has been validated by various studies [1,2]. However, since the end of the 20th century, various authors have criticized this model in studies that demonstrate a high customer attrition rate despite having high rates of satisfaction [3–5], calling for a shift in the paradigm where loyalty becomes the new strategic objective for business success through customer retention.

The tourism sector is also susceptible to this paradigm shift. Both public and private institutions, in addition to academics, are growing increasingly interested in learning about the factors that influence loyalty to a tourist destination [6,7] and, consequently, the intention to revisit the destination and recommend it to other travelers.

Loyalty has been considered by different perspectives: attitudinal loyalty, behavioral loyalty and composite loyalty, as well as, the integration of attitudinal and behavioral [8]. Composite loyalty is often operationalized as behavioral intention which includes revisit intention and intention to recommend [9,10]. From now onwards, we use the term loyalty to refer to composite loyalty and we will use, alternatively, both terms—loyalty and behavioral intention—as synonyms.

Recent studies have established that destination loyalty, or behavioral intention, is the cornerstone of a tourist destination's brand equity and a determining factor in its competitiveness [11]. Within this context, numerous studies have confirmed that satisfaction with the travel experience [12–15] and destination image [16–18] are antecedents of loyalty and, therefore, the desire to return to a destination and recommend it to others. There is also discussion about the magnitude and direction of the relationships between the different components of destination image (both cognitive and affective) and tourist loyalty [8,19,20].

However, there is very little information about the antecedents of tourists' destination loyalty in the scientific literature, especially considering that the cruise tourism sector is so important, although there are more studies that analyze the influence of image and/or cruisers' satisfaction on loyalty to the cruise line or their experience on board the ship [21–24] than the tourist destination itself [25–27]. There is also a lack of research regarding the relationship between the elements and attributes associated with destination image and the satisfaction and behavioral intention of travelers in general and cruisers in particular.

Considering the lack of empirical research that can explain these relationships, this article analyzes the antecedents of cruisers' loyalty to the tourist destination they are visiting and, specifically, the predictive power of the cognitive attributes of the tourist destination image on overall satisfaction and the influence of both on the behavioral intention of the cruisers visiting the city of Malaga (Spain).

Therefore, one key contribution of this article is to study, in the specific field of cruise tourism, the relationship between cognitive image and cruise passenger satisfaction and loyalty to the destination. Likewise, to the best of our knowledge, this is a novelty in which the attributes of the cognitive image of the destination are divided into five dimensions and their direct and indirect influence on satisfaction and loyalty are analyzed in the cruise tourism sector.

According to data from the Port Authority of Malaga [28] a total of 477,001 cruisers arrived at the Port of Malaga in 2019 on 288 ships, making it the second largest port in continental Spain in terms of visitors, after Barcelona. It is estimated that 11% of the tourists that visit Spain do so via cruise ships, generating a total revenue of 1.255 billion euros [29]. Following the United Nations World Tourism Organization [30], cruise travel has grown nearly 7%, for a total of 28.5 million travelers. Cruises in the Mediterranean have increased by 8%, for a total of over 4 million cruisers. These data justify the interest in studying a sector as important as cruise tourism and a destination as established as the city of Malaga (capital of the Costa del Sol).

In this study, destination loyalty is considered to be a dependent variable of variance-based structural equation modeling using partial least squares (PLS-SEM), while the different attributes of tourist destination cognitive image and overall satisfaction with the destination are predictor variables.

This article is organized in five sections, including the introduction. The second section offers a literature review regarding the perception of cruise passengers towards tourism, as well as the theoretical bases of the study and the relationship between image, satisfaction and loyalty. The third section includes the methodology and the data collected from the survey give to cruise passengers. The fourth section presents the results and the fifth section includes a discussion about these results, in addition to conclusions, limitations and proposed future lines of research.

2. Literature Review and Hypotheses

2.1. Attributes of the Cognitive Image of a Tourist Destination

Although the first generic studies about the importance of perceived image on human behavior were conducted in the 1950s [31,32], it was towards the end of the 20th century when researchers began to consider image as a determining factor in the individual perception of travelers and, consequently, in understanding their behavior and the tourist destination selection process [33–35].

Accordingly, measuring the image of a tourist destination [35–38], analyzing the changes in perception before, during and after the visit [39,40], determining the factors that influence image [41,42] and understanding the elements that compose said image [43–46], have been addressed by academics in numerous scientific articles. In the area of marketing, Hallmann et al. [47] demonstrated that the purchase decision process is simplified when the perception of a product or service is favorable and, therefore, their analysis is an effective marketing strategy, which plays a decisive role in the communication and commercialization of a destination as a tourism product.

There is a general consensus in the academic community that destination image is a multidimensional concept with at least two components: affective and cognitive [19,48], although some authors add a conative component as well [49,50].

The affective image of a destination refers to the emotions that individuals associate with the place they are visiting [41,43,51–54], while the cognitive image of a destination is related to the beliefs and knowledge a tourist has about the destination and arises from their assessment of the perceived attributes [43,55–57].

Conative image, on the other hand, is associated with an individual's behavior or true intention towards the destination, such as revisiting it, recommending it [56,58,59] or speaking positively about it [60]. Within this context, there are numerous researchers that associate it with destination loyalty or behavioral intention more than a component of destination image [12,19,55,61].

These dimensions can be studied separately in order to better understand their complexity as a whole [49,62–64], as the authors have done in this paper, focusing on the attributes of cognitive destination image for cruisers.

So, present research focuses on cognitive destination image. This election is justified, as indicated by Walmsley and Young [65], because the cognitive component is more descriptive, measurable and observable than the affective component and provides more specific information about the destination that is easier to interpret [42]. There are numerous empirical studies that analyze and support the cognitive component of destination image for its ability to characterize a destination, e.g., [43,66], and, also, using a multi-attribute approach [67–70].

The attributes that are most highly referenced by academics include environment image (the main indicators of which include the natural environment and climate) [71], green management [72,73], the atmosphere [74] and the singularity of cultural image and cultural heritage [56]. Other noteworthy attributes include infrastructure image, which mainly refers to available restaurant, shopping and accommodation services [43]. Similarly, various authors consider other attributes to be integral to cognitive destination image, such as attractions and entertainment image [12], accessibility image [75] and the perceived value image of services [76].

For the purposes of this study, we consider cognitive destination image from a multi attribute perspective, consisting of the aforementioned attributes we have adapted from Chi and Qu [12]: environment image, available infrastructure image, attractions and entertainment image, accessibility image and the perceived value image of services.

2.2. Impact of the Attributes of Cognitive Destination Image on Cruisers' Satisfaction and Loyalty

Numerous researchers have studied the influence of cognitive destination image on travelers' satisfaction [19,77–81], loyalty and behavioral intention [55,82] or on both concepts [12,16,17,83–87].

The correlation of destination image with satisfaction and loyalty has been studied to a lesser degree in the cruise tourism sector [22,27,88,89]. However, there are very few scientific studies that specifically analyze the relationship between cognitive image and these constructs, which can be considered in certain cases to be either a relevant, positive relationship [27,90] or of little significance [85].

Therefore, although there is research on the influence of a destination's overall image and some research on cognitive image, there are very few studies on the relationship of each of the integral attributes of the destination's cognitive image with travelers' loyalty and satisfaction and more so in the area of cruise tourism.

2.2.1. Impact of Environment Image

Various authors have analyzed the influence of environment image on travelers' satisfaction and loyalty. Jin et al. [91] define environment image as the relevant cognitive experience that influences the perception of a place when an individual is traveling to a specific destination. This perception affects their satisfaction and loyalty, especially in relation to their intention to revisit or recommend the destination.

Several studies [79,92,93] have demonstrated a positive influence of the destination's environment image on tourists' satisfaction. Research has also been conducted on its influence on the behavioral intention model associated with loyalty [94]. Ortegón-Cortázar and Royo-Vela [95] also demonstrated the indirect and positive influence of natural environment on behavioral intention through satisfaction.

Among the specific literature about cruise tourism, authors such as Sanz-Blas and Carvajal-Trujillo [96] or Toudert and Bringas-Rábago [27] have confirmed the direct relationship of the environment image construct with satisfaction on the one hand and destination loyalty on the other. Others, however, DiPietro and Peterson [25] or Silvestre et al. [26] have focused their research on the relationship of the latent variable (environment image) with the cruiser's intention to return to the destination and recommend it.

Keeping in mind all of these considerations, we postulate the following hypotheses:

H1. *Environment image has a direct, positive, influence on satisfaction.*

H2. *Environment image has a direct, positive, influence on loyalty.*

H3. *Satisfaction mediates between environment image and loyalty.*

2.2.2. Impact of Attractions and Entertainment Image

As indicated by Albaity and Melhem [16], the attractions and entertainment image is a determining factor in the selection of a tourist destination. This perception also affects the general image of the destination, influencing other attributes and the decision-making process [97]. Other authors such as Zhang [98] include this attribute in leisure experience.

The influence of a destination's attractions and entertainment on travelers' satisfaction has been adequately studied in the scientific literature [93,99]. Similarly, the influence of this cognitive attribute on loyalty and behavioral intention has also been considered in multiple studies [92,100,101].

In the cruise tourism sector, from the best of our knowledge, no one has studied the direct influence of attractions and entertainment image on cruisers' satisfaction or their future behavioral intention (as loyalty) in regard to the destination. However, there are studies that measure indirectly how attractions and entertainment can influence satisfaction such as Sanz-Blas and Carvajal-Trujillo [96] and Sanz-Blas et al. [89] include this attribute in a second order construct (destination image) and demonstrate its influence on cruisers' satisfaction.

Furthermore, the items related to attractions and entertainment image (which the authors have included in the dimension entitled tourism resources) have a low degree of relevance in relation to the destination image construct as compared to other attributes such as infrastructure image and environment image. However, Toudert and Bringas-Rábago [27] consider that this construct has the greatest influence on cruisers' satisfaction for the destination Baja California.

We have also found few studies that indirectly analyze the influence of attractions and entertainment on destination loyalty (revisiting and recommending it). Thus, Silvestre et al. [26] have included what we call attractions and entertainment in a construct entitled satisfaction with the city and the visit, concluding that there is no significant direct relationship between this construct (which includes attractions and entertainment image) and cruisers' destination loyalty. Additionally, Lee [102] tested a significant positive relation between leisure experience and behavioral intention through destination perceived value.

Therefore, we have developed the following hypotheses based on the literature review:

H4. *Attractions and entertainment image has a direct, positive, influence on satisfaction.*

H5. *Attractions and entertainment image has a direct, positive, influence on loyalty.*

H6. *Satisfaction mediates the relationship between attractions and entertainment image and loyalty.*

2.2.3. Impact of Infrastructure Image

Accommodations and restaurant services consist of a series of products and services, making it a much more complex task to present an image that meets travelers' expectations [103]. It is important to study the role of this infrastructure on destination image since general satisfaction in the tourism sector is strongly related to infrastructure image, specifically restaurant and accommodation services, among other variables [104]. A study by Simarmata et al. [105] highlights that tourism infrastructure image (travel agencies, accommodations and restaurants) has a significant influence on visitors' satisfaction. In the area of cruise tourism, various authors have confirmed that infrastructure is a decisive component of destination image [41,96]. However, Puh [80] does not consider infrastructure to play a decisive role in the formation of said image and, consequently, on satisfaction.

In regard to loyalty, various studies have concluded that if the image of the services and infrastructure available at a destination are positive, this could stimulate word-of-mouth recommendations, as well as travelers choosing to revisit the same destination in the future [12,14,106]. Similarly, regarding cruise tourism, a positive experience at the port of call has a decisive influence on travelers' behavioral intention to revisit a specific port and even to recommend visiting that destination to people they know [26]. Furthermore, Nasir et al. [107] studied the indirect relationship between the infrastructure and loyalty to destination

Consequently, based on the relevant literature, we propose the following hypotheses:

H7. *Infrastructure and services image has a direct, positive, influence on satisfaction.*

H8. *Infrastructure and services image has a direct, positive, influence on loyalty.*

H9. *Satisfaction mediates between infrastructure and service image, and loyalty.*

2.2.4. Impact of Perceived Value Image

Perceived value is defined by Zeithaml [108] as a consumer's overall assessment regarding the utility of a product or service, which is based on their perception of what is received and what is given. According to Lovelock [109], this concept is the balance between the perceived benefits and perceived costs.

Various studies have highlighted the importance of the relationship of the quality-value-loyalty chain [76,110] specifically determined that quality is an antecedent of perceived value, while satisfaction is a consequence of perceived value, which results in an attitude of loyalty. Many authors have studied this relationship in the tourism sector [15,111–118]. In particular, a tourist's perceived value of the quality of the trip is a construct that is directly related to their satisfaction with said trip [119,120]. In turn, satisfaction has an indirect relationship with loyalty [10,121]. Specifically, in the area of cruise tourism, various authors have studied the price-quality relationship (a component of our perceived value attribute) and its positive effect on satisfaction [26,88,122].

Similarly, other authors have demonstrated that perceived value in relation to loyalty has a positive impact on a tourist's decision to return to a destination [114,123]. According to studies by various authors, this relationship is also true in the area of cruise tourism [124,125].

Based on the aforementioned literature, we propose the following hypotheses:

H10. *Perceived value image has a direct, positive, influence on satisfaction.*

H11. *Perceived value image has a direct, positive, influence on destination loyalty.*

H12. *Satisfaction mediates between perceived value and loyalty.*

2.2.5. Impact of Accessibility Image

The accessibility of a destination is one of the attributes of cognitive image that determines tourists' satisfaction [79]. Many authors have considered the accessibility construct: making sure that the area has the appropriate signage [126], the importance of the location's accessibility factors for disabled tourists [127] and access to tourism information [128]. In regard to cruisers, the urban environment has a significant influence [27] on having a positive experience at the destination, which in turn influences the development of a positive image of the location visited. Furthermore, Wisker et al. [129] studied the indirect relationship between the accessibility and loyalty to destination.

Regarding its relationship with loyalty, tourists that positively value a destination's accessibility are more prone to travel during the off-season, which can help reduce the effects of seasonality as they have more time and, on average, stay longer at the destination, thereby spending more than average at said location. Furthermore, the more satisfied visitors are with the tourism available at a destination, the more loyal they are to said destination [75].

Consequently, we propose the following hypotheses:

H13. *Accessibility image has a direct, positive, influence on satisfaction.*

H14. *Accessibility image has a direct, positive, influence on destination loyalty.*

H15. *Satisfaction mediates between accessibility image and loyalty.*

2.3. Influence of Tourists' Satisfaction on Destination Loyalty

The relationship between satisfaction with the travel experience and destination loyalty or behavioral intention has been amply studied in the scientific literature. Satisfaction is considered to be both an antecedent [13–15] and a mediator of image and loyalty [12,83,86,130].

In addition to a rational preference, satisfaction often leads to an emotional relationship, which creates greater customer loyalty [131]. Loyalty also makes customers more prone to recommend a tourist service to the people they know [132]. In parallel, in the cruise tourism sector, cruisers' satisfaction is also a determining factor of loyalty. Silvestre et al. [26] determined that satisfaction with the city and the visit is a more decisive determining factor than satisfaction with establishments and services. Similarly, Brida and Coletti [133] concluded that the probability of cruise tourists revisiting a destination depends on their satisfaction with their experience at the destination. Along these lines, other researchers have concluded that general satisfaction has a positive influence on cruisers' loyalty, which is understood as their intention to return to the destination and recommend it [96,134]. Similarly, according to Satta et al. [135], there is a positive relationship between satisfaction with the destination and cruisers' word-of-mouth recommendations. Finally, Toudert and Bringas-Rábago [27] and Albaity and Melhem [16] have analyzed the influence of satisfaction as a mediator between a cruise's image and destination loyalty.

Therefore, in accordance with the points made above, we propose the following hypothesis:

H16. *Overall satisfaction has a direct, positive, significant influence on destination loyalty.*

3. Data Collection and Methods

3.1. Description of the Study Location: Malaga (Costa del Sol, Spain)

Malaga, the capital of the Costa del Sol (Andalusia, Spain) is a coastal city in the Mediterranean and tourism is the most important component of its economy. In 2019, the city had a population of 574,654 residents and received over 1.4 million tourists, making it the third largest tourist destination in Spain in terms of the annual number of visitors [136] and the second largest in terms of the number of cruisers in continental Spain, after Barcelona, with a total of 477,001 cruise passengers [29]. Consequently, Malaga was chosen for this study due to its importance in Spain's cruise tourism sector (Figure 1).

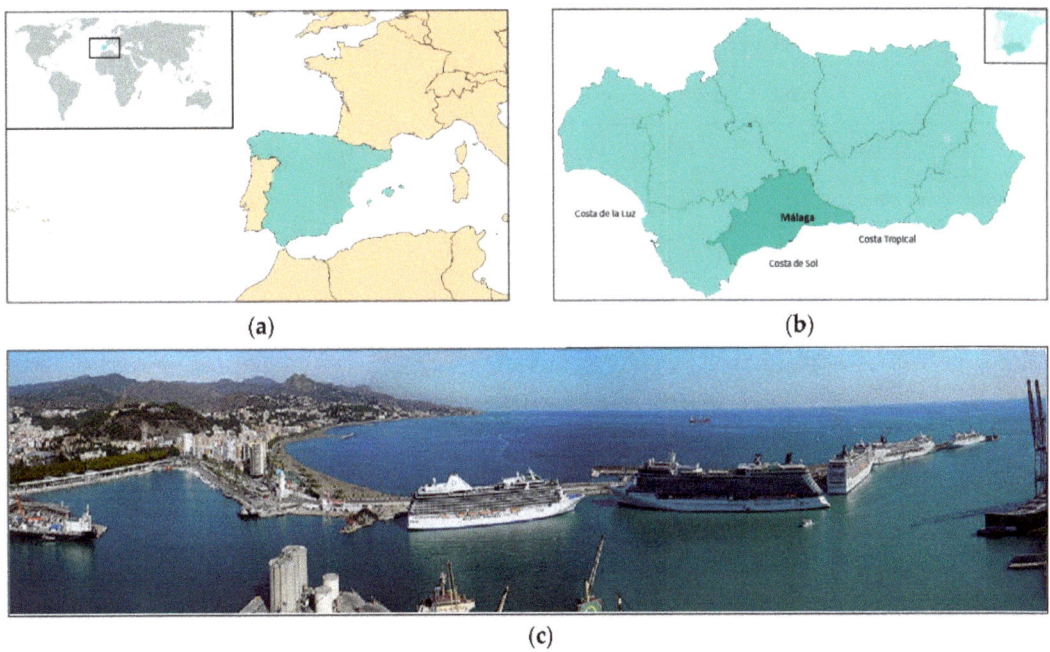

Figure 1. Location of Malaga (Andalusia, Spain) and photo of Port of Malaga. Source: Map (**a**) and map (**b**) from [137], map (**b**) own modification and photo (**c**) [138].

3.2. Sample, Data Collection and Construct Measures

A cross-sectional study was conducted with a random sample using raw data from a survey in which cruise tourists are the unit of analysis. The data were collected throughout the year 2019 from cruisers traveling on different ships. The questionnaire was handed out in person, in hard copy format, as passengers were boarding the ship after returning from visiting the city. Once the data were collected, a total of 457 valid surveys were considered (from a total of 470 collected) and there were no missing values (Table 1). The survey is included in the Supplementary Material.

Table 1. Descriptive sociodemographic.

	Range	Frequency	Percentage	Indicator	Category	Frequency	Percentage
Age	18–30	74	16.2	Sex	Man	217	47.5
	30–40	51	11.2		Woman	240	52.5
	40–50	59	12.9		Total	457	100.0
	50–60	86	18.8	N (sample size)			
	60–70	129	28.2	Valid	457		100.0
	70–90	58	12.7	Missing	0		0

Source: Own elaboration from SPSS.

A 7-point Likert scale was used to evaluate the degree of agreement or disagreement with each statement (1 = strongly disagree, 4 = neutral, 7 = strongly agree). The frequency and percentage of the respondents' ages and genders were included in the survey (see Table 1). At a 5% level of significance and a statistical power of 80%, with a maximum of 3 arrows pointing at a construct in the structural model and a minimum of 10% explained variance, according to this author, the minimum sample size is 300 [139] and, therefore, the sample more than fulfills this requirement. Table 2 shows the items included in the survey and their descriptive measures.

Table 2. Survey items about the city of Malaga (Andalusia, Spain).

Indicators or Items	No.	Mean	Me	Min	Max	Standard Deviation	Excess Kurtosis	Skewness
1st construct: 1. Environment								
1.1. Pleasant weather	1	5.796	6	1	7	1.43	1.394	−1.279
1.2. Clean and tidy environment	2	6.039	6	2	7	1.19	0.513	−1.139
1.3. Friendly and helpful local people	3	5.937	6	1	7	1.194	0.943	−1.101
1.4. Safe and secure environment	4	5.921	6	1	7	1.237	0.597	−1.082
1.5. Picturesque views	5	6.035	7	1	7	1.185	0.189	−1.018
2nd construct: 2. Attractions and entertainment								
2.1. Wide arrays of shows/exhibitions	6	4.521	4	1	7	1.075	0.903	1.33
2.2. Wide variety of entertainment	7	4.519	4	2	7	1.059	0.885	1.345
2.3. Tempting cultural events	8	4.593	4	1	7	1.15	0.27	1.118
3rd construct: 3. Infrastructure and service								
3.1. Wide selection of restaurants/cuisine	9	5.508	6	1	7	1.312	−0.837	−0.373
3.2. Wide variety of shopping options	10	5.597	6	1	7	1.274	−0.391	−0.537
3.3. Wide choice of accommodations	11	4.648	4	1	7	1.144	0.185	0.949
3.4. Reasonable price for food and accommodation	12	5.186	5	1	7	1.287	−0.737	−0.079
4th construct: 4. Perceived value								
4.1. Good value for money	13	5.263	5	1	7	1.325	−0.694	−0.265
4.2. Reasonable price for attractions and activities	14	5.026	5	1	7	1.323	−0.793	0.168
4.3. Good bargain shopping	15	4.814	4	1	7	1.429	−0.399	0.046
4.4. Appropriate signposting	16	5.449	6	1	7	1.382	−0.523	−0.501
5th construct: 5. Accessibility								
5.1. Easy access to city center	17	5.917	6	1	7	1.265	1.022	−1.146
5.2. Disabled access	18	4.814	4	1	7	1.261	−0.382	0.498
5.3. Tourism information points	19	5.56	6	1	7	1.3	−0.386	−0.552
6th construct: 6. Satisfaction with destination								
6.1. Overall Satisfaction (1 item)	20	6.088	6	3	7	1.008	0.514	−1.05
7th construct: 7. Loyalty with destination								
7.1. Intention to revisit it	21	5.737	6	1	7	1.415	1.024	−1.171
7.2. Intention to recommend	22	6.057	6	1	7	1.142	1.696	−1.304

Note: N = Total participants = 457; Me = Median. Source: Own elaboration from [140].

Infrastructure, perceived value and accessibility have been validated as measurement instruments by Chi and Qu [12]. Environment and, also, attractions and entertainment has been validated by means of individual factor analysis for each construct with SPSS software prior to import the database to the software SmartPLS. Different studies have used the overall satisfaction as measure, e.g., [12,79,141,142]. In addition, as we mentioned in the introduction section, loyalty can be measured through the variables used to measure behavioral intention, the return intention and the recommendations to friends and relatives [8–10,12,61,80,143–146].

3.3. Data Analysis

In our case, the descriptive analysis of the items in the measurement scales confirmed that the data does not follow a normal distribution based on skewness as the data exceeds the recommended thresholds of ± 1 [147].

For this study, we opted to use variance-based structural equation modeling estimated with partial least squares (PLS-SEM). These models are intended to predict and explain the variance of the endogenous construct under study (loyalty) by means of different latent variables that are predictors (cognitive image and satisfaction attributes).

Theoretical concepts of behavioural investigation are usually represented as latent variables according to Benitez et al. [148]. Following Dijkstra and Henseler [149], the use of PLSc is a recommended method to consistently estimates model in case the common factor model is true; and it is equally recommended in comparison with the covariance based structural equation modelling. Additionally, Dijkstra and Henseler conclude that the PLSc approach report less problems in the framework of non-recursive linear reflective model [149]. Among other reasons, these aforementioned facts justify the selection of PLSc. Later, we have computed the model with partial least squares structural equation modeling with the software SmartPLS version 3.3 [140] using PLSc to test the hypotheses.

Two phases englobe the assessment of the PLS-SEM: Firstly, the evaluation of the measurement model and, secondly, the assessment of the structural model. Concerning the mediation, Nitzl et al. [150] is followed. To end with this section and based on the literature review described earlier, we propose a confirmatory model with a nomogram (Figure 2) showing the relationships between the constructs, as well as the sign and direction of said relationships.

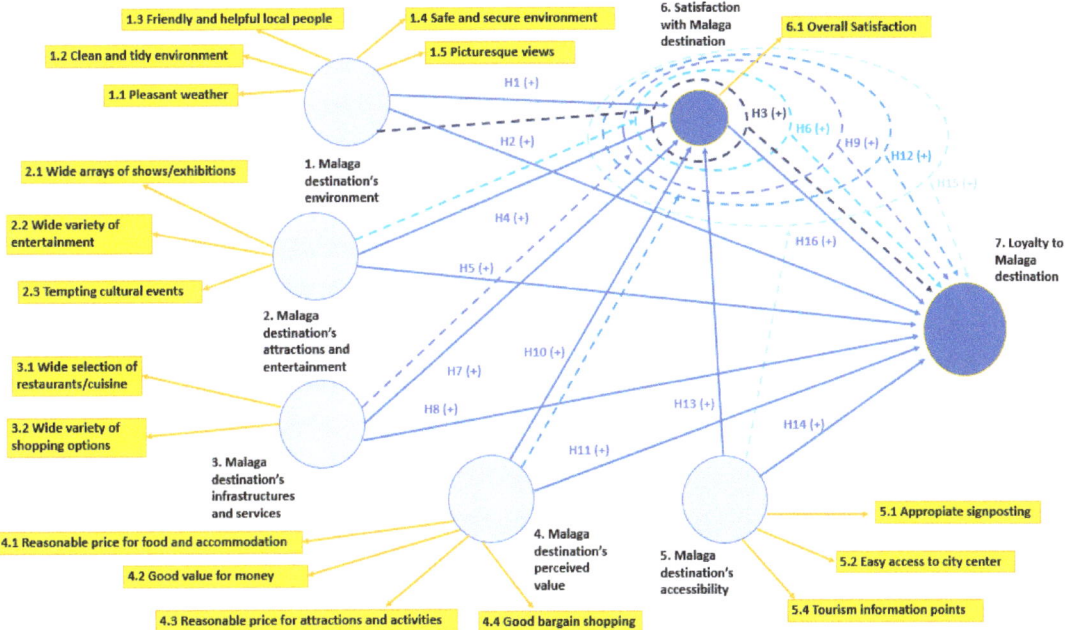

Figure 2. Confirmatory model. Source: Own elaboration using [140].

4. Results

4.1. Measurement Model

Following the recommendations based on the functionality of SmartPLS, PLSc was selected since the constructs are common factors. The evaluation of the measurement model of reflective constructs with its corresponding items included various assessments [151,152] item reliability, internal consistency reliability, convergent validity and discriminant validity.

Concerning the items removed from the measurement models, firstly, it should be noted the item 5.3 of construct accessibility was removed; the outer loading was below 0.6 and after removing from the model, AVE has increased [153,154]. Secondly, it was removed from infrastructure the indicator 3.3 variety of accommodation as we realize that cruise passengers do not spend nights in destination. The rest items from first construct environment with outer below 0.7, but near of this value, were not removed due to content validity.

Firstly, in regard to item reliability, the external load must be greater than or equal to 0.6 for an exploratory model and 0.7 for a confirmatory model (Tables 3 and A1 of Appendix A for more description). Secondly, internal consistency reliability was evaluated using composite reliability, [155,156], Dijkstra–Henseler RhoA and Cronbach's alpha [157], both of which must be greater than 0.7. Convergent validity was measured using the Average Variance Extracted (AVE) [158], which must be greater than 0.5. The aforementioned threshold requirements were all met in our case, as shown in Table 3.

Table 3. Item reliability, internal consistency reliability and convergent validity.

Constructs Related to Malaga Destination	CA > 0.7	RhoA > 0.7	CR > 0.7	AVE > 0.7	Factorial Loadings (Min.; Max.)
1. Environment	0.836	0.843	0.838	0.510	(0.533; 0.823) *
2. Attractions and entertainment	0.915	0.919	0.917	0.787	(0.736; 0.962) *
3. Infrastructure and service	0.822	0.824	0.822	0.699	(0.735; 0.904) *
4. Perceived value	0.859	0.861	0.860	0.606	(0.663; 0.851) *
5. Accessibility	0.807	0.808	0.807	0.583	(0.681; 0.843) *
6. Satisfaction with destination (1 item)	1	1	1	1	(1.000)
7. Loyalty with destination	0.832	0.871	0.846	0.736	(0.680; 1.000) *

Note: CA: Cronbach's alpha; RhoA: Dijkstra–Henseler rho (also denotes by ρ_A), CR: Composite reliability; AVE: Average variance extracted. Obtained from the * 95% Confidence Intervals. Source: [140].

Multiple criteria can be used to analyze discriminant validity, which measures how different the constructs are from each other. In this case, two criteria were used: first, the criterion of Fornell and Larcker [158], which consists of comparing the correlations between constructs with the square root of the AVE and the correlations between constructs and any of said correlations exceeding the AVE; second, the Heterotrait–Monotrait ratio of correlations (HTMT), which should not exceed 0.90 [159]. All of the established threshold requirements for the measurement model evaluation were met in this study, according to the calculated values shown in Table 4.

Table 4. Discriminant validity: Fornell and Lacker versus Heterotrait–Monotrait Ratio criteria.

Construct	1. Environment	2. Attractions and Entertainment	3. Infrastructures and Services	4. Perceived Value	5. Accessibility	6. Satisfaction	7. Loyalty
1. Environment	0.714	0.357	0.589	0.541	0.681	0.646	0.667
2. Attractions and entertainment	0.357	0.887	0.462	0.472	0.387	0.262	0.299
3. Infrastructures and services	0.592	0.460	0.836	0.555	0.590	0.408	0.425
4. Perceived value	0.539	0.470	0.556	0.778	0.608	0.472	0.467
5. Accessibility	0.679	0.386	0.589	0.607	0.763	0.538	0.508
6. Satisfaction	0.644	0.262	0.409	0.472	0.538	1.000	0.791
7. Loyalty	0.658	0.297	0.422	0.460	0.507	0.784	0.858

Source: [140]. Note: The square of AVE in the diagonal. Constructs' inter-correlations under diagonal and HTMT values over the diagonal.

4.2. Structural Model

In this case, all of the bootstrap exact fit tests (the standardized root mean square Residual, SRMR, the unweighted least squares discrepancy, DULS, and the geodesic discrepancy, DG) of the assessed model were computed, providing no significant results ($p > 0.005$); thus, the model fit is not satisfactory. However, other scholars disagree with these requirements according to Chin et al. [160].

The structural model was evaluated according to the recommendations made in various studies [147,161,162]. The assessment of the internal model includes: (1) collinearity between constructs; (2) the model's predictive power using the coefficient of determination (R^2) and the Stone Geisser Q^2 coefficient [163,164], (3) evaluation of effect size f^2; and (4) statistical significance and relevance of the path coefficients.

4.2.1. Collinearity Assessment

In order to evaluate the degree of collinearity with the constructs preceding the endogenous construct, the software calculates the reciprocal of the tolerance (TOL); that is, the variance inflation factor (VIF = 1/TOL). In the context of PLS-SEM, TOL values that are less than or equal to 0.20, equivalent to VIF values greater than or equal to 5, mean that there is a potential problem of collinearity [152]. Nevertheless, other authors use other thresholds, considering VIF values greater than 3.3 [165], as well as a value of 5 [166]. After calculating the VIF, the results indicate that there are no potential problems of collinearity between the latent variables (see Table 5).

Table 5. Variance inflation factor (VIF).

Constructs' VIF	6. Satisfaction	7. Loyalty
1. Environment	2.109	2.560
2. Attractions and entertainments	1.388	1.389
3. Infrastructure and services	1.929	1.932
4. Perceived value	1.896	1.938
5. Accessibility	2.293	2.329
6. Satisfaction		1.804

Source: [140].

4.2.2. Predictive Power

Two measurements were used to quantify the predictive power of this structural model: the coefficient of determination R^2 and Stone–Geisser Q^2 criteria. First, the coefficient of determination measures the model's predictive precision and $R^2 \geq 0.1$ is recommended [167]. However, another criterion has been established, considering that values of 0.1, 0.25 and 0.5 are weak, moderate and substantial, respectively [147]. The results are greater than the minimum threshold $R^2 \geq 0.1$, with loyalty values at $R^2 = 0.658$ and satisfaction at $R^2 = 0.446$ (next to substantial); therefore, our model has a substantial predictive power in relation to R^2 (Table 6). The general rule indicates that a cross-validated redundancy of $Q^2 > 0.5$ is considered to be a predictive model [139]. In this case, the Q^2 criterion by Stone–Geisser is close to the established threshold in the case of loyalty (Table 6).

Table 6. Variance explained and cross-validated redundancy index.

Dependent Construct	R^2	Q^2
6. Satisfaction	0.446 *	0.378
7. Loyalty	0.658 *	0.466

Source: [140]. Note. * $p < 0.05$.

4.2.3. Effect Size f^2

The change produced in the value of R^2 by eliminating an exogenous latent variable from the structural model can be used to indicate whether the omitted construct has a substantial impact on the endogenous construct. This change is measured by the effect size (f^2). The construct's effect size measures the impact of the exogenous variable in the structural model's predictive power. Values in the range: $0.02 \leq f^2 < 0.15$, $0.15 < f^2 \leq 0.35$ and $f^2 \geq 0.35$ indicate weak, moderate and strong effects, respectively, according to Cohen [168]. In our case, all of the f^2 values fluctuate between 0.00 and 0.606. The effect size of cruise tourists' satisfaction when explaining loyalty is strong (0.606), the f^2 of environment image when explaining satisfaction is moderate (0.214). Moreover, the f^2 of perceived value when explaining satisfaction (0.022) are weak likewise environment on loyalty (0.071). The rest of the effects are essentially non-existent or without any effect (0.000 to 0.016), whether for the image of the rest of the latent variables when explaining satisfaction or the image of the rest of the constructs when explaining loyalty (Table A2).

4.2.4. Path Coefficients, Significance and Relevance

In bootstrapping, subsamples are created with randomly drawn observations (with replacement) from the original set of data, with the same sample size as the study's original sample (457 cases). To ensure the stability of the results, there must be a large number of subsamples. The significance of the normalized path coefficients was obtained through a bootstrap resampling procedure of 5000 samples of the same number of observations (457 cases), which is considered to be the ideal number of bootstrap samples [169,170]. The standardized path coefficients reported in the path diagram (Figure 3) are significant at a level of 5%.

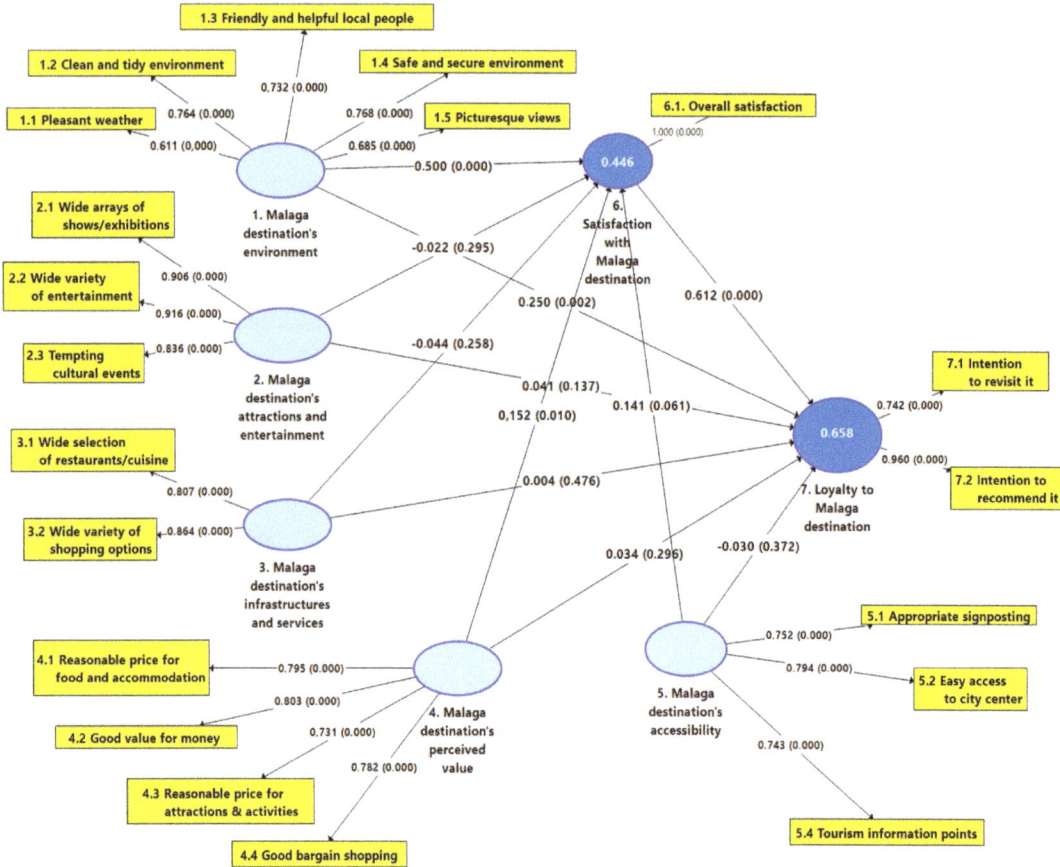

Figure 3. Tested model. Note: p-values in brackets. Source: Own elaboration from [140].

The relevance of the normalized path coefficients was obtained from the comparison between said coefficients, keeping in mind that the greater the value of these coefficients, greater the relevance. The direct path coefficients of the constructs of environment image and perceived value are significant and positive (Table 7) in relation to satisfaction, with environment image being the most relevant of the two image sub-dimensions with a value of 0.500, followed by perceived value. However, the relationships between attractions and entertainment image, infrastructure and services and accessibility with satisfaction are not significant. In regard to loyalty, only the constructs of environment image and satisfaction have significant direct effects, with satisfaction having the most substantial effect (0.500) after satisfaction in loyalty (0.612). Furthermore, we included the variance account for (VAF) equals to the proportion that indirect effect represents from total effect (also in percentage). It is a supplementary and secondary criterion: VAF below 20% is considered no mediation; VAF between 20% and 80% is partial mediation and greater than 80% full mediation [147].

Table 7. Direct, indirect and total effect concerning to model for Malaga destination.

HYPOS	Direct Path	Coefficient (Percent.Boot.95%CI)	Sig.	
supported	H1 Environment directly affects satisfaction.	0.500 (0.392; 0.617)	*	
supported	H2 Environment directly affects loyalty.	0.250 (0.103; 0.394)	*	
rejected	H4 Attractions and entertainments image directly affects satisfaction.	−0.022 (−0.090; 0.042)	NS	
rejected	H5 Attractions and entertainments image directly affects loyalty.	0.041 (−0.018; 0.103)	NS	
rejected	H7 Infrastructure and service image directly affects satisfaction.	−0.044 (−0.156; 0.064)	NS	
rejected	H8 Infrastructure and service image directly affects loyalty.	0.004 (−0.103; 0.106)	NS	
supported	H10 Perceived value image directly affects satisfaction.	0.152 (0.046; 0.254)	*	
rejected	H11 Perceived value image directly affects loyalty.	0.034 (−0.072; 0.141)	NS	
rejected	H13 Accessibility image directly affects satisfaction.	0.141 (−0.014; 0.287)	NS	
rejected	H14 Accessibility image directly affects loyalty.	−0.030 (−0.186; 0.121)	NS	
supported	H16 Overall satisfaction affects loyalty.	0.612 (0.508; 0.714)	*	
	Specific Indirect Path	Coefficient (Percent.Boot.95%CI)	Sig.	VAF
supported	H3 Environment indirectly affects loyalty through satisfaction.	0.306 (0.230; 0.396)	*	55.07% partial mediation
rejected	H6 Attractions and entertainments affects loyalty indirectly through satisfaction.	−0.013 (−0.057; 0.025)	NS	–
rejected	H9 Infrastructure and service image affects loyalty indirectly through satisfaction.	−0.027 (−0.096; 0.040)	NS	–
supported	H12 Perceived value image affects loyalty indirectly through satisfaction.	0.093 (0.028; 0.159)	*	73.30% partial mediation
rejected	H15 Accessibility image affects loyalty indirectly through satisfaction.	0.086 (−0.008; 0.182)	NS	–

Table 7. Cont.

	Total = Direct + Indirect	Coefficient (Percent.Boot.95%CI)	Sig.
Direct effect	H1 Environment influences on satisfaction.	0.500 (0.392; 0.617)	*
Direct and indirect	H2 Environment influences on loyalty.	0.556 (0.410; 0.707)	*
No effect	H4 Attractions and entertainments influences on satisfaction.	−0.022 (−0.090; 0.042)	NS
No effect	H5 Attractions and entertainments influences on loyalty.	0.028 (−0.042; 0.101)	NS
No effect	H7 Infrastructure and service influences on satisfaction.	−0.044 (−0.156; 0.064)	NS
No effect	H8 Infrastructure and service influences on loyalty.	−0.023 (−0.153; 0.102)	NS
Direct effect	H10 Perceived value influences on satisfaction.	0.152 (0.046; 0.254)	*
Direct and indirect	H11 Perceived value influences on loyalty.	0.127 (0.011; 0.240)	*
No effect	H3 Accessibility influences on satisfaction.	0.141 (−0.014; 0.287)	NS
No effect	H14 Accessibility influences on loyalty.	0.056 (−0.121; 0.228)	NS
Direct effect	H16 Satisfaction influences on loyalty.	0.612 (0.508; 0.714)	*

Note: HYPOS, hypotheses, Perc.Boot.CI, Percentile bootstrap confidence intervals, Sig.: Significant, NS: Not significant. *: $p < 0.05$. Variance account for (VAF). Source: Own from [140].

Considering that the direct effects of the other constructs of the destination's cognitive image are not significant in relation to loyalty, it is interesting to study the mediating role of satisfaction on said constructs and classify all these relationships. In the case of attractions and entertainment, infrastructure and services and, finally, in accessibility, there is no mediation; that is, they have no effect on loyalty, as both the direct and indirect effects are not statistically significant. Additionally, satisfaction is a partial collaborative—complementary—mediator between environment image and loyalty likewise perceived value and loyalty is (Table 8). Whether the direct or indirect path coefficients are significant or not is repeated in Table 8, summarizing the comparisons under consideration. Furthermore, Figure 3 offers a comprehensive visualization of the path model for the estimated model.

Table 8. Mediation analysis of satisfaction.

Path	Sig.	Sign	Hypotheses	Result of Satisfaction as Mediator between Each Image's Sub-Dimension and Loyalty
Environment directly affects loyalty.	Yes	(+)	H3: Satisfaction mediates between environment and loyalty.	H3 supported. Type of mediation: partial and positive. It is a collaborative-complementary-partial mediator.
Environment affects indirectly to loyalty through satisfaction.	Yes	(+)		
Conclusion: Satisfaction mediates between environment and loyalty through satisfaction, partially and positively.				
Attraction and entertainment directly affects loyalty.	No		H6: Satisfaction mediates between attraction and entertainment and loyalty.	H6 rejected. No effect, no mediation.
Attraction and entertainment indirectly affects loyalty.	No			
Conclusion: Satisfaction does not mediate between attractions and entertainments and loyalty. Attraction and entertainment have no effect on loyalty (neither direct nor indirect)				
Infrastructure and services directly affects loyalty.	No		H9: Satisfaction mediates between infrastructure and services and loyalty.	H9 rejected. No effect, no mediation.
Infrastructure and services indirectly affects loyalty.	No			
Conclusion respect to infrastructure and services: Satisfaction mediates between infrastructure and services and loyalty through satisfaction. Infrastructure and services have no effect on loyalty (neither direct nor indirect)				
Perceived value directly affects loyalty.	Yes	(+)	H12: Satisfaction mediates between perceived value and loyalty.	H12 supported. Type of mediation: partial and positive. It is a collaborative-complementary-partial mediator.
Perceived value indirectly affects loyalty.	Yes	(+)		
Conclusion: Satisfaction mediates between perceived value and loyalty through satisfaction, partially and positively.				
Accessibility directly affects loyalty.	No		H15: Satisfaction mediates between accessibility and loyalty.	H15 rejected. No effect, no mediation.
Accessibility indirectly affects loyalty.	No			
Conclusion: Satisfaction does not mediate between accessibility and loyalty. Accessibility has no effect on loyalty (neither direct nor indirect).				

Note: Sig.: Significant at level 5% based on one-side test. Source: Own elaboration.

5. Discussions and Conclusions

5.1. Theoretical Implications

This study presents a model that explains the influence of the cognitive attributes of tourist destination image on destination loyalty, as well as the mediating role of cruisers' overall satisfaction between each of these attributes and loyalty. One of the most relevant elements of this research as compared to other prior studies is the consideration of the individual influence of each of the attributes of cognitive destination image on both cruisers' satisfaction and loyalty. The few prior studies on cruise tourism offer overall results of destination image as a second-order construct, providing generic conclusions about its influence on satisfaction and loyalty [22,27,88,89]. However, this study allows us to answer which specific constructs of cognitive image are determining factors of cruisers' satisfaction and which constructs influence destination loyalty.

In comparison with Chi and Qu's research [12], our study includes the image destination in sub-dimensions in the model; meanwhile, Chi and Qu considered a construct of the overall image destination. Similarly, we also analyze direct and indirect effects of image through overall satisfaction. Specifically, our model is a sub-model of these authors with the novelty of considering the image divided into different dimensions in order to detect which constructs affects more to the satisfaction and the loyalty. Finally, this survey was conducted during the whole year; therefore, seasonality may not restrict the research findings to summer tourists.

With regards to Chen and Phou [66], they did not assess the destination image direct relation to loyalty with destination, only indirectly through the satisfaction. Similarly, we do assess the cognitive dimension of the destination image; however, they did this as an overall image, while in our study, the destination image is divided into five dimensions, as we have mentioned above.

From the results of this study, we can conclude that a destination's environment image is the attribute of cognitive image that has the greatest influence on cruisers' overall satisfaction, through a direct, positive, substantial relationship. This conclusion is in line with the study by Sanz-Blas and Carvajal-Trujillo [96], which demonstrates that the travel environment is the component of destination image that plays the most decisive role in cruisers' satisfaction in Valencia (Spain). However, in a study about the destination Ensenada (Baja California), Toudert and Bringas-Rabago [27] consider that although environment image has a significant direct effect on satisfaction, it has less of an influence than other cognitive components such as a destination's attractions and entertainment and the urban environment. These coincidences may be due to the fact that both destinations Valencia and Malaga are Mediterranean ports and, therefore, present a similar conception of environment image as compared to destinations at other latitudes, such as Ensenada. It also may be possible that cruisers in the Mediterranean have different expectations and that factors such as climate and picturesque views are more decisive in the construct of satisfaction than in other types of cruises, depending on their destinations. Ultimately, maintaining a pleasant, distinguishable environment should satisfactorily meet cruisers' motivations and expectations, as suggested by Qu and Ping [171].

Attributes such as the destination's attractions and entertainment image, or the image of infrastructure such as restaurants, hotels and shops, do not have an influence on cruisers' satisfaction or destination loyalty. These results corroborate the results obtained by Sanz-Blas et al. [89] and, partially, by Toudert and Bringas-Rábago [27], since the latter authors conclude that attractions are especially relevant.

On the other hand, cognitive attributes of the destination's image such as accessibility and the perceived value of services do have a direct influence on cruisers' satisfaction and also have an indirect influence on loyalty, to a lesser degree, although it has a weaker effect than environment image. These conclusions are in line with the results of a study conducted by Meng et al. [88], which were obtained from the opinions of passengers on the Star Cruise line.

Another important finding of this article is that environment image not only has an influence on cruisers' satisfaction, but that it also has a positive, direct, significant relationship with destination loyalty. These results are in accordance with the findings of DiPietro and Peterson [25], who also concluded that positive experiences with environment image contribute to cruisers revisiting and recommending a destination. However, these results differ from those obtained by Silvestre et al. [26] in regard to a similar construct that does not influence cruisers' loyalty.

Perceived value does not have a direct influence on loyalty; however, it does have an indirect effect through satisfaction, which acts as a total mediator in this case, although the indirect effect is weak. While these results are in line with the findings of various authors [124,125], they differ from those obtained by Murphy et al. [114] in regard to the direct relationship between the perceived value of tourism services and loyalty. Similarly, accessibility only has an indirect influence on loyalty through the mediation of satisfaction. This may be due to the fact that cruisers' assessment of accessibility depends on a positive perception of overall satisfaction.

Along these lines, the conclusion that cruisers' overall satisfaction has a decisive influence on loyalty confirms what has been established by prior studies, both in the tourism sector in general [13–15] and in cruise tourism in particular [96,134].

Finally, the results of this study corroborate the mediating role of satisfaction between image and loyalty, in accordance with the results obtained by other authors in both a general context [12,84,86,130] and within the specific context of cruises [16,27].

5.2. Managerial Implications

The attributes of cognitive image that have a significant influence on cruisers' satisfaction and loyalty (environment image, perceived value and accessibility) fall under the responsibility of destination management organizations such as city councils, business owners, trade associations and other public and private institutions. These institutions must proactively manage the variables that can produce satisfactory experiences for cruisers and help them develop a bond, both cognitive and emotional, when visiting the city.

The confirmation that overall satisfaction influences loyalty enables management organizations to develop more efficient loyalty strategies for their respective destinations. Therefore, if the goal is to get cruisers to return to the destination or for them to recommend it to the people they know, it is of the utmost importance to focus on issues of cleanliness, picturesque locations, safety and hospitality. These variables (which pertain to the attribute environment image) have the greatest influence on cruisers' satisfaction and loyalty, both directly and indirectly.

Meeting the expectations that cruisers associate with a destination's image is also crucial in this process of producing satisfaction and, consequently, loyalty. Otherwise, not only does it limit the likelihood of cruisers revisiting a destination, it also means that they would not give positive feedback about the destination to the people they know. If cruisers were to revisit the destination on another cruise, it would actively contribute to moderating the significant seasonality of hotels in the city of Malaga.

We believe that it is crucial to involve not only relevant institutions, but also the destination's local residents through appropriate communication and loyalty strategies for these kinds of initiatives, for example, through challenges, public greetings or promotional messages regarding the importance of welcoming cruise tourists and their significance in the local economy. These initiatives should be integrated in the destination's marketing and strategic plans.

5.3. Limitations and Suggestions for Future Research

This study reveals various limitations in addition to laying the groundwork for future research. One limitation to keep in mind is that the data refers to a specific port in the Mediterranean, which has specific, unique characteristics that are different from other ports of a similar scale. Similarly, the data were collected just before the emergence of COVID-19 and its consequences for cruise ships and their respective destinations.

It would, therefore, be important for future studies to expand the field of research, conducting a comparison with other destination ports that have other characteristics that differ from the Mediterranean. It would also be interesting to apply panel data using information post-COVID-19, in addition to the prior data that were collected. Lastly, we recommend expanding the cognitive attributes of the destination's image, considering the specific health risks associated with the disease and the destination's healthcare system.

Supplementary Materials: The following are available online at https://www.mdpi.com/article/10.3390/math9111218/s1, Table S1: Survey items about the city of Malaga (Andalusia, Spain).

Author Contributions: Conceptualization, E.M.S.-T.; G.B.-G.; methodology, software, validation, formal analysis, M.D.B.-M.; investigation, M.D.B.-M.; E.M.S.-T.; G.B.-G.; E.C.-R.; resources, G.B.-G.; E.C.-R.; data curation, G.B.-G.; E.C.-R.; writing—original draft preparation M.D.B.-M.; E.M.S.-T.; G.B.-G.; writing—review and editing, M.D.B.-M.; E.M.S.-T.; G.B.-G.; supervision, M.D.B.-M.; E.M.S.-T.; G.B.-G. All authors have read and agreed to the published version of the manuscript.

Funding: This research received no external funding.

Institutional Review Board Statement: Not applicable. All subjects gave their informed consent for inclusion before they participated in the study. The study was conducted in accordance with the Declaration of Helsinki.

Informed Consent Statement: Informed consent was obtained from all subjects involved in the study.

Data Availability Statement: Data available on request due to restrictions privacy. The data are not publicly available due to further researching.

Acknowledgments: We would like to thank two anonymous reviewers for helpful comments on this paper.

Conflicts of Interest: The authors declare no conflict of interest.

Appendix A

Table A1. Outer loadings.

Constructs Indicators	Outer Loadings	t Statistics	*p*-Value	5.0%	95.0%
1. Environment					
1.1. Pleasant weather	0.611	13.572	0.000	0.533	0.681
1.2. Clean and tidy environment	0.764	21.962	0.000	0.704	0.818
1.3. Friendly and helpful local people	0.732	19.363	0.000	0.669	0.793
1.4. Safe and secure environment	0.768	22.219	0.000	0.709	0.823
1.5. Picturesque views	0.685	14.857	0.000	0.604	0.757
2. Attractions and entertainment					
2.1. Wide arrays of shows/exhibitions	0.906	26.209	0.000	0.847	0.962
2.2. Wide variety of entertainment	0.916	31.524	0.000	0.866	0.962
2.3. Tempting cultural events	0.836	14.684	0.000	0.736	0.919
3. Infrastructure and service					
3.1. Wide selection of restaurants/cuisine	0.807	25.841	0.000	0.754	0.857
3.2. Wide variety of shopping options	0.864	33.260	0.000	0.819	0.904
3.3. Wide choice of accommodations	–	–	–	–	–
4. Perceived value					
4.1. Reasonable price for food and accommodation	0.795	23.608	0.000	0.735	0.847
4.2. Good value for money	0.803	25.353	0.000	0.749	0.851
4.3. Reasonable price for attractions and activities	0.731	18.390	0.000	0.663	0.793
4.4. Good bargain shopping	0.782	18.252	0.000	0.710	0.850

Table A1. *Cont.*

Constructs Indicators	Outer Loadings	t Statistics	*p*-Value	5.0%	95.0%
5. Accessibility					
5.1. Appropriate signposting	0.752	23.573	0.000	0.699	0.804
5.2. Easy access to city center	0.794	25.280	0.000	0.741	0.843
5.3. Disabled access	–	–	–	–	–
5.4. Tourism information points	0.743	20.796	0.000	0.681	0.799
6. Satisfaction with destination (1 item)					
6.1. Overall satisfaction	1.000			1.000	1.000
7. Loyalty with destination					
7.1. Intention to revisit it	0.742	21.059	0.000	0.680	0.798
7.2. Intention to recommend	0.960	38.125	0.000	0.918	1.002

Source: Own elaboration from [140].

Table A2. Effect size f^2.

Constructs Related to Malaga Destination	f^2	*p*-Value	Type
Environment on satisfaction	0.214	0.002	Moderate
Environment on loyalty	0.071	0.112	Weak
Attractions and entertainments on satisfaction	0.001	0.439	
Attractions and entertainments on loyalty	0.004	0.333	
Infrastructure and services on satisfaction	0.002	0.413	
Infrastructure and services on loyalty	0.000	0.499	
Perceived value on satisfaction	0.022	0.126	Moderate
Perceived value on loyalty to destination	0.002	0.443	
Accessibility on satisfaction with destination	0.016	0.243	
Accessibility on loyalty to destination	0.001	0.476	
Satisfaction with destination on loyalty	0.606	0.002	Substantial

Source: Own elaboration from [140].

References

1. Pickle, H.; Abrahamson, R.; Porter, A. Customer satisfaction and profit in small business. *J. Retail.* **1970**, *46*, 38.
2. Gnagi, P.; Bischofberger, J. *Customer Satisfaction as the Way to Business Success*; Tekstil: Zagreb, Croatia, 1995.
3. Jones, T.O.; Sasser, W.E. Why satisfied customers defect. *Harv. Bus. Rev.* **1995**, *76*, 88–99. [CrossRef]
4. Oliver, R.L. Whence consumer loyalty. *J. Mark.* **1999**, *63*, 33–44. [CrossRef]
5. Reichheld, F.F. Learning from customer defections. *Harv. Bus. Rev.* **1996**, *74*, 56–69.
6. Su, H.; Cheng, K.F.; Huang, H.H. Empirical study of destination loyalty and its antecedent: The perspective of place attachment. *Serv. Ind. J.* **2011**, *31*, 2721–2739. [CrossRef]
7. Chiu, W.; Zeng, S.; Cheng, P.S.T. The influence of destination image and tourist satisfaction on tourist loyalty: A case study of Chinese tourists in Korea. *Int. J. Cult. Tour. Hosp. Res.* **2016**, *10*, 223–234. [CrossRef]
8. Zhang, X.; Fu, L.; Cai, L.A.; Lu, L. Destination image and tourist loyalty: A meta-analysis. *Tour. Manag.* **2014**, *40*, 213–223. [CrossRef]
9. Baldinger, A.A.; Rubionson, J. Brand loyalty: Link between attitudes and behaviors. *J. Mark. Resk.* **1996**, *36*, 22–34.
10. Chen, C.F.; Tsai, D. How destination image and evaluative factors affect behavioral intentions. *Tour Manag.* **2007**, *28*, 1115–1122. [CrossRef]
11. Krishna, A.; Schwarz, N. Sensory perception, embodiment, and grounded cognition: Implications for consumer behavior. *J. Consum. Psychol.* **2014**, *24*, 453–459. [CrossRef]
12. Chi, C.G.Q.; Qu, H. Examining the structural relationships of destination image, tourist satisfaction and destination loyalty: An integrated approach. *Tour. Manag.* **2008**, *29*, 624–636. [CrossRef]
13. Cronin, J.; Brady, M.; Hult, G. Assessing the effects of quality, value, and customer satisfaction on consumer behavioral intentions in service environments. *J. Retail.* **2000**, *76*, 193–218. [CrossRef]
14. Oppermann, M. Tourism destination loyalty. *J. Travel Res.* **2000**, *39*, 78–84. [CrossRef]
15. Petrick, J.F.; Morais, D.D.; Norman, W.C. An examination of the determinants of entertainment vacationers' intentions to revisit. *J. Travel Res.* **2001**, *40*, 41–48. [CrossRef]
16. Albaity, M.; Melhem, S.B. Novelty seeking, image, and loyalty. The mediating role of satisfaction and moderating role of length of stay: International tourists' perspective. *Tour. Manag. Perspect.* **2017**, *23*, 30–37. [CrossRef]

17. Lee, S.W.; Xue, K. A model of destination loyalty: Integrating destination image and sustainable tourism. *Asia Pac. J. Tour. Res.* **2020**, *25*, 393–408. [CrossRef]
18. Lordanova, E. Tourism destination image as an antecedent of destination loyalty: The case of Linz, Austria. *Eur. J. Tour. Res.* **2017**, *16*, 214–232.
19. Del Bosque, I.R.; San Martín, H. Tourist satisfaction a cognitive-affective model. *Ann. Tour. Res.* **2008**, *35*, 551–573. [CrossRef]
20. Lv, X.; McCabe, S. Expanding theory of tourists' destination loyalty: The role of sensory impressions. *Tour. Manag.* **2020**, *77*, 104026. [CrossRef]
21. Chua, B.L.; Lee, S.; Han, H. Consequences of cruise line involvement: A comparison of first-time and repeat passengers. *Int. J. Contemp. Hosp. Manag.* **2017**, *29*, 1658–1683. [CrossRef]
22. Han, H.; Hyun, S.S. Role of motivations for luxury cruise traveling, satisfaction, and involvement in building traveler loyalty. *Int. J. Hosp. Manag.* **2018**, *70*, 75–84. [CrossRef]
23. Han, H.; Eom, T.; Chung, H.; Lee, S.; Ryu, H.B.; Kim, W. Passenger repurchase behaviours in the green cruise line context: Exploring the role of quality, image, and physical environment. *Sustainability* **2019**, *11*, 1985. [CrossRef]
24. Hwang, J.; Hyun, S.S. Perceived firm innovativeness in cruise travelers' experience and perceived luxury value: The moderating effect of advertising effectiveness. *Asia Pac. J. Tour. Res.* **2016**, *21* (Suppl. S1), S101–S128. [CrossRef]
25. DiPietro, R.B.; Peterson, R. Exploring cruise experiences, satisfaction, and loyalty: The case of Aruba as a small-island tourism economy. *Int. J. Hosp. Tour. Adm.* **2017**, *18*, 41–60. [CrossRef]
26. Silvestre, A.L.; Santos, C.M.; Ramalho, C. Satisfaction and behavioural intentions of cruise passengers visiting the Azores. *Tour. Econ.* **2008**, *1*, 169–184. [CrossRef]
27. Toudert, D.; Bringas-Rábago, N.L. Impact of the destination image on cruise repeater's experience and intention at the visited port of call. *Ocean Coast. Manag.* **2016**, *130*, 239–249. [CrossRef]
28. Puertos del Estado. Ministerio de Transporte, Movilidad y Agenda Urbana, 2020. Available online: http://www.puertos.es/es-es/estadisticas/Paginas/estadistica_mensual.aspx (accessed on 11 August 2020).
29. Puertos del Estado. Ministerio de Transporte, Movilidad y Agenda Urbana. *Los Puertos Españoles Superan los 9 Millones de Cruceristas en 2017*. 2018. Available online: http://www.puertos.es/es-es/Paginas/Noticias/Fitur2018.aspx (accessed on 11 August 2020).
30. United Nations World Tourism Organization. Available online: https://www.e-unwto.org/loi/wtobarometereng?expanded=d2010 (accessed on 14 July 2020).
31. Boulding, K.E. General systems-theory—The skeleton of science. *Manag. Sci.* **1956**, *2*, 197–208. [CrossRef]
32. Martineau, P. Sharper focues for the corporate image. *Harv. Bus. Rev.* **1958**, *36*, 49–58.
33. Chon, K.S. Traveler destination image modification process and its marketing implications. In *Proceedings of the 1990 Academy of Marketing Science (AMS) Annual Conference*; Springer: Cham, Switzerland, 2015; pp. 480–482.
34. Crompton, J.L.; Ankomah, P.K. Choice set propositions in destination decisions. *Ann. Tour. Res.* **1993**, *20*, 461–476. [CrossRef]
35. Echtner, C.M.; Ritchie, J.B. The measurement of destination image: An empirical assessment. *J. Travel Res.* **1993**, *31*, 3–13. [CrossRef]
36. Chen, J.S.; Hsu, C.H.C. Measurement of Korean tourists' perceived images of overseas destinations. *J. Travel Res.* **2000**, *38*. [CrossRef]
37. Driscoll, A.; Lawson, R.; Niven, B. Measuring tourists' destination perceptions. *Ann. Tour. Res.* **1994**, *21*, 499–511. [CrossRef]
38. Govers, R.; Go, F.M.; Kumar, K. Virtual destination image a new measurement approach. *Ann. Tour. Res.* **2007**, *34*, 977–997. [CrossRef]
39. Phelps, A. Holiday destination image—The problem of assessment: An example developed in Menorca. *Tour. Manag.* **1986**, *7*, 168–180. [CrossRef]
40. Prayag, G.; Hosany, S.; Muskat, B.; Del Chiappa, G. Understanding the relationships between tourists' emotional experiences, perceived overall image, satisfaction, and intention to recommend. *J. Travel Res.* **2017**, *56*, 41–54. [CrossRef]
41. Beerli, A.; Martin, J.D. Factors influencing destination image. *Ann. Tour. Res.* **2004**, *31*, 657–681. [CrossRef]
42. Jeong, Y.; Kim, S. A study of event quality, destination image, perceived value, tourist satisfaction, and destination loyalty among sport tourists. *Asia Pac. J. Mark. Logist.* **2019**, 940–960. [CrossRef]
43. Baloglu, S.; McCleary, K.W. US international pleasure travelers' images of four Mediterranean destinations: A comparison of visitors and nonvisitors. *J. Travel Res.* **1999**, *38*, 144–152. [CrossRef]
44. Dann, G.M. Tourists' images of a destination-an alternative analysis. *J. Travel Res.* **1996**, *5*, 41–55. [CrossRef]
45. Gallarza, M.G.; Saura, I.G.; García, H.C. Destination image: Towards a conceptual framework. *Ann. Tour. Res.* **2002**, *29*, 56–78. [CrossRef]
46. Ruan, W.Q.; Li, Y.Q.; Liu, C.H.S. Measuring tourism risk impacts on destination image. *Sustainability* **2017**, *9*, 1501. [CrossRef]
47. Hallmann, K.; Zehrer, A.; Müller, S. Perceived destination image: An image model for a winter sports destination and its effect on intention to revisit. *J. Travel Res.* **2015**, *54*, 94–106. [CrossRef]
48. Yacout, O.M.; Hefny, L.I. Use of Hofstede's cultural dimensions, demographics, and information sources as antecedents to cognitive and affective destination image for Egypt. *J. Vacat. Mark.* **2015**, *21*, 37–52. [CrossRef]
49. Agapito, D.; Oom do Valle, P.; da Costa Mendes, J. The cognitive-affective-conative model of destination image: A confirmatory analysis. *J. Travel Tour. Mark.* **2013**, *30*, 471–481. [CrossRef]

50. Woosnam, K.M.; Stylidis, D.; Ivkov, M. Explaining conative destination image through cognitive and affective destination image and emotional solidarity with residents. *J. Sustain. Tour.* **2020**, *28*, 917–935. [CrossRef]
51. Chen, J.S.; Uysal, M. Market positioning analysis: A hybrid approach. *Ann. Tour. Res.* **2002**, *29*, 987–1003. [CrossRef]
52. Gartner, W.C. Image formation process. *J. Travel Tour. Mark.* **1994**, *2*, 191–216. [CrossRef]
53. Kim, H.; Richardson, S.L. Motion picture impacts on destination images. *Ann. Tour. Res.* **2003**, *30*, 216–237. [CrossRef]
54. Walmsley, D.J.; Jenkins, J.M. Appraisive images of tourist areas: Application of personal constructs. *Aust. Geogr.* **1993**, *24*, 1–13. [CrossRef]
55. Bigné Alcañiz, E.; Sánchez García, I.; Sanz Blas, S. The functional-psychological continuum in the cognitive image of a destination: A confirmatory analysis. *Tour. Manag.* **2009**, *30*, 715–723. [CrossRef]
56. Pike, S.; Ryan, C. Destination positioning analysis through a comparison of cognitive, affective, and conative perceptions. *J. Travel Res.* **2004**, *42*, 333–342. [CrossRef]
57. Xie, K.L.; Lee, J.S. Toward the perspective of cognitive destination image and destination personality: The case of Beijing. *J. Travel Tour. Mark.* **2013**, *30*, 538–556. [CrossRef]
58. Bigné, J.E.; Sánchez, M.I.; Sánchez, J. Tourism image, evaluation variables and after purchase behaviour: Inter-relationship. *Tour. Manag.* **2001**, *22*, 607–616. [CrossRef]
59. Konecnik, M.; Gartner, W.C. Customer-based brand equity for a destination. *Ann. Tour. Res.* **2007**, *34*, 400–421. [CrossRef]
60. Baker, D.A.; Crompton, J.L. Quality, satisfaction and behavioral intentions. *Ann. Tour. Res.* **2000**, *27*, 785–804. [CrossRef]
61. Li, R.X.; Petrick, J.F. Revisiting the commitment-loyalty distinction in a cruising context. *J. Leis. Res.* **2010**, *42*, 67–90. [CrossRef]
62. Kim, D.; Perdue, R.R. The effects of cognitive, affective, and sensory attributes on hotel choice. *Int. J. Hosp. Manag.* **2013**, *35*, 246–257. [CrossRef]
63. Kim, S.; Yoon, Y. The hierarchical effects of affective and cognitive components on tourism destination image. *J. Travel Tour. Mark.* **2003**, *14*, 1–22. [CrossRef]
64. Li, M.; Cai, L.A.; Lehto, X.Y.; Huang, J.Z. A missing link in understanding revisit intention—The role of motivation and image. *J. Travel Tour. Mark.* **2010**, *27*, 335–348. [CrossRef]
65. Walmsley, D.; Young, M. Evaluative images and tourism: The use of personal constructs to describe the structure of destination images. *J. Travel Res.* **1998**, *36*, 65–69. [CrossRef]
66. Chen, C.F.; Phou, S. A closer look at destination: Image, personality, relationship and loyalty. *Tour. Manag.* **2013**, *36*, 269–278. [CrossRef]
67. Chon, K.S. Tourism destination image modification process: Marketing implications. *Tour. Manag.* **1991**, *12*, 68–72. [CrossRef]
68. Court, B.; Lupton, R.A. Customer portfolio development: Modeling destination adopters, inactives, and rejecters. *J. Travel Res.* **1997**, *36*, 35–43. [CrossRef]
69. Echtner, C.M.; Ritchie, J.B. The meaning and measurement of destination image. *J. Tour. Stud.* **1991**, *2*, 2–12.
70. Gartner, W.C.; Shen, J. The impact of Tiananmen Square on China's tourism image. *J. Travel Res.* **1992**, *30*, 47–52. [CrossRef]
71. Deng, J.; King, B.; Bauer, T. Evaluating natural attractions for tourism. *Ann. Tour. Res.* **2020**, *29*, 422–438. [CrossRef]
72. Tilikidou, I. The effects of knowledge and attitudes upon Greeks' pro-environmental purchasing behaviour. *Corp. Soc. Responsib. Environ. Manag.* **2007**, *14*, 121–134. [CrossRef]
73. Webb, D.J.; Mohr, L.A.; Harris, K.E. A re-examination of socially responsible consumption and its measurement. *J. Bus. Res.* **2008**, *61*, 91–98. [CrossRef]
74. Baloglu, S.; Mangaloglu, M. Tourism destination images of Turkey, Egypt, Greece, and Italy as perceived by US-based tour operators and travel agents. *Tour. Manag.* **2001**, *22*, 1–9. [CrossRef]
75. Smith, M.; Amorim, E.; Soares, C. O turismo acessível como vantagem competitiva: Implicações na imagem do destino turístico. *PASOS Rev. Tur. Patrim. Cult.* **2013**, *11*, 97–103. [CrossRef]
76. Gallarza, M.G.; Saura, I.G. Value dimensions, perceived value, satisfaction and loyalty: An investigation of university students' travel. *Tour. Manag.* **2006**, *27*, 437–452. [CrossRef]
77. Allameh, S.M.; Pool, J.; Jaberi, A.; Salehzadeh, R.; Asadi, H. Factors influencing sport tourists' revisit intentions: The role and effect of destination image, perceived quality, perceived value and satisfaction. *Asia Pac. J. Mark. Logist.* **2015**, *27*, 191–207. [CrossRef]
78. Lee, B.; Lee, C.K.; Lee, J. Dynamic nature of destination image and influence of tourist overall satisfaction on image modification. *J. Travel Res.* **2014**, *53*, 239–251. [CrossRef]
79. Prayag, G.; Ryan, C. Antecedents of tourists' loyalty to Mauritius: The role and influence of destination image, place attachment, personal involvement, and satisfaction. *J. Travel Res.* **2012**, *51*, 342–356. [CrossRef]
80. Puh, B. Destination image and tourism satisfaction: The case of a Mediterranean destination. *Mediterr. J. Soc. Sci.* **2014**, *5*, 538–544. [CrossRef]
81. Veasna, S.; Wu, W.Y.; Huang, C.H. The impact of destination source credibility on destination satisfaction: The mediating effects of destination attachment and destination image. *Tour. Manag.* **2013**, *36*, 511–526. [CrossRef]
82. Barroso Castro, C.; Martín Armario, E.; Martín Ruiz, D. The influence of market heterogeneity on the relationship between a destination's image and tourists' future behaviour. *Tour. Manag.* **2007**, *28*, 175–187. [CrossRef]
83. Assaker, G.; Hallak, R. Moderating effects of tourists' novelty-seeking tendencies on destination image, visitor satisfaction, and short-and long-term revisit intentions. *J. Travel Res.* **2013**, *52*, 600–613. [CrossRef]

84. Assaker, G.; Vinzi, V.E.; O'Connor, P. Examining the effect of novelty seeking, satisfaction, and destination image on tourists' return pattern: A two factor, non-linear latent growth model. *Tour. Manag.* **2011**, *32*, 890–901. [CrossRef]
85. Prayag, G. Tourists' evaluations of destination image, satisfaction, and future behavioral intentions—The case of Mauritius. *J. Travel Tour. Mark.* **2009**, *26*, 836–853. [CrossRef]
86. Song, Z.; Su, X.; Li, L. The indirect effects of destination image on destination loyalty intention through tourist satisfaction and perceived value: The bootstrap approach. *J. Travel Tour. Mark.* **2013**, *30*, 386–409. [CrossRef]
87. Kim, J.H. The impact of memorable tourism experiences on loyalty behaviors: The mediating effects of destination image and satisfaction. *J. Travel Res.* **2018**, *57*, 856–870. [CrossRef]
88. Meng, S.M.; Liang, G.S.; Yang, S.H. The relationships of cruise image, perceived value, satisfaction, and post-purchase behavioral intention on Taiwanese tourists. *Afr. J. Bus. Manag.* **2011**, *1*, 19–29. [CrossRef]
89. Sanz-Blas, S.; Buzova, D.; Carvajal-Trujillo, E. Familiarity and visit characteristics as determinants of tourists' experience at a cruise destination. *Tour. Manag. Perspect.* **2019**, *30*, 1–10. [CrossRef]
90. Hung, K.; Petrick, J.F. Why do you cruise? Exploring the motivations for taking cruise holidays, and the construction of a cruising motivation scale. *Tour. Manag.* **2011**, *32*, 386–393. [CrossRef]
91. Jin, N.P.; Lee, H.; Lee, S. Event quality, perceived value, destination image, and behavioral intention of sports events: The case of the IAAF World Championship, Daegu, 2011. *Asia Pac. J. Tour. Res.* **2011**, *18*, 849–864. [CrossRef]
92. Chen, C.C.; Lai, Y.H.R.; Petrick, J.F.; Lin, Y.H. Tourism between divided nations: An examination of stereotyping on destination image. *Tour. Manag.* **2016**, *55*, 25–36. [CrossRef]
93. Cheng, T.M.; Wu, H.C. How do environmental knowledge, environmental sensitivity, and place attachment affect environmentally responsible behavior? An integrated approach for sustainable island tourism. *J. Sustain. Tour.* **2015**, *23*, 557–576. [CrossRef]
94. Paraskevaidis, P.; Andreotis, K. Values of souvenirs as commodities. *Tour. Manag.* **2015**, *48*, 1–10. [CrossRef]
95. Ortegón-Cortázar, L.; Royo-Vela, M. Nature in malls: Effects of a natural environment on the cognitive image, emotional response, and behaviors of visitors. *Eur. Res. Manag. Bus. Econ.* **2019**, *25*, 38–47. [CrossRef]
96. Sanz-Blas, S.; Carvajal-Trujillo, E. Cruise passengers' experiences in a Mediterranean port of call. The case study of Valencia. *Ocean Coast. Manag.* **2014**, *102*, 307–316. [CrossRef]
97. Lee, H.E. Does a server's attentiveness matter? Understanding intercultural service encounters in restaurants. *Int. J. Hosp. Manag.* **2015**, *50*, 134–144. [CrossRef]
98. Zhang, X. Study on the influence of ecotourism environmental image on leisure experience and tourism satisfaction. *Ekoloji* **2019**, *28*, 1251–1257.
99. Zaheer, N.; Trkman, P. An information sharing theory perspective on willingness to share information in supply chains. *Int. J. Logist. Manag.* **2017**, *28*, 417–443. [CrossRef]
100. Coudounaris, D.N.; Sthapit, E. Antecedents of memorable tourism experience related to behavioral intentions. *Psychol. Mark.* **2017**, *34*, 1084–1093. [CrossRef]
101. Mak, A.H.N. Online destination image: Comparing national tourism organisation's and tourists' perspectives. *Tour. Manag.* **2017**, *60*, 280–297. [CrossRef]
102. Lee, J.A. A study on the effects of leisure experience, leisure negotiation, perceived value & behavioral intention in rural areas. *J. Hotel Res.* **2016**, *15*, 247–267.
103. Pizam, A.; Ellis, T. Customer satisfaction and its measurement in hospitality enterprises. *Int. J. Contemp. Hosp. Manag.* **1999**, 326–339. [CrossRef]
104. Lounsbury, J.W.; Hoopes, L.L. An investigation of factors associated with vacation satisfaction. *J. Leis. Res.* **1985**, *17*, 1–13. [CrossRef]
105. Simarmata, J.; Yuliantini, Y.; Keke, Y. The Influence of Travel Agent, Infrastructure and Accommodation on Tourist Satisfaction. In *Proceedings of the International Conference on Tourism, Gastronomy, and Tourist Destination (ICTGTD 2016), Djakarta, Indonesia, 14–15 November 2016*; Advances in Economics, Business and Management Research; Atlantis Press: Paris, France, 2016; Available online: https://www.atlantis-press.com/proceedings/ictgtd-16 (accessed on 24 May 2021).
106. Yoon, Y.; Uysal, M. An examination of the effects of motivation and satisfaction on destination loyalty: A structural model. *Tour. Manag.* **2005**, *26*, 45–56. [CrossRef]
107. Nasir, M.; Mohamad, M.; Ghani, N.; Afthanorhan, A. Testing mediation roles of place attachment and tourist satisfaction on destination attractiveness and destination loyalty relationship using phantom approach. *Manag. Sci. Lett.* **2020**, *10*, 443–454. [CrossRef]
108. Zeithaml, V.A. Consumer perceptions of price, quality and value: A means-end model and synthesis of evidence. *J. Mark.* **1988**, *52*, 2–22. [CrossRef]
109. Lovelock, C.H. *Service Marketing*; Prentice Hall International: Hoboken, NJ, USA, 2000.
110. Parasuraman, A.; Grewal, D. The impact of technology on the quality-value-loyalty chain: A research agenda. *J. Acad. Mark. Sci.* **2000**, *28*, 168–174. [CrossRef]
111. Babin, B.J.; Kim, K. International students' travel behavior: A model of the travel-related consumer/dissatisfaction process. *J. Travel Tour. Mark.* **2001**, *10*, 93–106. [CrossRef]
112. Kashyap, R.; Bojanic, D.C. A structural analysis of value, quality, and price perceptions of business and leisure travelers. *J. Travel Res.* **2000**, *39*, 45–51. [CrossRef]

113. McDougall, G.H.; Levesque, T. Customer satisfaction with services: Putting perceived value into the equation. *J. Serv. Mark.* **2000**, *14*, 392–410. [CrossRef]
114. Murphy, P.; Pritchard, M.P.; Smith, B. The destination product and its impact on traveller perceptions. *Tour. Manag.* **2000**, *21*, 43–52. [CrossRef]
115. Oh, H. Price fairness and its asymmetric effects on overall price, quality, and value judgments: The case of an upscale hotel. *Tour. Manag.* **2003**, *24*, 387–399. [CrossRef]
116. Petrick, J.F.; Backman, S.J. An examination of the construct of perceived value for the prediction of golf travelers' intentions to revisit. *J. Travel Res.* **2002**, *41*, 38–45. [CrossRef]
117. Ravald, A.; Grönroos, C. The value concept and relationship marketing. *Eur. J. Mark.* **1996**, 19–30. [CrossRef]
118. Tarn, J.L. The effects of service quality, perceived value and customer satisfaction on behavioral intentions. *J. Hosp. Leis. Mark.* **1999**, *6*, 31–43. [CrossRef]
119. Chen, C.F.; Chen, F.S. Experience quality, perceived value, satisfaction and behavioral intentions for heritage tourists. *Tour. Manag.* **2010**, *31*, 29–35. [CrossRef]
120. Lee, S.; Jeon, S.; Kim, D. The impact of tour quality and tourist satisfaction on tourist loyalty: The case of Chinese tourists in Korea. *Tour. Manag.* **2011**, *32*, 1115–1124. [CrossRef]
121. Bajs, I. Tourist perceived value, relationship to satisfaction and behavioural intentions: The example of the Croatian tourist destination Dubrovnik. *J. Travel Res.* **2013**, *54*, 1–13. [CrossRef]
122. Petrick, J.F. Segmenting cruise passengers with price sensitivity. *Tour. Manag.* **2005**, *26*, 753–762. [CrossRef]
123. Quintal, V.A.; Polczynski, A. Factors influencing tourists' revisit intentions. *Asia Pac. J. Mark. Logist.* **2010**, 554–578. [CrossRef]
124. Forgas-Coll, S.; Palau-Saumell, R.; Sanchez-Garcia, J.; Garrigos-Simon, F.J. Comparative analysis of american and spanish cruise passengers' behavioral intentions. *RAE Rev. Adm. Empresas.* **2016**, *56*, 87–100. [CrossRef]
125. Lobo, A.C. Enhancing luxury cruise liner operators' competitive advantage: A study aimed at improving customer loyalty and future patronage. *J. Travel Tour. Mark.* **2008**, *25*, 1–12. [CrossRef]
126. Taplin, R.H. The influence of competition on visitor satisfaction and loyalty. *Tour. Manag.* **2013**, *36*, 238–246. [CrossRef]
127. Israeli, A.A. A preliminary investigation of the importance of site accessibility factors for disabled tourists. *J. Travel Res.* **2002**, *41*, 101–104. [CrossRef]
128. Li, Y.; Hu, C.; Huang, C.; Duan, L. The concept of smart tourism in the context of tourism information services. *Tour. Manag.* **2017**, *58*, 293–300. [CrossRef]
129. Wisker, Z.L.; Kadirov, D.; Nizar, J. Marketing a destination brand image to Muslim tourists: Does accessibility to cultural needs matter in developing brand loyalty? *J. Hosp. Tour. Res.* **2020**. [CrossRef]
130. Wang, C.Y.; Hsu, M.K. The relationships of destination image, satisfaction, and behavioral intentions: An integrated model. *J. Travel Tour. Mark.* **2010**, *27*, 829–843. [CrossRef]
131. Lee, K.Y.; Lee, C.K.; Lee, S.K.; Babin, B.J. Festivalscapes and patrons' emotions, satisfaction, and loyalty. *J. Bus. Res.* **2008**, *61*, 56–64. [CrossRef]
132. Zeithaml, V.A.; Berry, L.L.; Parasuraman, A. The behavioral consequences of service quality. *J. Mark.* **1996**, *60*, 31–46. [CrossRef]
133. Brida, J.G.; Coletti, P. Tourists' intention of returning to a visited destination: Cruise ship passengers in Cartagena de Indias, Colombia. *Tour. Mar. Environ.* **2012**, *8*, 127–143. [CrossRef]
134. Brida, J.G.; Pulina, M.; Riaño, E.; Zapata-Aguirre, S. Cruise visitors' intention to return as land tourists and to recommend a visited destination. *Anatolia* **2012**, *23*, 395–412. [CrossRef]
135. Satta, G.; Parola, F.; Penco, L.; Persico, L. Word of mouth and satisfaction in cruise port destinations. *Tour. Geogr.* **2015**, *17*, 54–75. [CrossRef]
136. Instituto Nacional de Estadística (National Statistics Institute of Spain). Cifras Oficiales de Población Resultantes de la Revisión del Padrón Municipal a 1 de Enero. Resumen por Capitales de Provincia. 2019. Available online: https://www.ine.es/jaxiT3/Tabla.htm?t=2911&L=0 (accessed on 6 June 2020).
137. Sánchez-Teba, E.M.; Benítez-Márquez, M.D.; Romero-Navas, T. Residents' negative perceptions towards tourism, loyalty and happiness: The case of Fuengirola, Spain. *Sustainability* **2019**, *11*, 6841. [CrossRef]
138. Puertomalaga.com. Photo (c) Figure 1. Panoramic View of the Port of Malaga (with Authorization from the Port of Malaga). Available online: https://www.puertomalaga.com/wp-content/themes/puerto/images/panoramica-puerto.jpg (accessed on 24 April 2021).
139. Chin, W.W. How to write and report PLS analyses. In *Handbook of Partial Least Squares: Concepts, Methods and Applications*; Esposito Vinzi, V., Chin, W.W., Henseler, J., Wang, H., Eds.; Springer Handbooks of Computational Statistics Series; Springer: Berlin/Heidelberg, Germany, 2010; Volume II, pp. 655–690.
140. Ringle, C.M.; Wende, S.; Becker, J.M. *Software "SmartPLS 3" (Version Professional 3.3.3, Update 2021)*; Boenningstedt, Germany, 2015. Available online: http://www.smartpls.com (accessed on 24 May 2021).
141. Bloemer, J.; Ruyter, K. On the relationship between store image, store satisfaction and store loyalty. *Eur. J. Mark.* **1998**, *32*, 499–513. [CrossRef]
142. Bolton, R.N.; Lemon, K.N. A dynamic model of customers' usage of services: Usage as an antecedent and consequence of satisfaction. *J. Mark. Res.* **1999**, *36*, 171–186. [CrossRef]

143. Antón, C.; Camarero, C.; Laguna-Garcia, M. Towards a new approach of destination loyalty drivers: Satisfaction, visit intensity and tourist motivations. *Curr. Issues Tour.* **2017**, *20*, 238–260. [CrossRef]
144. Campo-Martínez, S.; Garau-Vadell, J.B.; Martínez-Ruiz, M.P. Factors influencing repeat visits to a destination: The influence of group composition. *Tour. Manag.* **2010**, *31*, 862–870. [CrossRef]
145. Moore, S.A.; Rodger, K.; Taplin, R. Moving beyond visitor satisfaction to loyalty in nature-based tourism: A review and research agenda. *Curr. Issues Tour.* **2015**, *18*, 667–683. [CrossRef]
146. Vareiro, L.; Ribeiro, J.C.; Remoaldo, P.C. What influences a tourist to return to a cultural destination? *Int. J. Tour. Res.* **2019**, *21*, 280–290. [CrossRef]
147. Hair, J.F., Jr.; Hult, G.T.M.; Ringle, C.M.; Sarstedt, M. *A Primer on Partial Least Squares Equation Modeling (PLS-SEM)*, 2nd ed.; Sage Publication, Inc.: Thousand Oaks, CA, USA, 2017.
148. Benitez, J.; Henseler, J.; Castillo, A.; Schuberth, F. How to perfom and report an impactful analysis using partial least squares: Guidelines for confirmatory and explanatory IS research. *Inf. Manag.* **2020**, *57*, 103168. [CrossRef]
149. Dijkstra, T.K.; Henseler, J. Consistent partial least squares path modeling. *Manag. Inf. Syst.* **2015**, *39*, 297–316. Available online: https://www.jstor.org/stable/26628355 (accessed on 23 April 2021).
150. Nitzl, C.; Roldan, J.L.; Cepeda, G. Mediation analysis in partial least squares path modeling: Helping researchers discuss more sophisticated models. *Ind. Manag. Data Syst.* **2016**, *116*. [CrossRef]
151. Cepeda-Carrion, G.; Cegarra-Navarro, J.G.; Cillo, V. Tips to use partial least squares structural equation modelling (PLS-SEM) in knowledge management. *J. Knowl. Manag.* **2019**, *23*, 67–89. [CrossRef]
152. Hair, J.F.; Risher, J.J.; Sarstedt, M.; Ringle, C.M. When to use and how to report the results of PLS-SEM. *Eur. Bus. Rev.* **2019**, *31*, 2–24. [CrossRef]
153. Hair, J.F.; Hult, G.T.; Ringle, C.M.; Sarstedt, M.; Castillo-Apraiz, J.; Cepeda Carrion, G.; Roldan, J.L. *Manual de Partial Least Squares Structural Equation Modeling (PLS-SEM)*; Omnia Science Omnia Publisher SL: Barcelona, Spain, 2019.
154. Hair, J.F., Jr.; Sarstedt, J.; Hopkins, L.; Kuppelwieser, V.G. Partial least squares structural equation modeling (PLS-SEM). *Eur. Bus. Rev.* **2014**, *26*, 106–121. [CrossRef]
155. Werts, C.E.; Linn, R.L.; Jöreskog, K.G. Intraclass reliability estimates: Testing structural assumptions. *Educ. Psychol. Meas.* **1974**, *34*, 25–33. [CrossRef]
156. Nunnally, J.C.; Bernstein, I.H. *Psychometric Theory*, 3rd ed.; McGraw-Hill, Inc.: New York, NY, USA, 1994.
157. Cronbach, L.J. Coefficient alpha and internal structure of test. *Psychometrika* **1951**, *16*, 297–334. [CrossRef]
158. Fornell, C.; Larcker, D.F. Evaluating structural equation models with unobservable variables and measurement error. *J. Mark. Res.* **1981**, *18*, 39–50. [CrossRef]
159. Henseler, J.; Ringle, C.R.; Sarstedt, M. A new criterion for assessing discriminant validity in variance-based structural equation modeling. *J. Acad. Mark. Sci.* **2015**, *43*, 115–135. [CrossRef]
160. Chin, W.; Cheah, J.H.; Liu, Y.; Ting, H.; Lim, X.J.; Cham, T.H. Demystifying the role of causal-predictive modeling using partial least squares structural equation modeling in information systems research. *Ind. Manag. Data Syst.* **2020**, *120*, 2161–2209. [CrossRef]
161. Ali, F.; Rasoolimanesh, S.M.; Sarstedt, M.; Ringle, C.M.; Ryu, K. An assessment of the use of partial least squares structural equation modeling (PLS-SEM) in hospitality research. *Int. J. Contemp. Hosp. Manag.* **2018**, *30*, 514–538. [CrossRef]
162. Marin-Garcia, J.; Alfalla-Luque, R. Key issues on partial least squares (PLS) in operations management research: A guide to submissions. *J. Ind. Eng. Manag.* **2019**, *12*. [CrossRef]
163. Stone, M. Cross-validatory choice and assessment of statistical predictions (with discussion). *J. R. Stat. Soc. Ser. B Stat. Methodol.* **1976**, *38*, 102. [CrossRef]
164. Geisser, S. The predictive sample reuse method with applications. *J. Am. Stat. Assoc.* **1975**, *70*, 320–328. [CrossRef]
165. Diamantopoulos, A.; Siguaw, J.A. Formative versus reflective indicators in organizational measure development: A comparison and empirical illustration. *Br. J. Manag.* **2006**, *17*, 263–282. [CrossRef]
166. Hair, F.J.; Black, C.W.; Babin, J.B.; Anderson, E.R. *Multivariate Data Analysis*, 7th ed.; Pearson Education Limited: Essex, UK, 2014.
167. Falk, R.F.; Miller, N.B. *A Primer for Soft Modeling*; University of Akron Press: Akron, OH, USA, 1992.
168. Cohen, J. *Statistical Power Analysis for the Behavioral Sciences*; Lawrence Erlbaum: Mahwah, NJ, USA, 1988.
169. Ringle, C.M.; Sarstedt, M.; Mitchell, R.; Gudergan, S.P. Partial least squares structural equation modeling in HRM research. *Int. J. Hum. Resour. Manag.* **2018**, 1–27. [CrossRef]
170. Streukens, S.; Leroi-Werelds, S. Bootstrapping and PLS-SEM: A step-by-step guide to get more out of your bootstrap results. *Eur. Manag. J.* **2016**, *34*, 618–632. [CrossRef]
171. Qu, H.; Ping, E.W.Y. A service performance model of Hong Kong cruise travelers' motivation factors and satisfaction. *Tour. Manag.* **1999**, *2*, 237–244. [CrossRef]

Article

Using Higher-Order Constructs to Estimate Health-Disease Status: The Effect of Health System Performance and Sustainability

Alicia Ramírez-Orellana *, María del Carmen Valls Martínez and Mayra Soledad Grasso

Department of Economics and Business, University of Almería, 04120 Almeria, Spain; mcvalls@ual.es (M.d.C.V.M.); mayragrasso21@gmail.com (M.S.G.)
* Correspondence: aramirez@ual.es; Tel.: +34-950-01-5696

Abstract: This article aims to provide information to public agencies and policymakers on the determinants of health systems and their relationships that influence citizens' health–disease status. A total of 61 indicators for each of 17 Spanish autonomous communities were collected from the Spanish Ministry of Health, Social Services, and Equality between 2008 and 2017. The applied technique was partial least squares structural equation modeling (PLS-SEM). Concerning health–disease status, an influence of sustainability and performance on the health system was hypothesized. The findings revealed that health system sustainability had a negative effect on health–disease status, measured in terms of disease incidence. However, the relationship between health system performance and health–disease status is positive. Furthermore, health system performance mediates the relationship between sustainability and health–disease status. According to our study, if we consider the opposite poles that make up the definition of health–disease status (well-being and disease), this concept is defined more by the incidence of the negative aspect.

Keywords: National Health Services; health–disease status; health system performance; health system sustainability; health policy; healthcare quality; partial least squares structural equation modeling (PLS-SEM)

Citation: Ramírez-Orellana, A.; Valls Martínez, M.d.C.; Grasso, M.S. Using Higher-Order Constructs to Estimate Health-Disease Status: The Effect of Health System Performance and Sustainability. *Mathematics* **2021**, *9*, 1228. https://doi.org/10.3390/math9111228

Academic Editor: Christoph Frei

Received: 26 March 2021
Accepted: 25 May 2021
Published: 27 May 2021

Publisher's Note: MDPI stays neutral with regard to jurisdictional claims in published maps and institutional affiliations.

Copyright: © 2021 by the authors. Licensee MDPI, Basel, Switzerland. This article is an open access article distributed under the terms and conditions of the Creative Commons Attribution (CC BY) license (https://creativecommons.org/licenses/by/4.0/).

1. Introduction

All countries seek to grow economically. Undoubtedly, this is reflected in improvements in the standard of the population's living. For its part, the population's health plays a fundamental role in its economic prosperity [1]. Health has a direct impact on the economy and economic growth [2]. Both the prevention of diseases and their treatment are necessary to reduce disease burden [3]. These activities will be directed by a health system for which the government is responsible [4]. The government will need tools to continuously evaluate and monitor the health system if its objective is for it to work properly [4,5]. Inadequate or inefficient health expenditure could slow down the economic growth of the entire country [1]. Having quality information when making decisions about health policies improves health, well-being, and patient satisfaction [5]. At the organizational level, one of the inputs for improving the health care system's efficiency, effectiveness, and equity is the health information systems. The use of health information systems leads to the achievement of administrative efficiency, maximizing the value of resources as an outcome [6]. Managing all the data that health workers routinely record enables gathering information on vital statistics, public health programs, reportable diseases, and mortality. The purpose of the health information system is to promote the development of an information culture where those responsible for health use information operatively for optimal planning and decision making to provide health services based on knowledge [7].

According to the World Health Organization (WHO), "Health is a state of complete physical, mental and social well-being, and not only the absence of diseases or illnesses" [8].

The measurement of health–disease status can be performed from the perspective of diagnostic morbidity (based on empirical data on diseases in the population) or from the perceived morbidity's perspective (based on self-perception of health–disease status [9]). Self-perception of the state of health is not the same in both sexes.

All health systems aim to improve citizens' health [10–15], respond to patient expectations, and equitably distribute the financial burden [12,16,17]. It is essential to know the needs in each region of the country and allocate resources accordingly in order to improve health–disease status [9]. For its part, responsiveness is a crucial element in patient satisfaction [15] and includes several concepts, such as confidentiality, autonomy, prompt attention, and access to social support networks, among others [6,16]. Finally, when discussing an equitable distribution of the financial burden, we refer to the fact that each household should pay the health system somewhat based, to a certain degree, on their income [4,18]. We can also refer to equity in terms of provision of the service, which aims to benefit each user based on their particular needs [6,19].

A quality health system will provide an excellent service when and where patients require it [20]. If the system malfunctions (poorly managed, poorly structured, ineffective, or poorly structured financially), it will not deliver its full potential, its costs will rise, and health outcomes will worsen [6]. In other words, it will not be able to fulfill its ultimate goal. For this reason, it is vitally important to manage these systems and evaluate their performance [21]. Periodically carrying out an efficiency analysis is a productive tool to investigate the potential for improvement in a hospital's resource use [22,23]. Nevertheless, the contemporary approach to measuring performance includes a cost analysis of services, quality, and patient satisfaction [24].

For its part, the measurement and evaluation of patient satisfaction are considered key points to work on if what is intended is the continuous improvement of the health system and its consequent excellence [20,21]. Patients' opinions are among of the main elements with which satisfaction is measured [22]. This feedback will serve as the basis for analyzing the health system and working towards its improvement [23,25,26]. Maintaining an excellent healthcare system has never been cheap. Furthermore, updating based on continuous technological advances nowadays requires even more effort than ensuring the system works efficiently. Managing quality will improve the quality of the services provided and reduce costs [10,26].

Today, companies worldwide are concerned with reporting on their sustainability. Through sustainability reports, corporations explain their planning in its economic, environmental, and social aspects [27]. When we speak of health system sustainability, we refer to the management of resources and expenses that are carried out, and the degree to which the health system's use is capable of meeting current needs without compromising the satisfaction of future needs [13,28]. It will be necessary to maintain the best possible cost-effectiveness ratio to meet this criterion. In other words, resources should be allocated to those interventions that provide the maximum improvement in health per monetary unit [3,22]. Additionally, for the organization to be sustainable over time, quality of service must exist [29]. The difficulty in measuring quality, in these terms, is that we have to look to the future and design a service that meets the needs of tomorrow [29].

On many occasions, health indicators are used to strategically direct resources and expenses [5]. When these indicators are comparable between countries, relevant and significant data can be extracted to improve them and identify good and bad practices [30]. Public health indicators contribute to transparency and good governance [31]. For example, in Europe, the ECHI (European Community Heath Indicator) is used, which functions as a hub of information and notifications on health at the European level [5].

The WHO established that governments have to guarantee the availability of health services to their citizens [6] in order to improve health status, meet patient expectations, and comply with the financial equity criteria [12]. One of the factors that most influence patient satisfaction is the health system's ability to comply with clinical requirements. The latter depend on the facilities' ability, for example, to provide laboratory reports on time, and to maintain the availability of required blood groups [26].

As we have already indicated, improving health is the main objective of the entire health system, but this should not be limited to physical diseases or symptoms. It is crucial that we extend ourselves to evaluating and treating depressive symptoms [31]. Moreover, those responsible should not be limited solely to the clinical aspects. For example, good management and administration of the health system will also improve society's health [32].

Previous research has studied the performance and sustainability of the health system. However, it is not common to find studies on its effect on the population's health as a complete health system, designed in the form of a nomological network and integrated by different explanatory subsystems of the health–disease status of citizens. Some studies use individual variables as isolated pieces with influence on a single non-latent dependent variable [33]. In this vein, we have not found investigations that use higher-order complex latent variables defined by several dimensions. Our study contributes to defining the boundaries of the health system, and highlighting the importance of the sustainability and performance subsystems as drivers of the levels of well-being, morbidity, and mortality of the population, that is, of the health–disease status. Moreover, we provide those responsible for managing the health system with information on the efficient and effective use of resources that does not compromise future needs and affect the population's health–disease status. In addition, our model offers policymakers information on the determining variables of the health system and the correlations between them to serve as an instrument for effective decision making. The rest of the article is structured as follows: First, we carry out the literature review and pose the hypotheses. Secondly, we describe the research methodology, and, after, the results are gathered and presented. Finally, we discuss the principal findings of the research and the conclusions.

Literature Background and Hypotheses

The economic development of a country depends on many factors, and one of them is the health of its inhabitants. A healthy population will always be more productive. To achieve this, it will be essential that the country has an effective and efficient health system [30,34]. Thus, countries should develop programs and policies to protect and improve the population's health [35] and reduce inequalities in health access [5]. In this sense, studying the quality-to-price ratio is increasingly crucial [18]. In Europe, health systems face increasing costs, as the population is aging and, therefore, making greater use of them [14,30]. The elderly are using the health care system more frequently, and the medical treatments they use are more expensive [36]. Innovations in health are imminent to ensure a healthy life [14,37]. Nevertheless, this is also costly and complex due to the system's dynamism [14].

The ultimate goal should be to promote and improve the population's quality of life [14], minimizing the risk of mortality [14,38]. Then, the health system's improvement will increase the population's quality of life and, therefore, reduce the mortality rate [39,40]. Mortality and morbidity ratios are used to measure the health–disease status of the population [12]. Both are associated with physical and psychological states [38,41]. The mortality rate is lower in women, which generates a higher incidence of morbidity [3,36]. A study on Spanish citizens' health status determined that neither in-hospital mortality nor morbidity are significant factors in establishing perceived health status [9]. The life expectancy of women is higher than that of men [3,42]. The difference in life expectancy between men and women can be influenced by male sex exposure to risk factors or occupational risk, or other risky behaviors [3]. Otherwise, people who suffer from a chronic illness have a negative self-perception of their health [43]. Furthermore, it is the female sex who is prone to chronic diseases [42]. Women tend to self-perceive worse than men [9].

The expected result of good health system management is the long-term well-being of the patient. However, these results depend not only on the provision of a good-quality health service, but also the characteristics of the patient [44]. For example, maternity in adolescence increases morbidity and mortality in women and their children, since they are usually born with medical complications [45].

It is expected that the government will take the necessary measures to offer the population quality and sustainable health services. How health services are provided will affect the health status of the population [30,46]. On the one hand, quality can be measured according to different pillars, such as safety, patient satisfaction, effectiveness, and pertinence. On the other hand, sustainability can be studied according to the health system's level of use, allocations of resources, and volume of expenses.

A safe health system manages risks to minimize incidents [34,47], for example, evaluating the effectiveness of new medical treatments and medicines [33]. These factors can be measured by the number of hospital infections and the rate of adverse drug reactions. One of the dimensions of healthcare quality is safety [37,48,49], which is related to efficiency, since fewer interventions are less expensive. The literature indicates that safe care can be provided with minimal waste of resources [34]. Regarding this issue, the WHO emphasizes the need to understand healthcare complexity to ensure patients' safety [13]. The characteristics of the patient directly influence safety. The higher the complications, the lower the security. The factors that influence risk exposure are age, disease burden, and gender [34].

A patient satisfied with the medical attention received will pay more attention to the treatments and recommendations that the health personnel indicate, and consequently, they will have better health results [50,51]. On the contrary, a dissatisfied patient will not adhere to the recommendations of health professionals. Hence, resources will be wasted, medical care productivity will decrease, and morbidity and mortality rates will increase [52]. From another perspective, we could say that the patient's satisfaction affects their life expectancy, and this relationship is strong [10]. Today, people are more demanding about the services they receive. To achieve their satisfaction, it will be necessary for healthcare to be "patient-oriented", that is, depending on the individual needs of each patient [51,53]. Additionally, previous studies indicate that when patients are allowed to participate in medical treatment decisions, they are more satisfied [20,54,55]. Other factors that influence a patient's satisfaction are confidence in the health system's professionals [23,26,29,55,56], the physician's behavior [26,29,32,55,56], and the degree of patient follow-up [29,57]. In their study, Ricci-Cabello et al. found that those patients who had a pleasant experience in medical care reported better self-perceived health [51].

The health system's ineffectiveness can be measured by the readmission rate to hospitals, which causes higher costs for the system and more anguish to the patient [58]. Repeated hospitalization could signal a failure in the quality of the health system [48]. There must be a balance between a hasty medical discharge and a prolonged hospital stay due to not yet solving the patient's problem. This could increase the probability of contracting other diseases as a result of staying in the hospital, such as nosocomial diseases, infections, and pressure ulcers [59]. Low self-perceived health states are associated with a higher risk of readmission [58]. Moreover, when patients are depressed during hospitalization, the risk of being readmitted increases [38]. Previous research found that hospitals with a longer average length of stay are less efficient [18,48].

Pertinence could be associated with equity in the provision of services. Equity in providing services means that each patient is cared for according to their needs at the right time [6]. When the health system can provide adequate care at the right time, this prevents an increase in the severity of diseases and saves possible future expenses [34].

Some previous studies indicate that higher-income countries show better efficiency rates, while others reach ambiguous conclusions [30]. Higher per capita health spending is directly reflected in the efficiency of health systems [18]. Healthcare effectiveness can be defined as the health system's ability to achieve maximum expected results without increasing unexpected results [30].

Previously, the term "sustainability" referred only to environmental factors. Today, it is studied with a multidimensional approach. In the area of health, the health system will be sustainable when it takes care of the well-being of patients, health professionals, and the entire community, preserving resources [28]. In other words, we must provide the best possible health service to improve the patient's health status, with the lowest waste of resources possible [13,34]. Budget cuts in health matters are increasingly frequent [14,55,56], so it is increasingly important to focus on sustainability, that is, to offer services of excellence while being efficient in the use of resources and the application of expenses [30,35,60].

A sustainable health system must focus on prevention [35]. In the European Union, the leading cause of death is cardiovascular diseases whose risk factors (smoking, high body mass index, lack of physical activity, and blood pressure) are highly preventable [61]. When people do not take preventive measures (low cost/high value), they will only rely on emergency services (high price/less effective results) [46]. Previous research found that countries with higher healthcare expenditure per capita have more efficient hospitals [18]. On the other hand, others indicate that efficiency is not defined by the volume of resources assigned to health [2,14].

This research's principal objective is to examine the influence of the health system performance and health system sustainability on health–disease status. After a careful review of the literature, we formulated the following hypotheses:

Hypothesis 1 (H1). *Health system sustainability influences health–disease status.*

Hypothesis 2 (H2). *Health system performance influences health–disease status.*

Hypothesis 3 (H3). *Health system sustainability influences health system performance.*

Hypothesis 4 (H4). *Health system performance mediates the relationship between health system sustainability and health–disease status.*

The theoretical model that we propose in Figure 1 relates the following three latent variables or constructs:

- Health system sustainability
- Health system performance
- Health–disease status

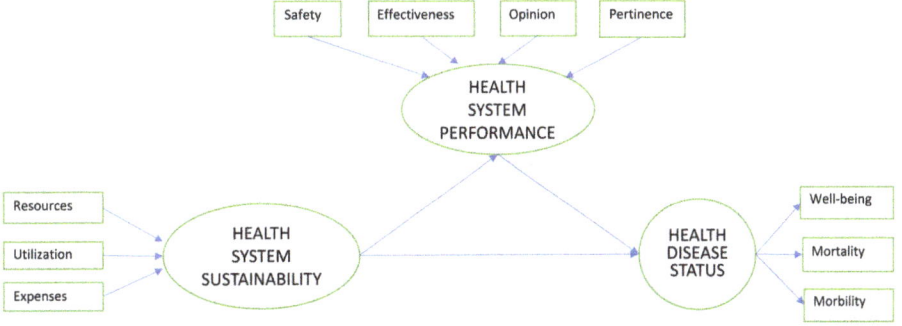

Figure 1. The theoretical model of health–disease status in Spain.

2. Research Methodology

In this section, we present the results of applying the algorithm of partial structural equations of higher-order constructs as an efficient solution for evaluating the health–disease status model in Spain.

2.1. PLS-SEM Analysis

PLS-SEM analysis come from two statistical traditions: linear regression and factor analysis. PLS-SEM models use theoretical concepts in the form of constructs or latent variables, such as unobserved variables, which are measured through indicators, data, or manifest variables [62]. Wold [63] was the author and developer of the PLS-SEM algorithm whose objective is to minimize the residual variances of the endogenous variable to be explained [64]. The basic PLS algorithm applies a two-stage method. In the first stage, the constructs' scores are iteratively estimated through a four-step procedure. The second stage computes the final estimates of coefficients (outer weights, loadings, and path coefficients) using the ordinary least squares method for each partial regression model [65].

The evaluation of a traditional PLS-SEM model requires firstly specifying the measurement model and secondly evaluating the structural model where the hypotheses are tested. In our case, there were two types of measurement relationships between indicators/items and constructs: reflective and formative [66]. Depending on the direction of the causal relationship between the latent variable and its indicators, a series of different criteria were verified according to the reflective or formative model (for more details, see [67–69]). Thus, in the case of reflective models, the causal relationship goes from the latent variable to the indicators, and in formative models the opposite. Reflective or A-mode models were assessed using four criteria: individual item reliability, construct reliability, convergent validity, and discriminant validity; formative or B-mode models were evaluated using the criteria of multicollinearity between items. With the specification of the measurement scale, it was possible to verify that the relationships among indicators and their constructs were valid and reliable, regardless of the measurement mode used. Once it was determined that the measure was valid and reliable, the structural analysis of the model was carried out. PLS-SEM used various criteria for structural validation, such as coefficients of determination (R^2), size of effects (f^2), and predictive validity (Q^2). The analysis of composites in PLS-SEM allows the calculation of latent variable scores as an exact linear combination of the indicators, which can be used to aggregate higher-order constructs [64,70]. Apart from being able to estimate mediation and moderation effects with multiple latent variables, PLS-SEM analysis allows analyzing of models with lower-order constructs (LOC) and higher-order constructs (HOC).

When using PLS composites, we consider the LOC as a mediator or aggregator between the indicators or dimensions, that is, the latent variable scores of the LOCs that constitute the HOC [71,72]. Therefore, we can build more parsimonious models [73] by grouping the relationships of sets of variables that make joint theoretical sense [74] and can be interpreted as a unit without losing the effect of each one of them separately. This is especially relevant as the number of variables increases, and the correlation between them and/or the sample size decreases. In such circumstances, multiple regression models without SEM can be strongly affected by net suppression conditions between variables with a high correlation between them [75]. In our particular case, the constructs that we wanted to examine were fairly complex and different from those first-order components in which constructs located on the same plane or level are considered. In this sense, constructs can be designed according to higher-order components (HOC). This type of model frequently requires higher-order structures to be examined, including various levels of components [68,76]. For example, the health system's quality represented by the health system performance construct in our model can be specified based on multiple abstraction grades. Mainly, quality can be constituted by various first-order components that separately identify numerous quality features. These may include safety and patient satisfaction through patient opinions, effectiveness, or relevance in the healthcare context. These first-order components or lower order components (LOC) make up the second-order component or higher-order components (HOC) of the quality of the system (health system performance), which presents a greater degree of abstraction.

Rather than modeling quality attributes as drivers of overall respondent quality in a unique level latent variable (see Figure 2), the higher-order model entails combining the lower order constructs into a single multidimensional construct. This modeling procedure is conducive to greater theoretical parsimoniousness and decreases the model's complexity, as shown in Figure 3.

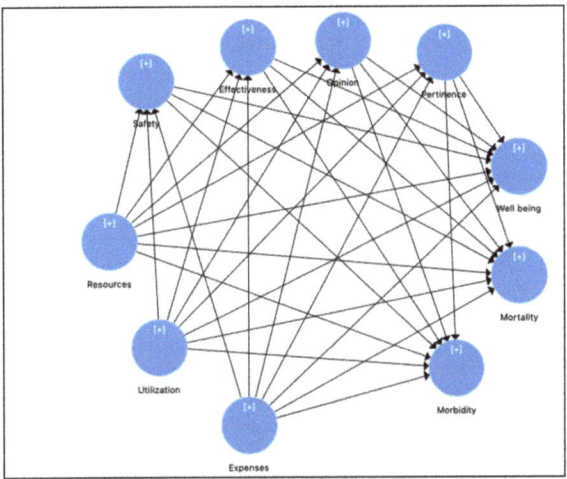

Figure 2. The first step: lower order components' measurement model.

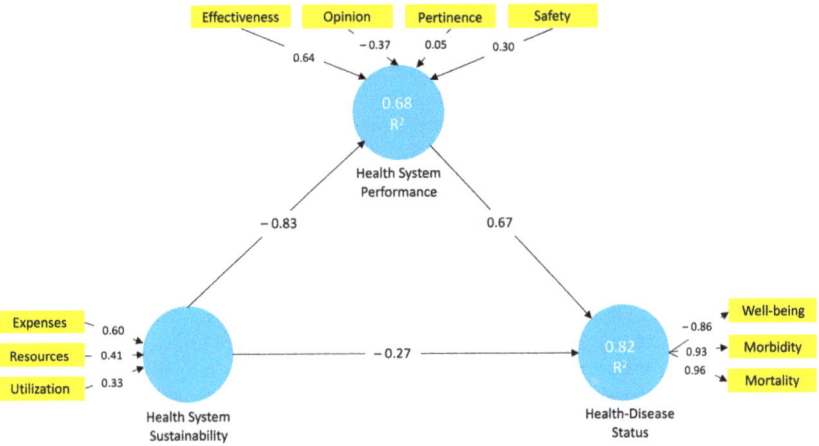

Figure 3. The second step: higher-order composites' structural model. Health system sustainability, health system performance, and health–disease status are higher-order constructs (HOC).

Researchers can choose between different approaches to identify the higher-order construct, with alternative approaches to repeated and two-step indicators being the most commonly used in the literature [77]. This work chose the two-step disjoint approach because it shows a better recovery of path parameters [78]. The disjointed approach was initially only based on evaluating the lower-order components' measurement model. These were directly related to all other constructs with which the higher-order construct is theoretically related (see Figure 2). That is, in this first step, we verified compliance with the criteria related to the measurement model of the PLS-SEM algorithm for the lower-order constructs. Thus, in the case of constructs in mode A, the criteria tested were individual item reliability, construct reliability, convergent validity, and discriminant validity; while for B-mode constructs, multicollinearity was verified [67]. During these checks, those eliminations of items that did not meet the criteria were made, subsequently providing the scores of the lower-order constructs. The scores' construct values were then saved, but only those of the lower-order constructs—in our case, the scores of the LOC effectiveness, safety, opinions, and pertinence to build the HOC health system's performance; resources, utilization, and expenses to form the HOC health system's sustainability; and the LOC well-being, mortality, and morbidity for the HOC health–disease status. In stage two, these scores are used as indicators to measure the corresponding higher-order construct. Therefore, we apply the PLS-SEM algorithm again in this second step, but exclusively for higher-order constructs with their lower-order dimensions as indicators. In this second step, the PLS-SEM algorithm was fully developed to evaluate both the measurement model and the structural model [78]. The criteria applied to verify the structural model were the inner model variance inflation factor, path coefficients, coefficient of determination, effect sizes, and predictive relevance.

According to Law et al. [79], a construct is higher-order or multidimensional when it refers to a set of different but related dimensions that must be treated as a single theoretical concept. This construct should not be confused with the one-dimensional construct or those multiple variables that manifest a relationship with each other but correspond to more than one theoretical concept. Consequently, a multidimensional construct is conceptualized based on its dimensions and, therefore, does not exist separately. Higher-order constructs constitute a holistic representation of a very complex reality, and their modeling increases the variance explained by the proposed model [80]. In addition, they reduce the number of relationships of the path model, as we can see in Figure 3, achieving greater model parsimony.

2.2. Specification of PLS-SEM Model

The specification of the higher-order model on Spanish health–disease status required defining the set of HOC constructs and the set of indicators related to the lower-order constructs. In this vein, the dimensions included in the health system performance's higher-order construct were the following:

Effectiveness (LOC): Effectiveness in health care refers to the degree to which an intervention—service, process, procedure, diagnostic test, or treatment—produces the desired result. It includes the following indicators: "birth of children from less-than-20-year-old women for every 100 births", "incidence of tetanus per 100,000 inhabitants", "incidence of hepatitis B per 100,000 inhabitants", and "incidence of mumps per 100,000 inhabitants".

Safety (LOC): This dimension refers to how the health system provides safe care and care to the patient. This involves minimizing the unnecessary risk of harm to the patient, which manifests itself in the absence of accidental injuries attributable to the provision of care or medical errors. Healthcare that promotes patient safety in the provision of care involves risk management; recording, analysis, and monitoring of incidents; and implementing solutions to minimize recurrence risk. This includes the following indicators: "reporting rate of suspected serious adverse reactions to medicines", "intrahospital mortality post-infarction per 100 highs from a heart attack", and "lower member amputation rate in diabetic people". In fact, incident reporting and monitoring are measured with these three indicators.

Opinion or Patient Satisfaction (LOC): One of the critical components of quality is the system's responsiveness to patient preferences, attitudes, and expectations. Patient-centered care is defined as one that establishes a good interrelationship between professionals and patients to ensure that decisions made regarding patients' care process take into account their needs, desires, and preferences, ensuring that these patients have the necessary training and support for effective participation. In a health system whose social legitimacy rests on reliability, satisfaction, and trust, this is understood as a significant quality component to generate a positive experience for patients and the population in contact with services. This includes indicators such as "degree of satisfaction of citizens with the functioning of the public health system".

Pertinence (LOC): The degree to which users receive the care they need, with the best use of resources according to available scientific evidence and side effects, is less than the potential benefits. This includes the following indicators: "laparoscopic cholecystectomy", "conservative breast cancer surgery", and "hip fracture patients with surgery in the first 48 h".

On the other hand, the dimensions included in the health system sustainability's higher-order construct were the following:

Expenses (LOC): Disbursement of goods and services intended to preserve, maintain, recover, or improve the population's health level. This includes indicators such as "percentage of health expenditure on primary care", "percentage of pharmacy expenditure", and "percentage of expenditure in specialized care", among others.

Utilization (LOC): Citizens take advantage of health services. This includes, among others, indicators such as "frequentation in specialized care inquiries (% SNS)" or "rates of surgical interventions (% SNS)".

Resources (LOC): High-quality healthcare requires the availability of sufficient resources to meet individual and population needs. The system's capacity refers to economic resources, infrastructure, equipment, human resources, medical devices, medicines, and health service technologies, including information and communication technologies. This includes, among others, indicators such as "medical staff in specialized care per 1000 inhabitants (% SNS)", "nursing staff in specialized care per 1000 inhabitants (% SNS)", "hospital beds in operation (% SNS)", and "posts in day hospitals per 1000 inhabitants (% SNS)".

Finally, the dimensions included in the health–disease status higher-order construct (HOC) were the following:

Well-being (LOC): Health well-being is measured through life expectancy, which is the average number of years a given absolute or total population lives in a certain period. This includes, among others, indicators such as "life expectancy at birth" and "life expectancy at 65 years".

Mortality (LOC): This is the proportion of people who die from the total population over a period of time, usually expressed as the proportion per one hundred thousand per year. This includes, among others, indicators such as "age-adjusted mortality rate from ischemic heart disease per 100,000 inhabitants", "age-adjusted mortality rate from cerebrovascular disease per 100,000 inhabitants", and "age-adjusted mortality rate from cancer per 100,000 inhabitants".

Morbidity (LOC): Morbidity is a sick state, disability, or poor health due to any cause. The term can refer to any form of disease or to the extent that a health condition affects the patient. This includes, among others, indicators such as "incidence of tuberculosis per 100,000 inhabitants", "incidence of new HIV diagnoses", and "adjusted hospitalization rate for acute myocardial infarction per 10,000 inhabitants (SNS)."

Likewise, the definitions of the individual indicators with their corresponding lower-order constructs and their modes of measurement are shown in Table 1 below.

Table 1. Composites and description of indicators.

Composites	Indicators	Description
Effectiveness (Mode B)	EF1	Birth of children from women less than 20 years old for each 100 births
	EF2	Incidence of tetanus per 100,000 inhab.
	EF3	Incidence of hepatitis B per 100,000 inhab.
	EF4	Incidence of mumps per 100,000 inhab.
Safety (Mode B)	SA1	Rate of suspected severe adverse effects rate to medication notified per 1,000,000 inhab.
	SA2	Intrahospital mortality of post-heart attack for every 100 discharges per a heart attack
	SA3	Amputation rate of the lower limb in diabetes patients
Opinion (Mode A)	O1	Level of satisfaction of citizens with the public health system
	O2 *	Level of satisfaction of citizens with their historical knowledge and the tracking of their health condition by their family doctor and pediatrician
	O3	Level of satisfaction of citizens with the information provided by their doctor about their health condition

Table 1. *Cont.*

Composites	Indicators	Description
Pertinence (Mode B)	PE1	Percentage of laparoscopic cholecystectomy
	PE2	Percentage of conservative breast cancer surgery
	PE3	Percentage of hip fracture patients with surgery in the first 48 h
Expenses (Mode B)	EX1	Percentage of health expenditure in primary care
	EX2	Percentage of health expenditure in pharmacy
	EX3	Public health expenditure per covered population
	EX4 *	Percentage of health expenditure in specialized care
	EX5	Percentage of health expenditure on salaries
	EX6	Percentage of health expenditure on intermediate consumption
	EX7 *	Percentage of health expenditure on public–private contract
	EX8	Percentage of health expenditure on internship training
Utilization (Mode B)	U1	Consultations with specialist doctors (% NHS)
	U2 *	Hospitalizations (% NHS)
	U3	Surgical interventions (% NHS)
	U4 *	CT utilization (% NHS)
	U5 *	Use rate of nuclear magnetic resonance (% NHS)
	U6	Hemodialysis usage (% NHS)
	U7	Hemodynamic usage (%NHS)
Resources (Mode B)	RE1	Specialist doctors (% NHS)
	RE2 *	Specialized nursing (% NHS)
	RE3	Beds in operation (% NHS)
	RE4	Day hospital places (% NHS)
	RE5 *	Operating rooms (% NHS)
	RE6	CT equipment (% NHS)
	RE7 *	Nuclear magnetic resonance equipment (% NHS)
	RE8	Hemodialysis equipment (% NHS)
	RE9	Hemodynamic equipment (% NHS)
Well-being (Mode A)	WB1	Life expectancy at birth
	WB2	Life expectancy at 65 years
	WB3	Healthy life years at birth
	WB4	Healthy life years at the age of 65 years
Mortality (Mode B)	MT1 *	Ischemic heart disease mortality rate per 100,000 inhab.
	MT2	Cerebrovascular disease mortality rate per 100,000 inhab.
	MT3	Cancer mortality rate per 100,000 inhab.
	MT4	Chronic obstructive pulmonary disease mortality rate per 100,000 inhab.
	MT5	Pneumonia and influenza mortality rate per 100,000 inhab.
	MT6 *	Chronic liver disease mortality rate per 100,000 inhab.
	MT7	Diabetes mellitus mortality rate per 100,000 inhab.
	MT8	Unintentional accidents mortality rate per 100,000 inhab.
	MT9	Suicide mortality rate per 100,000 inhab.
	MT10	Alzheimer's mortality rate per 100,000 inhab.
Morbidity (Mode B)	MB1	Tuberculosis incidence
	MB2	New HIV diagnosis
	MB3	Diabetes in adult population
	MB4	Acute myocardial infarction hospitalization per 10,000 inhab. (NHS only)
	MB5	Cerebrovascular disease hospitalization per 10,000 inhab. (NHS only)
	MB6	Chronic obstructive pulmonary disease hospitalization per 10,000 inhab. (NHS only)
	MB7	Diabetes mellitus hospitalization per 10,000 inhab. (NHS only)
	MB8	Hypertensive disease hospitalization per 10,000 inhab. (NHS only)
	MB9 *	Congestive heart failure hospitalization per 10,000 inhab. (NHS only)
	MB10	Victims of traffic accidents
	MB11	Work accidents
	MB12	Frequency of work accidents

Source: Ministry of Health, Social Services, and Equality (MHSE), 2008–2017. * These indicators were not included in latent variables due to the multicollinearity criteria of PLS-SEM or item reliability.

2.3. Data and Sample

In the sample configuration, data from key indicators of Spain's national health system from 2008 to 2017 were used. The model was tested with a secondary dataset and used repeated cross-sectional data [81]. The Spanish Ministry of Health, Social Services, and Equality (MHSE) has a statistical portal with information about each autonomous community's average national health system key indicators. Of the total of 19 autonomous communities existing in Spain, the lack of data from two of them (Ceuta and Melilla) led to them being excluded, leaving the sample composed of 17 autonomous communities. Faul, Erdfelder, Buchner, and Lang [82] explain the minimum sample size required when we set an effect size f^2 of 0.15 and a significance level of 0.05, using the statistical program G * Power. Our results show a minimum size of 103 observations for a statistical power of 0.8. Therefore, the minimum sample size required of 103 observations is less than the 165 used.

The selection of sets of indicators is a procedure used by different supra- and international organizations that are beginning to use said sets of indicators or are in the process of preparing them. Among them, the European Commission works to obtain comparable information on health, the habits of the population related to health and diseases, and health systems. The objective of the commission is to have an integrated system of indicators, common at the European level, whose work scheme is based on the ECHI (European Community Health Indicators) project. At the Spanish level, the country has significantly developed its health information systems in order to obtain executive and multidimensional information. In Spain, this is known under the generic name of "key indicators of the SNS", which also serve as the basis for submitting the information to the ECHI project of the European Commission. The conceptual model of the European Core Health Indicators (ECHI) was adapted to the Spanish national health system's characteristics, which determined the relationships between the constructs. In this sense, Table 1 presents a summary of all the variables and indicators included in the model, their acronyms, and the data sources used.

Table 2. Hierarchical component of study.

Lower Order Composites	Higher-Order Composites
Effectiveness Safety Opinion Pertinence	Health system performance
Expenses Utilization Resources	Health system sustainability
Well-being Mortality Morbidity	Health–disease status

Source: Own elaboration.

The series of indicators used are grouped around their meaning, and some indicators are both secondary and primary care and/or exclusive to one of the two types, depending on the case. For example, the indicator "EX2—Percentage of health expenditure in pharmacy" includes the pharmaceutical expenses of both hospitals and primary health centers. However, for example, the indicator "EX1—Percentage of health expenditure in primary care" is exclusive to primary care centers, while the indicator "U2—Hospitalizations (% NHS)" is exclusive to secondary care centers, that is, hospitals. This means that the key indicators of the SNS used in this work include both information from secondary care data and information from primary care data.

Concerning the higher-order construct (see Table 2) health–disease status, 24 items were used grouped into three theoretical dimensions: mortality, morbidity, and well-being [12]. To measure quality or health system performance (HS performance), we follow Cinaroglu and Baser's [10] recommendations. A scale of 13 items initially grouped into four dimensions was used: effectiveness, opinion, safety, and pertinence. Finally, health system sustainability (HS sustainability) was measured with the scale proposed by Valls Martínez and Ramírez-Orellana [47], consisting of 24 items grouped into three dimensions: utilization, resources, and expenses. The second-order HS performance and health–disease status constructs were mixed type [76], and according to the results of confirmatory tetrad analysis, we considered the formative-formative type for HS sustainability.

3. Assessing PLS-SEM Results

This section presents the results of applying the disjoint two-step method to our higher-order component model. Initially, at the first step, the PLS algorithm was performed to evaluate the lower-order composites' measurement model. The second step evaluated both the measurement model and the structural model of the higher-order composites. The evaluation of the measurement model allowed us to check the validity and reliability of the proposed scales, before proceeding to evaluate the structural model (see Figure 3).

3.1. Evaluation of LOC Measurement Model

The lower-order composites measurement model was evaluated concerning the four criteria identified to meet the models' reliability and validity: individual item reliability, construct reliability, convergent validity, and discriminant validity.

3.1.1. Reflective Measurement Model

- Individual item reliability LOC

According to the latent variables' specifications (see Table 1), only the opinion and well-being constructs were measured in mode A. Therefore, we refined those items with load values lower than the reference value of 0.707 [83]. According to this criterion, the second item of the opinion composite (level of satisfaction of citizens with their historical knowledge and monitoring of their health status by the family doctor and pediatrician) was eliminated due to not exceeding the reference threshold.

- Construct Reliability LOC

The Cronbach alpha coefficient (α), the Dijkstra–Henseler (ρ_A) index, and the composite reliability statistics were calculated to check the construct reliability criterion [78]:

$$\text{Cronbach's } \alpha = \frac{N \cdot \bar{c}}{1+(N-1) \cdot \bar{c}}$$

$$\rho_A := (\hat{w}'\hat{w})^2 \cdot \frac{\hat{w}'[S-diag(S)]\hat{w}}{\hat{w}'[\hat{w}\hat{w}'-diag(\hat{w}\hat{w}')]\hat{w}'}$$

$$\rho_C = \frac{\left(\sum_{i=1}^{N} l_i\right)^2}{\left(\sum_{i=1}^{N} l_i\right)^2 + \sum_{i=1}^{N} \text{var}(e_i)},$$

where N is the number of lower-order components ($i = 1, 2, \ldots, N$); $\bar{c}$ is the average correlation between the lower-order components; $\hat{w}'$ is the higher-order constructs' estimated weight vector, and the number of lower-order constructs is the dimension of $\hat{w}$; S is the empirical covariance matrix of the lower-order components; l_i is the loading of the lower-order component i in a particular higher-order construct; and $\text{var}(e_i)$ is the measurement error's variance of the lower-order component i.

All three indicators share the same benchmark threshold of 0.7 [84,85], and this was met for the sample data (see Table 3).

Table 3. Construct reliability LOC.

Constructs	Cronbach Alpha	ρ_A	Composite Reliability
Opinion	0.774	1.210	0.884
Well-being	0.841	0.878	0.890

Source: Own elaboration.

- Convergent validity LOC

The convergent validity of the model's constructs was verified by analyzing the average variance extracted (AVE) [78]:

$$\text{AVE} = \frac{1}{N} \sum_{i=1}^{N} l_i^2.$$

The AVE values in this study were 0.793 for opinion and 0.672 for well-being. These results are adequate as the values should be above 0.5, according to Hair et al. [67].

- Discriminant validity LOC

To close the LOC measurement analysis in mode A, the discriminant validity was verified through the HTMT ratio of the higher-order constructs Y_i and Y_j developed by Henseler, Ringle, and Sarstedt [86]:

$$\text{HTMT} = \frac{\frac{1}{K_i K_j} \sum_{g=1}^{K_i} \sum_{h=1}^{K_j} r_{ig,jh}}{\left(\frac{2}{K_i(K_i-1)} \sum_{g=1}^{K_i-1} \sum_{h=g+1}^{K_i} r_{ig,jh} \cdot \frac{2}{K_j(K_j-1)} \sum_{g=1}^{K_j-1} \sum_{h=g+1}^{K_j} r_{jg,jh} \right)}$$

where K_i (respectively K_j) is the number of lower-order constructs considered as indicators of the higher-order construct Y_i (respectively Y_j); and $r_{ig,jh}$ is the correlations of the lower-order constructs within and across the higher-order constructs Y_i and Y_j. Observe that the numerator represents the average heterotrait–heteromethod correlation, and the denominator is the geometric mean of the average monotrait–heteromethod correlation of construct Y_i and the average monotrait–heteromethod correlation of construct Y_j.

The ratio should not exceed the threshold value of 0.85 or 0.90 [87]. In this study, the HTMT ratio had a value of 0.409, thus reaching discriminant validity.

The Fornell and Larcker [88] criterion was also used to measure discriminatory validity. This criterion explains that the amount of variance that a construct captures from its indicators (AVE) should be greater than the variance that such as construct shares with other constructs in the model (the squared correlation between the two constructs). To facilitate this assessment, the root square of the AVE of each latent variable should be greater than the correlations it has with the other latent variables in the model.

The values indicating that there is an adequate discriminatory validity according to the Fornell and Larcker criterion are shown in bold on the diagonal (see Table 4).

Table 4. Fornell and Larcker criterion LOC.

Constructs	EF	EX	MB	MT	O	PE	RE	SA	U	W–B
EF	n/a									
EX	−0.537	n/a								
MB	0.854	−0.71	n/a							
MT	0.846	−0.635	0.873	n/a						
O	−0.271	0.364	−0.452	−0.369	0.891					
PE	−0.186	0.342	−0.136	−0.033	0.167	n/a				
RE	−0.548	0.421	−0.627	−0.555	0.614	0.141	n/a			
SA	0.578	−0.622	0.673	0.599	−0.188	−0.199	−0.422	n/a		
U	−0.405	0.112	−0.382	−0.465	0.341	−0.104	0.403	−0.252	n/a	
W–B	−0.657	0.543	−0.665	−0.761	0.367	0.110	0.519	−0.405	0.431	0.820

Source: Own elaboration.

3.1.2. Formative Measurement Model

- Collinearity of mode B indicators' LOC

As the measurement mode A models have been evaluated in the previous subsections, it is now necessary to assess the formative measurement models, or mode B. To do this, within the two-step method in higher-order models, we examine the degree of collinearity of the indicators in mode B. If there is multicollinearity, we proceed to eliminate these items. For items EX4, EX7, U2, U4, U5, RE2, RE5, RE7, MT1, and MT6 (see Table 2), variance inflation factor (VIF) values equal to or greater than 5 were found, which indicated a multicollinearity problem, and they were eliminated from the model. The VIF of the k-th indicator is calculated as follows:

$$\text{VIF}_k = \frac{1}{1 - R_k^2},$$

where R_k^2 is the explained variance of the k-th regression. A high value of R_k^2 denotes that the variance of the k-th indicator can be explained by other items of the construct.

- Compute the LOC scores

Finally, the disjoint two-stage approach does not interpret the model estimates. According to the PLS algorithm, it proceeds to compute the lower-order constructs scores to use as new variables to measure the higher-order construct in stage two. The lower order components are linked to all other constructs that the higher-order construct is theoretically related to, as shown in Figure 2. In evaluating the HOC model, these scores are used as indicators of the higher-order construct [78].

3.2. Evaluation of HOC Measurement Model

3.2.1. Reflective Measurement Model

- Individual item reliability HOC

The reflective indicators' individual reliability is valued by examining the factorial loads (λ) or simple correlations of the measures or indicators with their respective construct. The indicators are reliable if $\lambda \geq 0.707$ [83]. Several researchers argue that this heuristic rule should not be as rigid in the early stages of scale development [64] or when scales apply to different contexts [89].

In the model, the values for loads conform to what is recommended (see Table 5); however, the health–disease status construct has a negative value for the well-being dimension (-0.860). This value means that the condition is satisfied, since the squared value of -0.86 is 0.74, so the variance is explained in 74%; therefore, it must maintain the well-being item.

Table 5. Individual item reliability HOC.

Constructs	Morbidity	Mortality	Well-Being
Health–disease Status	0.934	0.960	-0.860

Source: Own elaboration.

- Construct Reliability HOC

The measurements are the Cronbach alpha coefficient (α), the ρ_A index, and composite reliability.

Composite reliability is more appropriate than Cronbach alpha for PLS, as it does not assume that all indicators receive the same weight [84]. The value of 0.7 is suggested as an appropriate level for "modest" reliability in the early stages of research, and a stricter 0.8 or 0.9 for more advanced research stages. Dijkstra–Henseler index (ρ_A) was also evaluated and is considered to be a measure of consistent reliability [85].

As shown in Table 6, the Dijkstra–Henseler index (ρ_A) value meets the recommended threshold to conform with our evaluation. Dijkstra and Henseler [85] presented their index

ρ_A as an exact and consistent measure of construct reliability, since Cronbach's alpha is conservative in excess and composite reliability the opposite.

Table 6. Construct reliability HOC.

Construct	Cronbach Alpha	ρ_A	Composite Reliability
Health–disease Status	−0.876	0.926	0.696

Source: Own elaboration.

- Convergent validity HOC

Convergent validity implies that a set of indicators represents a single underlying construct, demonstrated by its one-dimensionality [90]. For average variance extracted (AVE) values, it is recommended that their values be equal to or greater than 0.50. In this case, the health–disease status with a 0.845 value of AVE is given validity.

- Discriminant validity HOC

Discriminant validity indicates the extent to which a given construct is different from other constructs. We measure it through the Fornell and Larcker criterion.

The values indicating that there is an adequate discriminatory validity according to the Fornell and Larcker criterion are shown in bold on the diagonal (see Table 7).

Table 7. Fornell and Larcker criterion HOC.

Constructs	Health–Disease Status	HS Performance	HS Sustainability
Health–disease St.	0.919		
HS Performance	0.890	n/a	
HS Sustainability	−0.821	−0.826	n/a

Source: Own elaboration.

3.2.2. Formative Measurement Model

The measurement model for mode B composites (HS performance and HS sustainability) was evaluated in terms of collinearity between indicators, significance, and relevance of external weights.

First, discarding indicators was carried out when the indicator exceeded the variance impact factor (VIF > 5). As a result of this process, all the HOC indicators remained without collinearity.

Second, the relevance of weights was analyzed. Figure 3 shows the relevance of indicators within their construction.

Thus, for the latent higher-order HS performance, the most positively relevant dimensions were effectiveness and safety. Additionally, opinion has negative relevance, while pertinence lacked weight within the system's quality with a weight of 0.05, very close to zero.

For the HS sustainability higher-order variable measured through its dimensions, it was established that expenses are the most weighted dimension, followed by resources and, finally, utilization. All three dimensions bring positive relevance to the construct.

Finally, to evaluate the significance, we can start bootstrapping with 10,000 subsamples to check if the external weights are significantly different from zero, that is, the minimum recommended by Hair, Ringle, and Sarstedt [91]. Since weights provide information on their contribution, they can be classified according to their respective composition [64]. Indicators with a non-significant weight but with significant loads of 0.50 or more were considered relevant [64]. Our results show that all the indicators' weights were significant, except pertinence (Table 8).

Table 8. Significance of weights.

Constructs	Original Sample	t	Loadings	Lo95	Hi95
Health System Sustainability					
Expenses	0.600 ***	9.974	0.810	[0.479;	0.714]
Resources	0.413 ***	7.978	0.798	[0.314;	0.517]
Utilization	0.328 ***	6.194	0.562	[0.220;	0.427]
Health System Performance					
Effectiveness	0.639 ***	14.307	0.902	[0.547;	0.723]
Opinion	−0.367 ***	7.616	−0.588	[−0.459;	−0.270]
Pertinence	0.051 ns	1.410	−0.189	[−0.018;	0.122]
Safety	0.298 ***	6.267	0.727	[0.205;	0.392]

ns not significant. * $p < 0.05$; ** $p < 0.01$; *** $p < 0.001$. Significance, t statistic, and 95% bias-corrected confidence interval performed by 10,000 replications bootstrapping procedure.

3.3. Evaluation of HOC Structural Model

Once the measures of the constructs were verified to be appropriate, the structural model was assessed.

3.3.1. Evaluation of Path Coefficients

Path coefficients and their significance are reported in Table 9 and Figure 3, with their 10,000 bootstrap resampling levels. In addition, Table 9 shows that the VIF of the constructs ranged from 1.000 to 3.152, suggesting that collinearity is not a problem. This study also evaluated quality by verifying that the Q^2 value is greater than 0.5, which shows a situation of high predictive relevance [67]. This suggests a good fit in model prediction.

Table 9. Full sample results.

Constructs	Path	t	p	Lo95	Hi95	f^2	VIF
Direct effects							
HSP→HS	0.667 ***	14.413	0.000	0.577;	0.760	0.766	3.152
HSS→HS	−0.821 ***	36.448	0.000	−0.864;	−0.775	0.125	3.152
			R^2: 0.816; Q^2: 0.672				
HSS→HSP	−0.826 ***	35.197	0.000	−0.873;	−0.781	2.152	1.000
			R^2: 0.683				
Indirect effect						VAF	
HSS→HSP→HS	−0.551 ***	13.219	0.000	−0.640	−0.475	67.31%	n/a

* $p < 0.05$; ** $p < 0.01$; *** $p < 0.001$. Significance, t statistic, and 95% bias-corrected confidence interval performed by 10,000 replication bootstrapping procedure. VIF—inner model variance inflation factor; VAF—variance accounted for.

Our results suggest that HS performance has a positive and significant impact on health–disease status at a level of 5%; the higher the quality of the system, the higher the health–disease status. Additionally, HS sustainability has a significant but negative impact on health–disease level, suggesting that the health system's higher sustainability lowers the rate of morbidity and mortality. Likewise, HS sustainability's influence on HS performance is negative and significant. In short, all model hypotheses that relate latent variables to each other are accepted (H1, H2, and H3).

We analyzed the mediation (H4) hypothesis, resulting in the indirect effects being significant [92]. The indirect effect of HS sustainability on health–disease status through HS performance was positive and significant (p-value 0.000), supporting H4 (Table 9). The direct effect was also significant, which indicated that the mediation effect was partial [93]; HS sustainability directly influenced health–disease status (H1), and indirectly influenced it through HS performance. The value of the variance accounted for (VAF) indicated that the mediated ratio was 67.31% of HS sustainability's total effect on health–disease status (see the indirect effect in Table 9):

$$\text{VAF} = \frac{\text{indirect effect}}{\text{total effect}}.$$

3.3.2. Assessment of the Coefficient of Determination (R^2)

The coefficient of determination (R^2) represents a measure of predictive power. It indicates the amount of variance of a construct explained by the predictor variables of that endogenous construct in the model. R^2 values range from 0 to 1; the higher the value, the more predictive capacity the model has for that variable.

The values of R^2 should be high enough for the model to reach a minimum level of explanatory power. Falk and Miller [94] suggest at least ≥ 0.10; Chin [64] states that 0.67 is substantial, 0.33 is moderate, and 0.19 is weak.

The health–disease status construct's predictive level, with a value of 0.842, can be considered more than substantial (see Table 9). The HS performance constructs with a value equal to 0.680 are also more than substantial because they exceed 67% and are close to 1 (see Table 9).

3.3.3. Review of Effect Sizes (f^2)

The effect sizes (f^2) value the degree to which an exogenous construct helps explain a certain endogenous construct in terms of R^2 [95]:

$$f^2 = \frac{R^2_{included} - R^2_{excluded}}{1 - R^2_{included}},$$

where R^2 is calculated including and excluding a specific predictor construct in the model.

A Cohen [95] heuristic rule for evaluating f^2 holds that $0.02 \leq f^2 < 0.15$ is a small effect; $0.15 \leq f^2 < 0.35$ is a moderate effect; and $f^2 \geq 0.35$ is a large effect.

The results in Table 10 show that the effect between the exogenous construct HS sustainability and its contribution to the endogenous construct health–disease status (0.125) was small and significant, while HS performance (2.152) had a large effect. In contrast, the HS performance construct with health–disease status, with a value of 0.766, had a significant and large effect.

Table 10. Effect sizes (f^2) and p-values.

HS Performance → Health–Disease Status	0.766	0.000
HS Sustainability → Health–Disease Status	0.125	0.010
HS Sustainability → HS Performance	2.152	0.000

Source: Own elaboration.

4. Discussion

The results of this study have important implications for hospital managers and policymakers. Healthcare officials and managers will have one more tool with which to establish the determinant factors for achieving their objective: to improve the population's health and quality of life. The findings revealed that health system sustainability had a negative effect on health–disease status, measured in terms of mortality and morbidity rates. However, the relationship between health system performance and health–disease status is positive.

We analyzed 61 indicators belonging to lower-order components that define three higher-order components. Data were obtained from the Spanish Ministry of Health Social Services, and Equality for the entire Spanish territory, except Ceuta and Melilla, for the period between 2008 and 2017. The applied technique was partial least squares structural equation modeling (PLS-SEM).

The health–disease status construct, composed of three lower-order components, was reflected in two components of disease incidence and mortality with loads in a positive sense, presenting the well-being dimension's inverse correlation with the value of the construct. Therefore, the model was further delimited by mortality and morbidity. In other words, the latent variable health–disease status was defined more by incidences of diseases than by health status in a positive sense. For example, a previous study discovered that the mortality rate increases when the person suffers from heart disease or cancer and, on the other hand, when the patient is hospitalized through an emergency unit [60]. A previous study indicated that injured people have a higher Charlson Comorbidity Index (CCI) than non-injured people, that is, pre-injury morbidity was higher [96].

A positive relationship between the constructs of health system performance and health–disease status was confirmed. The most relevant dimensions were effectiveness and safety, in this order, and lastly, pertinence with a non-significant influence. One of the most investigated components of effectiveness is the quality of the system, which, for example, can materialize in annual tests of hemoglobin A1C in diabetic patients and the use of aspirin in cases of myocardial infarction [6]. Moreover, opinion had a negative effect on performance. This clearly confirmed patient satisfaction as an indicator of quality care [97]. Despite this, the patient being in a state of discomfort might not be the best criterion when evaluating the health system [98].

Moreover, health system sustainability negatively influenced health–disease status, which shows that increases in expenditures, resources, and extent of use in the healthcare system improve the population's health, reducing mortality and morbidity or increasing well-being. The three dimensions analyzed have a positive influence on the formative construct. The weights inform us about the contribution of the indicators to the construct. The indicator with the most weight was expenses, followed by the allocation of resources, and lastly, the use and exploitation of health services by citizens. When public agencies provide an adequate allocation of resources according to the patients' needs, not only are effective, safe, and timely results offered—the efficiency of the system is improved [34]. It would be interesting for health systems to also invest their resources in prevention. For example, cardiovascular diseases, in many cases, and especially in young patients, are driven by behaviors that can be avoided, such as a sedentary lifestyle, smoking, poor diet, and alcohol consumption [99]. In another way, the literature shows that between 25% and 40% of cardiovascular diseases are attributable to work-related stress. For this reason, health systems policymakers should also address issues related to occupational health psychology, not only for mental morbidity, but also for other diseases that include the risk of death [41]. However, a study revealed that depression is a common factor in hospitalized patients, and when it is present, the risk of death after myocardial infarctions is higher [38]. On the other hand, hospital readmission is higher when it comes to cardiac patients [58].

Our research indicated that the health system's better performance would be reflected as a better health–disease status of the population, which is consistent with the bibliography, which considers that with greater effectiveness and safety in the health system, the patient will obtain the desired results in a safe way [30]. Moreover, according to another study, we can observe that efficiency showed a negative relationship with mortality rates because the treatment's efficiency allows better clinical results to be obtained [60]. Effectiveness is one of the health system's performance indicators and reflects the effect that its treatment and interventions have on the health–disease status of the population [10]. A study carried out with exclusive data from hospitals revealed that the most competent and efficient hospitals have lower mortality rates [60].

Maintaining a sustainable health system is the basis for improving people's health [2,28,34]. Hospitals that do not allocate their resources properly are more insecure, which means that they are more likely to have unwanted clinical events [34]. In that sense, the authorities must improve the services' quality and deliver services effectively and professionally [32]. The results showed an inverse relationship between sustainability and health–disease status. In other words, better spending, resource allocation, and use of the health system lowers the incidence of diseases and improves health, which also explains the negative influence of sustainability on performance, since adequate management of resources and expenses will lead to a better-quality system. Tumors are the second leading cause of death in women worldwide. In this sense, the health system must promote and be able to attend on time the periodic controls that are required [42]. A previous study found that when it comes to diseases such as diabetes or lung conditions, patients adhere less to treatments. Therefore, the health system administration should direct its efforts to persuade the population by communicating with and educating them about the need to control these diseases [52]. The direct and indirect effects of sustainability on health–disease status were confirmed. The mediating effect, through the performance construct, was 32.09% of the total effect.

A favorable health–disease status of the population will require decision making by public authorities regarding the right laws of health in accordance with the WHO. This will imply implementing an efficient financing system with sufficient budgetary allocation to stimulate the system's performance [18,19]. Thus, for example, spending on primary and specialized medicine are basic pillars, passing through the distribution of facilities that allow accessibility and use of the resources invested in the national health system [47]. All the budgetary allocation to cover expenses and resources must be done with balance, regardless of whether the healthcare offer is public and/or public–private arrangements, as is the case in some Spanish regions, hence the necessary regulation of the private–public provider mix. Regarding health system performance, the authorities must effectively attend to the composition of essential services packages to reduce health incidents. Another important factor in performance is having a good management and information system that allows data to be available at an opportune moment to make decisions that may affect the health system [6].

In this empirical study, the created model predicts the population's health–disease status as 84.2%, which is considered more than substantial. On the other hand, the performance construct explained 68% of the variance.

Although our research model uses the Spanish Ministry of Health, Social Service, and Equality data to verify our hypotheses, there remain some limitations. One limitation was due to the phenomenon of the invisibility of data [1] related to social care arrangements. In addition, our results are based solely on the Spanish territory, which opens up the possibility that the findings are specific only to this country. Future research should be focused on other countries. Furthermore, differences in the patient's gender, educational level, socioeconomic level, and other characteristics could yield interesting results in the future. Otherwise, future research could try to compare the performance of different secondary care centers, that is, hospitals, within the national territory, in order to verify if there are differences between autonomous communities (since in Spain it is the autonomous communities who have health competence).

5. Conclusions

Using the structural equation modeling approach, we developed a health–disease status model. The research reveals that a health system's administrators and government must pay their attention to continuously improving health system performance and health system sustainability to fulfill their ultimate goal, which is to enhance citizens' health–disease status.

The study's findings showed that patient health improves when the health system's performance is excellent, effective, and safe. Furthermore, patient health improves when the health system is sustainable over time, which implies that expenses, resources, and the use made of medical services are consistent with the needs of patients and do not compromise their future needs. Furthermore, health system performance mediates the relationship between sustainability and health–disease status. In other words, sustainability has a double effect (direct and indirect) on health–disease status.

Author Contributions: Conceptualization, A.R.-O.; methodology, A.R.-O.; validation, A.R.-O. and M.d.C.V.M.; formal analysis, A.R.-O. and M.S.G.; writing—original draft preparation, A.R.-O. and M.S.G.; writing—review and editing, A.R.O., M.d.C.V.M., and M.S.G.; supervision, A.R.-O. and M.d.C.V.M. All authors have read and agreed to the published version of the manuscript.

Funding: This research received no external funding.

Institutional Review Board Statement: Not applicable.

Informed Consent Statement: Not applicable.

Data Availability Statement: Data publicly available from the Spanish Ministry of Health.

Conflicts of Interest: The authors declare no conflict of interest.

References

1. Blouin, G.G. Data Performativity and Health: The Politics of Health Data Practices in Europe. *Sci. Technol. Hum. Values* **2020**, *45*, 317–341. [CrossRef]
2. Hejduková, P.; Kureková, L. National Health Systems' Performance: Evaluation WHO Indicators. *Procedia Soc. Behav. Sci.* **2016**, *230*, 240–248. [CrossRef]
3. Hanson, K. La Medición Del Estado de la Salud: Género, Carga de Morbilidad Y Establecimiento de Prioridades en el Sector Salud. *OPS Publ. Occas.* **2000**, *5*.
4. Kramers, P.G.N. The ECHI project. *Eur. J. Public Health* **2003**, *13*, 101–106. [CrossRef]
5. Verschuuren, M.; Gissler, M.; Kilpeläinen, K.; Tuomi-Nikula, A.; Sihvonen, A.-P.; Thelen, J.; Gaidelyte, R.; Ghirini, S.; Kirsch, N.; Prochorskas, R.; et al. Public health indicators for the EU: The joint action for ECHIM (European Community Health Indicators & Monitoring). *Arch. Public Health* **2013**, *71*, 12. [CrossRef]
6. Kruk, M.E.; Freedman, L.P. Assessing health system performance in developing countries: A review of the literature. *Health Policy* **2008**, *85*, 263–276. [CrossRef]
7. Cuggia, M.; Toubiana, L. Health Information Systems. *Yearb Med. Inform.* **2013**, *22*, 114–116. [CrossRef]
8. World Health Oragnization. 2020. Available online: https://www.who.int/es/about/who-we-are/constitution (accessed on 1 May 2021).
9. Castro-Vázquez, Á.; Espinosa-Gutiérrez, I.; Rodríguez-Contreras, P.; Santos-Iglesias, P. Relación entre el estado de salud percibido e indicadores de salud en la población española. *Int. J. Clin. Health Psychol.* **2007**, *7*, 883–898.
10. Cinaroglu, S.; Baser, O. Understanding the relationship between effectiveness and outcome indicators to improve quality in healthcare. *Total. Qual. Manag. Bus. Excel.* **2018**, *29*, 1294–1311. [CrossRef]
11. Murray, C.J.L.; Frenk, J. A framework for assessing the performance of health systems. *Bull. World Health Organ.* **2000**, *78*, 717–731.
12. Anderson, G.; Hussey, P.S. Comparing Health System Performance in OECD Countries. *Health Aff.* **2001**, *20*, 219–232. [CrossRef]
13. Faezipour, M.; Ferreira, S. A System Dynamics Perspective of Patient Satisfaction in Healthcare. *Procedia Comput. Sci.* **2013**, *16*, 148–156. [CrossRef]
14. Puertas, R.; Marti, L.; Guaita-Martinez, J.M. Innovation, lifestyle, policy and socioeconomic factors: An analysis of European quality of life. *Technol. Forecast. Soc. Chang.* **2020**, *160*, 120209. [CrossRef]
15. García-Alfranca, F.; Puig, A.; Galup, C.; Aguado, H.; Cerdá, I.; Guilabert, M.; Pérez-Jover, V.; Carrillo, I.; Mira, J.J. Patient Satisfaction with Pre-Hospital Emergency Services. A Qualitative Study Comparing Professionals' and Patients' Views. *Int. J. Environ. Res. Public Health* **2018**, *15*, 233. [CrossRef]
16. Rosenbusch, J.; Ismail, I.R.; Ringle, C.M. The agony of choice for medical tourists: A patient satisfaction index model. *J. Hosp. Tour. Technol.* **2018**, *9*, 267–279. [CrossRef]
17. Vogus, T.J.; McClelland, L.E. When the customer is the patient: Lessons from healthcare research on patient satisfaction and service quality ratings. *Hum. Resour. Manag. Rev.* **2016**, *26*, 37–49. [CrossRef]
18. Varabyova, Y.; Müller, J.-M. The efficiency of health care production in OECD countries: A systematic review and meta-analysis of cross-country comparisons. *Health Policy* **2016**, *120*, 252–263. [CrossRef]
19. Garcia-Lacalle, J.; Martin, E. Rural vs urban hospital performance in a 'competitive' public health service. *Soc. Sci. Med.* **2010**, *71*, 1131–1140. [CrossRef]
20. Chang, C.-W.; Tseng, T.-H.; Woodside, A.G. Configural algorithms of patient satisfaction, participation in diagnostics, and treatment decisions' influences on hospital loyalty. *J. Serv. Mark.* **2013**, *27*, 91–103. [CrossRef]
21. Ferreira, P.L.; Raposo, V.; Tavares, A.I. Primary health care patient satisfaction: Explanatory factors and geographic characteristics. *Int. J. Qual. Health Care* **2020**, *32*, 93–98. [CrossRef]
22. Manzoor, F.; Wei, L.; Hussain, A.; Asif, M.; Shah, S.I.A. Patient Satisfaction with Health Care Services; An Application of Physician's Behavior as a Moderator. *Int. J. Environ. Res. Public Health* **2019**, *16*, 3318. [CrossRef]
23. Aiken, L.H.; Sloane, D.M.; Ball, J.; Bruyneel, L.; Rafferty, A.M.; Griffiths, P. Patient satisfaction with hospital care and nurses in England: An observational study. *BMJ Open* **2021**, *8*, e019189. [CrossRef]

24. Pérez-Romero, S.; Gascón-Cánovas, J.J.; Salmerón-Martínez, D.; Parra-Hidalgo, P.; Monteagudo-Piqueras, O. Relevancia del contexto socioeconómico y sanitario en la satisfacción del paciente. *Gac. Sanit.* **2017**, *31*, 416–422. [CrossRef]
25. Draper, M.; Cohen, P.; Buchan, H. Seeking consumer views: What use are results of hospital patient satisfaction surveys? *Int. J. Qual. Health Care* **2001**, *13*, 463–468. [CrossRef]
26. Kamra, V.; Singh, H.; De, K.K. Factors affecting patient satisfaction: An exploratory study for quality management in the health-care sector. *Total. Qual. Manag. Bus. Excel.* **2016**, *27*, 1013–1027. [CrossRef]
27. Evangelinos, K.; Fotiadis, S.; Skouloudis, A.; Khan, N.; Konstandakopoulou, F.; Nikolaou, I.; Lundy, S. Occupational health and safety disclosures in sustainability reports: An overview of trends among corporate leaders. *Corp. Soc. Responsib. Environ. Manag.* **2018**, *25*, 961–970. [CrossRef]
28. Molero, A.; Calabrò, M.; Vignes, M.; Gouget, B.; Gruson, A.D. Sustainability in Healthcare: Perspectives and Reflections Regarding Laboratory Medicine. *Ann. Lab. Med.* **2021**, *41*, 139–144. [CrossRef]
29. Handayani, P.W.; Hidayanto, A.N.; Sandhyaduhita, P.I.; Kasiyah; Ayuningtyas, D. Strategic hospital services quality analysis in Indonesia. *Expert Syst. Appl.* **2015**, *42*, 3067–3078. [CrossRef]
30. lo Storto, C.; Goncharuk, A.G. Efficiency vs effectiveness: A benchmarking study on European healthcare systems. *Econ. Sociol.* **2017**, *10*, 102–115. [CrossRef]
31. Ruo, B.; Rumsfeld, J.S.; Hlatky, M.A.; Liu, H.; Browner, W.S.; Whooley, M.A. Depressive Symptoms and Health-Related Quality of Life. *JAMA* **2003**, *290*, 215–221. [CrossRef]
32. Hussain, A.; Asif, M.; Jameel, A.; Hwang, J.; Sahito, N.; Kanwel, S. Promoting OPD Patient Satisfaction through Different Healthcare Determinants: A Study of Public Sector Hospitals. *Int. J. Environ. Res. Public Health* **2019**, *16*, 3719. [CrossRef] [PubMed]
33. European Observatory on Health Systems and Policies. State of Health in the EU-Spain: Country Health Profile 2019. Available online: https://ec.europa.eu/health/sites/health/files/state/docs/2019_chp_es_english.pdf (accessed on 12 March 2021).
34. Ferreira, D.C.; Nunes, A.M.; Marques, R.C. Operational efficiency vs clinical safety, care appropriateness, timeliness, and access to health care. *J. Prod. Anal.* **2020**, *53*, 355–375. [CrossRef]
35. Rabar, D.; Fabić, M.G.; Petrlić, A. Financial performance–efficiency nexus in public health services: A nonparametric evidence-based approach. *Econ. Res. Ekonomska Istraživanja* **2020**, *33*, 3334–3355. [CrossRef]
36. Gutierrez-Robredo, L. Morbilidad en la poblacion mayor. El proceso de la transición epidemiológica. *Demos* **2001**, *14*, 8–9.
37. Díaz, R. Satisfacción del paciente: Principal motor y centro de los servicios sanitarios. *Rev. Calid. Asist.* **2002**, *17*, 22–29. [CrossRef]
38. Pederson, J.L.; Warkentin, L.M.; Majumdar, S.R.; McAlister, F.A. Depressive symptoms are associated with higher rates of readmission or mortality after medical hospitalization: A systematic review and meta-analysis. *J. Hosp. Med.* **2016**, *11*, 373–380. [CrossRef]
39. Purcărea, V.L.; Gheorghe, I.R.; Petrescu, C.M. The Assessment of Perceived Service Quality of Public Health Care Services in Romania Using the SERVQUAL Scale. *Procedia Econ. Financ.* **2013**, *6*, 573–585. [CrossRef]
40. Yang, H.; Guo, X.; Wu, T. Exploring the influence of the online physician service delivery process on patient satisfaction. *Decis. Support Syst.* **2015**, *78*, 113–121. [CrossRef]
41. Vieco-Gómez, G.F.; Abello Llanos, R. Psychosocial factors at work, stress and morbidity around the world. *Psicol. Desde Caribe* **2014**, *31*, 354–385. [CrossRef]
42. Llobet, C.V.; Banqué, M.; Fuentes, M.; Ojuel, J. Morbilidad diferencial entre mujeres y hombres. *Anuario Psicol.* **2008**, *39*, 9–22.
43. Carreras, M.; Puig, G.; Sánchez-Pérez, I.; Inoriza, J.M.; Coderch, J.; Gispert, R. Morbidity and self-perception of health, two different approaches to health status. *Gac. Sanit.* **2020**, *34*, 601–607. [CrossRef]
44. Berta, P.; Ingrassia, S.; Punzo, A.; Vittadini, G. Multilevel cluster-weighted models for the evaluation of hospitals. *Metron* **2016**, *74*, 275–292. [CrossRef]
45. Mendoza, L.A.; Arias, M.; Mendoza, L.I. Hijo de madre adolescente: Riesgos, morbilidad y mortalidad neonatal. *Rev. Chil. Obstet. Ginecol.* **2012**, *77*, 375–382. [CrossRef]
46. Kashian, R.; Lovett, N.; Xue, Y. Has the affordable care act affected health care efficiency? *J. Regul. Econ.* **2020**, *58*, 193–233. [CrossRef]
47. Martínez, M.D.C.V.; Ramírez-Orellana, A. Patient Satisfaction in the Spanish National Health Service: Partial Least Squares Structural Equation Modeling. *Int. J. Environ. Res. Public Health* **2019**, *16*, 4886. [CrossRef]
48. Horvat, A.; Filipovic, J. Healthcare system quality indicators: The complexity perspective. *Total. Qual. Manag. Bus. Excel.* **2020**, *31*, 161–177. [CrossRef]
49. Amin, M.; Nasharuddin, S.Z. Hospital service quality and its effects on patient satisfaction and behavioural intention. *Clin. Gov. Int. J.* **2013**, *18*, 238–254. [CrossRef]
50. Huynh, H.P.; Sweeny, K.; Miller, T. Transformational leadership in primary care: Clinicians' patterned approaches to care predict patient satisfaction and health expectations. *J. Health Psychol.* **2018**, *23*, 743–753. [CrossRef]
51. Ricci-Cabello, I.; Stevens, S.; Dalton, A.R.H.; Griffiths, R.I.; Campbell, J.L.; Valderas, J.M. Identifying Primary Care Pathways from Quality of Care to Outcomes and Satisfaction Using Structural Equation Modeling. *Health Serv. Res.* **2018**, *53*, 430–449. [CrossRef]
52. DiMatteo, M. Variations in patients' adherence to medical recommendations: A quantitative review of 50 years of research. *Med. Care* **2004**, *42*, 200–209. [CrossRef]

53. Hunter-Jones, P.; Line, N.; Zhang, J.J.; Malthouse, E.C.; Witell, L.; Hollis, B. Visioning a hospitality-oriented patient experience (HOPE) framework in health care. *J. Serv. Manag.* **2020**, *31*, 869–888. [CrossRef]
54. Macrae, H. "It's my body, my future": Older women's views of their interactions with physicians. *J. Women Aging* **2016**, *28*, 211–224. [CrossRef] [PubMed]
55. Gulías, E.; Díez, N.; Pereira, M. Los componentes de la satisfacción de los pacientes y su utilidad para la gestión hospitalaria. *Rev. Española Cienc. Política* **2013**, *32*, 161–181.
56. Fernández-Pérez, Á.; Sánchez, Á. Improving People's Self-Reported Experience with the Health Services: The Role of Non-Clinical Factors. *Int. J. Environ. Res. Public Health* **2020**, *17*, 178. [CrossRef]
57. Bible, J.E.; Shau, D.N.; Kay, H.F.; Cheng, J.S.; Aaronson, O.S.; Devin, C.J. Are Low Patient Satisfaction Scores Always Due to the Provider? *Spine* **2018**, *43*, 58–64. [CrossRef]
58. Vámosi, M.; Lauberg, A.; Borregaard, B.; Christensen, A.V.; Thrysoee, L.; Rasmussen, T.B.; Ekholm, O.; Juel, K.; Berg, S.K. Patient-reported outcomes predict high readmission rates among patients with cardiac diagnoses. Findings from the DenHeart study. *Int. J. Cardiol.* **2020**, *300*, 268–275. [CrossRef]
59. Ferreira, D.; Marques, R.C. Do quality and access to hospital services impact on their technical efficiency? *Omega* **2019**, *86*, 218–236. [CrossRef]
60. Martini, G.; Berta, P.; Mullahy, J.; Vittadini, G. The effectiveness–efficiency trade-off in health care: The case of hospitals in Lombardy, Italy. *Reg. Sci. Urban Econ.* **2014**, *49*, 217–231. [CrossRef]
61. Kilpeläinen, K.; Tuomi-Nikula, A.; Thelen, J.; Gissler, M.; Sihvonen, A.-P.; Kramers, P.; Aromaa, A. Health indicators in Europe: Availability and data needs. *Eur. J. Public Health* **2012**, *22*, 716–721. [CrossRef]
62. Williams, L.J.; Vandenberg, R.J.; Edwards, J.R. 12 Structural Equation Modeling in Management Research: A Guide for Improved Analysis. *Acad. Manag. Ann.* **2009**, *3*, 543–604. [CrossRef]
63. Wold, H. Soft modelling: Intermediate between traditional model building and data analysis. *Banach Cent. Publ.* **1980**, *6*, 333–346. [CrossRef]
64. Chin, W.W. The partial least squares approach to structural equation modeling. In *Modern Methods for Business Research*; Marcoulides, G., Ed.; Lawrence Erlbaum Associates: Mahwah, NJ, USA, 1998; pp. 295–336.
65. Roldán, J.; Sánchez-Franco, M. Variance-based structural equation modeling: Guidelines for using partial least squares in information systems research. In *Research Methodologies, Innovations and Philosophies in Software Systems Engineering and Information Systems*; Mora, M., Steenkamp, A., Johnston, L., Gamon, J., Eds.; IGI Global: Hershey, PA, USA, 2012; pp. 193–221.
66. Henseler, J.; Ringle, C.M.; Sarstedt, M. Testing measurement invariance of composites using partial least squares. *Int. Mark. Rev.* **2016**, *33*, 405–431. [CrossRef]
67. Hair, J.F.; Risher, J.J.; Sarstedt, M.; Ringle, C.M. When to use and how to report the results of PLS-SEM. *Eur. Bus. Rev.* **2019**, *31*, 2–24. [CrossRef]
68. Wetzels, M.; Odekerken-Schröder, G.; Van Oppen, C. Using PLS Path Modeling for Assessing Hierarchical Construct Models: Guidelines and Empirical Illustration. *MIS Q.* **2009**, *33*, 177–195. [CrossRef]
69. Chin, W.W. How to Write Up and Report PLS Analyses. In *Handbook of Partial Least Squares*; Esposito, V.V., Chin, W.W., Henseler, J., Wang, H., Eds.; Springer: Berlin/Heidelberg, Germany, 2009; pp. 655–690.
70. Richter, N.F.; Cepeda, G.; Roldán, J.L.; Ringle, C.M. European management research using partial least squares structural equation modeling (PLS-SEM). *Eur. Manag. J.* **2016**, *34*, 589–597. [CrossRef]
71. Bollen, K.A.; Bauldry, S. Three Cs in measurement models: Causal indicators, composite indicators, and covariates. *Psychol. Methods* **2011**, *16*, 265–284. [CrossRef]
72. Müller, T.; Schuberth, F.; Henseler, J. PLS path modeling—A confirmatory approach to study tourism technology and tourist behavior. *J. Hosp. Tour. Technol.* **2018**, *9*, 249–266. [CrossRef]
73. Ringle, C.M.; Sarstedt, M.; Mitchell, R.; Gudergan, S.P. Partial least squares structural equation modeling in HRM research. *Int. J. Hum. Resour. Manag.* **2020**, *31*, 1617–1643. [CrossRef]
74. Grace, J.B.; Bollen, K.A. Representing general theoretical concepts in structural equation models: The role of composite variables. *Environ. Ecol. Stat.* **2008**, *15*, 191–213. [CrossRef]
75. Alfalla-Luque, R.; Machuca, J.A.; Marin-Garcia, J.A. Triple-A and competitive advantage in supply chains: Empirical research in developed countries. *Int. J. Prod. Econ.* **2018**, *203*, 48–61. [CrossRef]
76. Polites, G.L.; Roberts, N.; Thatcher, J. Conceptualizing models using multidimensional constructs: A review and guidelines for their use. *Eur. J. Inf. Syst.* **2012**, *21*, 22–48. [CrossRef]
77. Hair, J.F.; Hult, G.; Ringle, C.; Sarstedt, M. *A Primer on Partial Least Squares Structural Equation Modeling (Pls-SeM)*, 2nd ed.; Sage Publications: Thousand Oaks, CA, USA, 2017.
78. Sarstedt, M.; Hair, J.F.; Cheah, J.-H.; Becker, J.-M.; Ringle, C.M. How to Specify, Estimate, and Validate Higher-Order Constructs in PLS-SEM. *Australas. Mark. J.* **2019**, *27*, 197–211. [CrossRef]
79. Law, K.S.; Wong, C.-S.; Mobley, W.H. Toward a Taxonomy of Multidimensional Constructs. *Acad. Manag. Rev.* **1998**, *23*, 741. [CrossRef]
80. Edwards, J.R. Multidimensional Constructs in Organizational Behavior Research: An Integrative Analytical Framework. *Organ. Res. Methods* **2001**, *4*, 144–192. [CrossRef]

81. Roemer, E. A tutorial on the use of PLS path modeling in longitudinal studies. *Ind. Manag. Data Syst.* **2016**, *116*, 1901–1921. [CrossRef]
82. Faul, F.; Erdfelder, E.; Buchner, A.; Lang, A.-G. Statistical power analyses using G*Power 3.1: Tests for correlation and regression analyses. *Behav. Res. Methods* **2009**, *41*, 1149–1160. [CrossRef] [PubMed]
83. Carmines, E.; Zeller, R. *Reliability and Validity Assessment*; SAGE Publications Inc.: London, UK, 1979.
84. Garson, G. *Partial Least Squares: Regression & Structural Equation Models*; Statistical Publishing Associates: Asheboro, NC, USA, 2016; pp. 1–262.
85. Dijkstra, T.K.; Henseler, J. Consistent partial least squares path modeling. *MIS Q.* **2015**, *39*, 297–316. [CrossRef]
86. Henseler, J.; Ringle, C.M.; Sarstedt, M. A new criterion for assessing discriminant validity in variance-based structural equation modeling. *J. Acad. Mark. Sci.* **2015**, *43*, 115–135. [CrossRef]
87. Franke, G.; Sarstedt, M. Heuristics versus statistics in discriminant validity testing: A comparison of four procedures. *Internet Res.* **2019**, *29*, 430–447. [CrossRef]
88. Fornell, C.; Larcker, D.F. Evaluating Structural Equation Models with Unobservable Variables and Measurement Error. *J. Mark. Res.* **1981**, *18*, 39–50. [CrossRef]
89. Barclay, D.; Higgins, C.; Thomson, R. The partial least squares approach to causal modeling personal computer adoption and use as an illustration. *Technol. Stud.* **1995**, *2*, 285–309.
90. Henseler, J.; Ringle, C.M.; Sinkovics, R.R. The use of partial least squares path modeling in international marketing. In *New Advances in International Marketing*; Emerald Publishing Limited: Bingley, UK, 2009; Volume 20, pp. 277–319.
91. Hair, J.F.; Ringle, C.M.; Sarstedt, M. PLS-SEM: Indeed a Silver Bullet. *J. Mark. Theory Pract.* **2011**, *19*, 139–152. [CrossRef]
92. Hayes, A.F. Beyond Baron and Kenny: Statistical Mediation Analysis in the New Millennium. *Commun. Monogr.* **2009**, *76*, 408–420. [CrossRef]
93. Cepeda, G.; Nitzl, C.; Roldán, J. Mediation analyses in partial least squares structural equation modeling: Guidelines and empirical examples. In *Partial Least Squares Path Modeling: Basic Concepts, Methodological Issues and Applications*; Latan, H., Noonan, R., Eds.; Springer: Cham, Switzerland, 2017; pp. 173–195.
94. Falk, R.; Miller, N. *A Primer for Soft Modeling*; University of Akron Press: Akron, OH, USA, 1992.
95. Cohen, J. Set Correlation and Contingency Tables. *Appl. Psychol. Meas.* **1988**, *12*, 425–434. [CrossRef]
96. Cameron, C.M.; Purdie, D.M.; Kliewer, E.V.; McClure, R.J. Differences in prevalence of pre-existing morbidity between injured and non-injured populations. *Bull. World Health Organ.* **2005**, *83*, 345–352.
97. Leino-Kilpi, H.; Vuorenheimo, J. Patient satisfaction as an indicator of the quality of nursing care. *Nord. J. Nurs. Res.* **1992**, *12*, 22–28. [CrossRef]
98. Duggirala, M.; Rajendran, C.; Anantharaman, R. Patient-perceived dimensions of total quality service in healthcare. *Benchmarking Int. J.* **2008**, *15*, 560–583. [CrossRef]
99. Urrego, C.; Romero, M.; Murcia, Z.; Medina, C.; Marulanda, J.; Zerón, H. Evaluación de factores de riesgo asociados a enfermedad cardiovascular en jóvenes universitarios de la Localidad Santafé en Bogotá, Colombia. *Nova* **2016**, *14*, 35–45. [CrossRef]

Article

Integrating and Controlling ICT Implementation in the Supply Chain: The SME Experience from Baja California

Rubén Jesús Pérez-López [1], Jesús Everardo Olguín-Tiznado [2], Jorge Luis García-Alcaraz [3,4,*], María Mojarro-Magaña [1], Claudia Camargo-Wilson [2] and Juan Andrés López-Barreras [5]

[1] Department of Industrial Engineering, Technological Institute of Ciudad Guzmán, Ciudad Guzmán 49100, Mexico; ruben.pl@cdguzman.tecnm.mx (R.J.P.-L.); maria.mm@cdguzman.tecnm.mx (M.M.-M.)
[2] Faculty of Engineering, Architecture, and Design, Autonomous University of Baja California, Ensenada 22860, Mexico; jeol79@uabc.edu.mx (J.E.O.-T.); ccamargo@uabc.edu.mx (C.C.-W.)
[3] Department of Industrial Engineering and Manufacturing, Autonomous University of Ciudad Juárez, Ciudad Juárez 32310, Mexico
[4] Division of Research and Postgraduate Studies, Tecnológico Nacional de México/Instituto Tecnológico de Ciudad Juárez, Ciudad Juárez 32500, Mexico
[5] Faculty of Engineering and Chemical Sciences, Autonomous University of Baja California, Tijuana 22390, Mexico; jalopez@uabc.edu.mx
* Correspondence: jorge.garcia@uacj.mx or jorge.ga01@itcj.edu.mx; Tel.: +52-656-6884843 (ext. 5430)

Abstract: It is mentioned that companies' competition is currently more associated with supply chains (SC) than production processes since sometimes logistics costs represent up to 70% of the total production cost in a product. To improve efficiency in SC, companies are implementing information and communication technologies (ICT). This paper reports a structural equation model that incorporates four latent variables related to ICT applied in SC: technological innovation, information management, and information availability as independent variables, and operating benefits gained as a dependent variable. These variables are related using six hypotheses that are validated using information obtained from 80 responses to a survey applied to small and medium-sized enterprises in Baja California (Mexico). The partial least squares technique is used to validate the hypotheses in the structural equation model. Findings indicate that technological innovation is the basis for the successful implementation of ICT and its application guarantees greater information availability and efficient management, leading to obtaining operating benefits in SC.

Keywords: structural equation model; information and communication technology; ICT integration

1. Introduction

Companies seek to optimize supply chain (SC) operations, and innovation is essential to ensure survival in the global market. Therefore, there is currently a frequent growing use of computer packages to support the control in a production system. Moreover, empirical evidence shows that most small and medium-sized enterprises (SMEs) implementing innovation activities will improve their production processes and, consequently, the financial income.

For such reason, information systems that generate knowledge that guide organizations towards a vision of "up-to-date information and communication technology (ICT) tools" should be investigated. These ICTs enable SC members to improve product flow, services, and information in real-time [1]. ICTs have evolved together with a complexity value chain (VC) of products or services offered by companies driven by economic and social changes and will increase in pace as they are applied in production systems [2].

ICTs provide mechanisms for companies to collect, store, access, share, analyze, and control information, improving SC performance [3]. Therefore, ICT facilitates the decision-making process to maximize business and SC profitability. However, social benefits are

also obtained, increasing information and knowledge exchange, accident reduction rates, and better motivation in human resources.

From the beginning, David et al. [4] indicated that ICT integrated into SC offers a reduction in cycle times and inventories, minimizes the whiplash effect, and allows collaboration among SC partners, among others. Here, ICT integration can be defined as the degree to which information systems are related to the company's internal functions and with SC members [3,5]. According to Llach and Alonso-Almeida [6], ICTs increase the efficiency of the organization's internal processes and the SC integration, facilitate the activities redesign, improving time, practicality, and accuracy.

These benefits are obtained since ICT facilitates the electronic data interchange (EDI) in real-time and the decision-making process in SC [7]. Therefore, using ICT in production processes, combined with human skills, generates *Operating benefits* in the SC, such as improving the process execution, management, and materials availability, decreasing information cost, improving quality by providing access to information, and increasing competitiveness [6].

In the last three decades, Mexico has had a boom in ICT adoption. Several types of research have studied this phenomenon in SC, associated with integration and *Information management* through causal models; other investigations analyze the critical factors for its implementation, such as *Technological innovation, Information availability*, and *Information management* [8,9].

Although Hallikas, Korpela, Vilko and Multaharju [1] mention that integrated process activities using ICT, information gathering, and sharing in SC must be studied, they recognize that there is not enough information on ICT control and management in SC. In addition, there is little empirical evidence on the contribution of ICT in SC performance in SMEs, so there is a need to improve the understanding of the impact of using ICT in SC, where the lag of technology has left a wide gap.

One SME sector with ICT widely applied is the maquiladora industry (MI), which is defined as a subsidiary company with its headquarters in a country other than the one in which it is established. Specifically, there are 5171 companies of this type in Mexico, generating 2,702,116 direct jobs and are based mainly on the country's northern part [10]. Those MI are close to the United States of America as the primary world market and take advantage of skilled and low-cost labor and tariff preferences due to free trade agreements between the two countries.

Specifically, in Baja California state (Mexico), there are 931 MIs, which generate 358,889 direct jobs, with Tijuana and Mexicali being the most representative cities, with 72,605 and 252,902 direct jobs, respectively [11]. This MI is characterized by importing its raw materials and exporting its finished products. Hence, communication with customers and suppliers outside the MI is intense due to their geographical characteristics and internal departments. Therefore, for SMEs, ICT is essential in their production and administrative processes for information flow horizontally and vertically.

However, even given the importance of IM in Baja California, few studies have focused on studying the ICT implementation. The most representative research by Pérez-López et al. [12] focuses on investigating the relationship between the planning process and the operational benefits obtained. Other studies in the region focus on studying ICT in other sectors, such as education [13], restaurants [14], or the integration of a cluster specialized in ICT [15].

Hence, this article aims to analyze three critical success factors in ICT implementation in MI in Baja California (Mexico): *Technological innovation, Information availability*, and *Information management* as independent variables, and to determine their impact on *Operating benefits* obtained in the SC as the dependent variable.

2. Literature Review and Hypotheses

2.1. Technological Innovation (TIN)

Technological innovation is defined as a kind of goods or services. Due to its novelty or degree of improvement, it will benefit end-users and meet their needs [16]. In addition, these innovations in the implementation and adoption processes develop new knowledge through internal engineering and development departments and the interaction of the company with external partners, such as society, government, clients, and suppliers [17].

For companies, innovation is vital to increase productivity and raise the competitiveness of their processes, so there is a need to promote and encourage ICTs to perform business. Orji et al. [18] express that innovation has a positive effect on SC economic performance.

2.2. Information Availability (IAV)

Information is essential for decision-making in any business. The availability, speed, accuracy, and visibility of this information are vital, and its analysis is currently part of the administrative tasks in SC [1]. Dutta et al. [19] point out that companies have integrated ICT devices to acquire, control, coordinate, and manage information from their operational processes, facilitating the decision-making process in Indian manufacturing companies. Therefore, *Technological innovation* used in SC allows having *Information availability* to link activities in production processes [20], so the following hypothesis is proposed:

Hypothesis 1 (H1). *Technological innovation in SC has a direct and positive effect on Information availability in SMEs.*

2.3. Information Management (IMA)

Information is commonly used for decision-making. That is, it is managed to improve customer relationship management (CRM). *IMA* has been developed in recent years, combined with strategy and technology to create, improve customer and supplier relationships, which maximizes the value generated, trust, and cooperation between partners. Thus, *Information management* is a competitive advantage for companies with more dynamic processes and supports a quick response to uncertainty [21]. However, this *Information management* depends on the levels of *Technological innovation* that the company has, so the following hypothesis is proposed:

Hypothesis 2 (H2). *Technological innovation in SC has a direct and positive effect on Information management performed in SMEs.*

Madonsela [22] states that investors trust that *Information management* with the use of ICT guarantees return on investment (ROI), while Llach and Alonso-Almeida [6] state that human ability is increased. However, *Information availability* allows partners to share operational, tactical, and strategic information to decrease uncertainty in SC, hence its importance. Some authors mention that the alliance between *Information availability* and *Information management* is fundamental for the success of companies [23]. In addition, some state that *Information management* for mutual growth among organizations depends on the accessibility to information, so the following hypothesis is proposed:

Hypothesis 3 (H3). *Information availability in SC has a direct and positive effect on Information management in SMEs.*

2.4. Operating Benefits (OBE)

Integrating and using ICT in SC to interact and coordinate activities generates *Operating benefits* in the short and long term [7], which means that logistics and transport operations are places to share and transfer data, thus improving operational performance.

Haj and Dhiaf [20] conclude that managers should be aware of the benefits of integrating ICT and performance improvement in their operations. In this sense, ICT is identified as the innovation to perform information exchange among companies, depending on the companies' technological and partners' integration level. Moreover, Madonsela [22] indicates that companies that invest and use innovative ICT grow faster and are more productive and profitable. Therefore, the following hypothesis is proposed:

Hypothesis 4 (H4). *Technological innovation in SC has a direct and positive effect on the Operating benefits obtained by SMEs.*

Obtaining information from the SC is not enough. It must be available for the SC members to facilitate decision-making and generate operational performance in terms of time savings in task executions, reduction of errors by handling information, and, above all, lower costs [1]. Kumar et al. [24] agree that ICT integration in production processes and *Information availability* add value to the product or service, affecting operations, customers, suppliers, and internal collaborators, so the following hypothesis is proposed:

Hypothesis 5 (H5). *Information availability in SMEs has a direct and positive effect on Operating benefits.*

Petrick, Maitland and Pogrebnyakov [7] indicate that *Information management* in the company's operational areas produces benefits and allows supervision, management, and operations control. Likewise, Partanen et al. [25] mention that most industries implement ICT to generate competitive advantage, which leads to business development and management of the organization's information system. At the same time, one of the leading *Operating benefits* offered by *Information management* within the SC comes from coordinated decision making, as it is the key to finding a skillful SC [25–27]. Therefore, the following hypothesis is proposed:

Hypothesis 6 (H6). *Information management in SC directly and positively affects the Operating benefits obtained in SMEs.*

Figure 1 graphically represents the hypotheses regarding the analysis variables mentioned above.

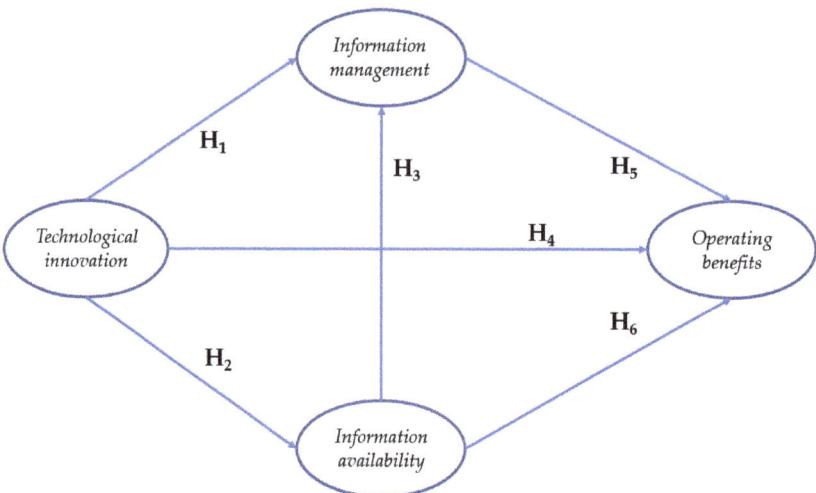

Figure 1. Proposed hypotheses.

3. Methodology

3.1. Questionnaire Preparation

A questionnaire was designed to obtain information about using ICT in SC activities in SMEs in Baja California, Mexico. Databases such as Elsevier, SpringerLink, Emerald, and EBSCOhost were consulted for generating the questionnaire. The scientific articles selection related to ICT was made using these keywords: ICT integration, ICT adoption, and ICT in SC. A list with 145 articles was obtained and reviewed but was reduced to only 55. In addition, only articles written in English and published in indexed journals were analyzed.

The questionnaire was integrated into six sections, and the first section sought to obtain demographic information associated with the respondent, industrial sector, respondent's profile, and age. From the second to the fourth section, information on the planning, execution, and control of ICT in the SC were obtained; the fifth and sixth sections were related to the benefits, with 91 items. However, in this report, only items corresponding to the control stage were analyzed, where the following latent variables were studied:

1. *Technological innovation*, with four items [17,20]
2. *Information availability*, with eight items [1,5,20,26]
3. *Information management*, with eight items [6,21,22,25]
4. *Operating benefits*, with five items [1,7,20,22,25,26]

The list of items belonging to each latent variable appears in the descriptive analysis of the items (Results section), but they were omitted here due to space restrictions.

3.2. Application and Data Capture

The questionnaire was uploaded to the Google forms platform for its application. The items must be answered using a five-point Likert scale. One indicates that the activity is not performed or that no benefits are obtained. Five means that this activity is always performed or that the benefit is always obtained [28]. Several authors have used that scale in productivity and manufacturing research in recent years [29,30].

An email was sent to managers identified in Baja California (Mexico) laboring for SMEs, explaining the study's objective and inviting them to participate. The platform was available from January to April 2020 to collect responses.

3.3. Debugging of the Database

An Excel file was downloaded from the platform, and questionnaires with duplicate responses were identified and discarded for future analysis. Missing and extreme values were determined and replaced by the median [31]; likewise, the standard deviation of each case (questionnaire answered) was obtained for non-compromised responders identification, and those with values lower than 0.5 were discarded from the analysis.

3.4. Validation of Variables

Latent variables were validated according to the following indexes [32]: R-squared and adjusted R-squared to measure parametric predictive validity, and values greater than 0.2 were accepted; Q-squared to measure nonparametric predictive validity, and values greater than zero, and similar to R-squared were taken; Cronbach's alpha and composite reliability index to measure internal validity, and values larger than 0.7 were accepted; average variance extracted (AVE) for convergent validity and values greater than 0.5 were accepted; variance inflation factor (VIF) to measure collinearity, and values lower than 3.3 were taken.

3.5. Descriptive Analysis of the Sample

A descriptive analysis was performed with the debugged database in the SPSS 23® statistical software due to its wide acceptance, friendly interphase, and use in scientific reports [33,34]. This analysis was performed with the demographic information obtained which makes it possible to characterize the sample using crosstabs.

3.6. Descriptive Analysis of Items

Central tendency and dispersion measures for items in latent variables were obtained. The median was obtained to measure central tendency, given that data are received on an ordinal scale as an assessment. High median values indicate that activities are always performed, or that benefits are always obtained, while low values indicate the opposite. The arithmetic mean was not used because the values are not on an interval scale [30].

As a measure of dispersion for every item, the interquartile range (IR) was obtained. A high value indicates low consensus among the surveyed persons about the value in an item, while low values indicate high agreement.

3.7. Structured Equation Modelling (SEM)

To validate hypotheses in Figure 1, the WarpPLS 5.0® software was used based on partial least squares (PLS), recommended for small samples by Kock [32]. Before interpreting the structural model, the following model efficiency indices with a significance level of 0.05 were evaluated [30]: average path coefficient (APC) to know the average predictive model validity; average R-squared and average adjusted R-squared (AARS) are measures of the explanatory power of the model; block average variance inflation index (AVIF) and VIF full collinearity index (AFVIF), which measures the collinearity between items in a latent variable, the ideal value should be less than 3.3; Tenenhaus index (GoF), which measures data fit to the model, the appropriate value should be greater than 0.36.

To validate the relationships or hypotheses, the direct effects between latent variables were measured (see Figure 1), estimating a standardized regression index β and tested using the following null hypothesis: $\beta_i = 0$ versus the alternative hypothesis: $\beta_i \neq 0$. If the hypothesis test indicates that $\beta_i \neq 0$, then there is statistical evidence to declare a relationship between two variables.

The relationship between the latent variables was measured by these three types of effects [30,35]: direct effects indicate the impact between latent variables and validate the hypotheses proposed; indirect effects between variables occur through a third variable called mediator, and at least one mediator can appear. Total effects are the sum of direct and indirect effects in each relationship analyzed.

3.8. Sensitivity Analysis

A sensitivity analysis was performed to determine the situations in certain scenarios of latent variable occurrence. In this case, since PLS uses standardized values for estimations, a probability less than minus one in a latent variable is considered low scenario $P(X_i < -1)$. In contrast, a probability greater than one is a high scenario $P(X_i > 1)$ [32]. Thus, conditional probabilities for the following scenarios were estimated for each relationship: $P(Z_d > 1)/P(Z_i > 1)$, $P(Z_d > 1)/P(Z_i < -1)$, $P(Z_d < -1)/P(Z_i > 1)$ and $P(Z_d < -1)/P(Z_i < -1)$; where, Z_d represents the standardized value for a dependent variable, and Z_i is for the independent variable. Moreover, probabilities of find variables co-occurring in high and low scenarios were reported, whereby we obtain: $P(Z_i > 1) \cap P(Z_d > 1)$, $P(Z_i > 1) \cap P(Z_d < -1)$, $P(Z_i < -1) \cap P(Z_d > 1)$ and $P(Z_i < -1) \cap P(Z_d < -1)$.

4. Results

4.1. Descriptive Analysis of the Sample

After debugging the database, 80 valid surveys were obtained from different SMEs established in Baja California, Mexico. Table 1 illustrates the descriptive analysis, indicating the industrial sector and the respondent's position. We observed that manufacturing industries represent 45% of participation, and the sectors of the food, textile, computer equipment, and electronic accessories industries represent 35%; the other sectors only hold 20%. Regarding the job position, supervisors represent 44%, general managers with 30%, and production managers with 26%.

Table 1. Industrial sector and job position.

Sector	General Managers	Production Manager	Supervisor	Total
Manufacturing industries	8	11	17	36
Food industries	4	1	2	7
Textile industries	2	5	0	7
Computer equipment manufacturing	1	1	5	7
Electronic accessories manufacturing	2	1	4	7
Plastic industries	1	2	3	6
Metal product industries	2	0	3	5
Printing and related industries	2	0	1	3
Non-metallic mineral industries	1	0	0	1
Furniture, mattresses, and blinds industries	1	0	0	1
Total	24	21	35	80

Table 2 shows the years of experience in the job and the gender. Results indicate that 37% of responders had less than two years of experience, followed by 33% of responders with 2 to 4 years, and 24% of those with more than five years. With regards to gender, women participated with 32%, while men participated with 68%.

Table 2. Years on the position.

| Sex | Time | | | | |
	<2 Years	2–5 Years	5–10 Years	>10 Years	Total
Male	18	21	7	8	54
Female	12	6	3	5	26
Total	30	27	10	13	80

4.2. Variables Validation

Table 3 shows the values obtained from the variable validation process. According to the methodology proposed, the model has parametric predictive validity because R-squared and adjusted R-squared are greater than 0.2. Overall, the variables comply with all the required validation indices (please see the cut-off value column).

4.3. Descriptive Analysis

Table 4 shows the median and interquartile range for items analyzed in latent variables. The second column illustrates the median. It is observed that six items have values greater than four, indicating that, based on the respondent's perception, those activities are essential and frequently performed in the company, or the benefits are commonly gained. Two values that are larger than four are associated with *Information availability*, and four are related to the *Operating benefits* gained. However, the lowest median value is related to items *search for and renew the most modern information technology*, and this means that SMEs have insufficient resources for renewing ICT frequently.

Table 3. Validation indexes for latent variables.

Indices	Cut-Off Value	Latent Variables			
		TIN	IAV	IMA	OBE
R-squared	≥0.2		0.61	0.529	0.757
Adjusted R-squared	≥0.2		0.605	0.517	0.748
Reliability index	≥0.7	0.951	0.920	0.889	0.956
Cronbach's Alpha	≥0.7	0.939	0.895	0.750	0.908
Average variance extracted	≥0.5	0.733	0.659	0.800	0.915
Variance inflation index	≤3.3	3.231	2.832	2.378	3.816
Q-squared	≥0.2		0.611	0.529	0.758

Table 4. Data descriptive analysis.

Items	Median	Interquartile Range
Technological innovation		
Keep a robust data network with suppliers and customers to monitor and evaluate the exchange of information	3.745	1.748
Efficient use of data exchange	3.714	1.695
Provide maintenance to the information system	3.558	1.705
Search for and renew the most modern information technology	3.533	1.921
Information availability for or from		
Customer order follow-up	4.164	1.413
Fulfillment and delivery order management	4.018	1.478
From commercial customers	3.904	1.507
Customer demand management	3.887	1.502
Inventory management materials	3.830	1.490
Suppliers and customers in product development processes	3.745	1.621
Suppliers	3.694	1.710
Management of supplier activities and relationships	3.569	1.526
Information management		
The Internet capacity of the organization	3.977	1.620
They are providing high-quality services	3.920	1.553
Purchase and sales systems	3.863	1.575
The internal computer network system	3.830	1.637
The planning and scheduling of the organization's activities	3.808	1.567
The various internal information systems	3.745	1.483
The software used in the information system	3.725	1.635
Electronic market trends	3.634	1.967
Warehouse management systems	3.609	1.850
Operating benefits		
Customer response flexibility	4.123	1.609
Flexibility of the systems to meet customer needs	4.091	1.630
Cost competitiveness	4.038	1.745
Strengthening the relationship with customers	4.000	1.629
Shorter order cycles	3.878	1.820

The third column illustrates the interquartile range. The highest IR value is 1.967, meaning that responders probably had doubts interpreting the electronic market trends. The lowest value is 1.397, referring to customer order follow-up; it indicates that there was a lot of agreement and consensus among respondents for the value in that item.

4.4. Model Efficiency Indices

The model efficiency indices appear in Table 5. The average path coefficient has a *p*-value lower than 0.001, indicating that the model has predictive validity. Similarly, the average R-squared (ARS) and adjusted R-squared (AARS) have an associated *p*-value lower than 0.001, indicating that the model has sufficient explanatory power.

Table 5. Model efficiency indices.

Indices	Results	Cut-Off Value
Average path coefficient (APC)	(APC) = 0.422, $p < 0.001$	$p \leq 0.05$
Average R-squared (ARS)	(ARS) = 0.632, $p < 0.001$	$p \leq 0.05$
Average adjusted R-squared (AARS)	(AARS) = 0.623, $p < 0.001$	$p \leq 0.05$
Average block VIF (AVIF)	(AVIF) = 2.435	Acceptable if ≤ 3.3
Average full collinearity VIF (AFVIF)	(AFVIF) = 3.064	Acceptable if ≤ 3.3
Tenenhaus GoF (GoF)	(GoF) = 0.701	Better if ≥ 0.36

Moreover, the average block variance inflation factor index (AVIF) and the VIF full collinearity index (AFVIF) have values lower than 3.3, concluding that there are no collinearity problems among latent variables analyzed. Similarly, the Tenenhaus index (GoF) shows that model has a good data fit since its value is greater than 0.36. Figure 2 illustrates the β, the *p*-value associated with relationships among latent variables as direct effects or hypotheses, and the R^2.

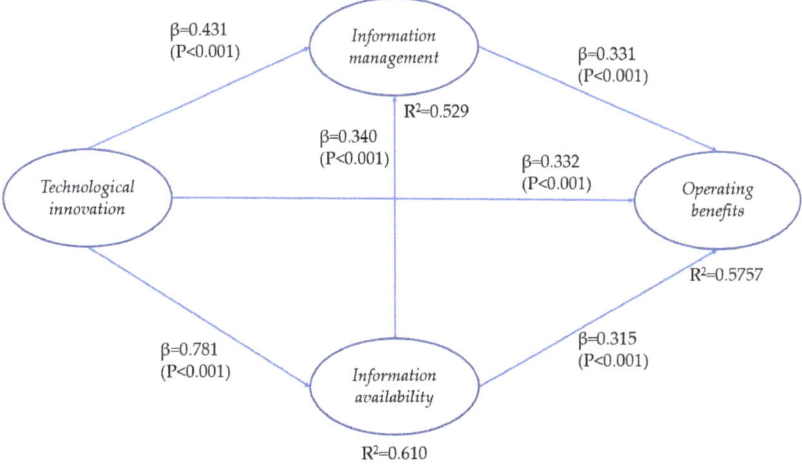

Figure 2. Structural equation model.

4.5. Direct Effects

Table 6 summarizes the results of direct effects between latent variables, indicating the associated *p*-value. According to the *p*-values associated with every β estimated, it is concluded that all relationships tested must be accepted (see the last column named decision). For example, the following conclusion can be done for H1: there is enough statistical evidence to declare that *Technological innovation* in SC has a direct and positive effect on *Information availability* they possess in SMEs since when the first latent variable increases its standard deviation by one unit, the second one goes up by 0.781 units.

Table 6. Direct effects.

Hi	Independent Variable	Dependent Variable	β (p-Value)	Decision
H1	Technological innovation	Information availability	0.781 (<0.001)	Accept
H2	Technological innovation	Information management	0.431 (<0.001)	Accept
H3	Information availability	Information management	0.340 (<0.001)	Accept
H4	Technological innovation	Operating benefits	0.332 (<0.001)	Accept
H5	Information availability	Operating benefits	0.315 (<0.001)	Accept
H6	Information management	Operating benefits	0.331 (<0.001)	Accept

4.6. Sum of Indirect Effects

The sum of indirect effects among latent variables appears in Table 7, where two of them are statistically significant, and one is not. For example, the indirect relationship between *Technological innovation* and *Information management* given by *Information availability* as a mediator variable has β = 0.265, with *p*-value < 0.001, and is statistically significant. It is the same case for the relationship between *Technological innovation* and *Operating benefits*, which is given through *Information availability* and *Information management* as mediator variables and has β = 0.477, and *p*-value < 0.001, being statistically significant. On the contrary, the indirect relationship of *Information availability* to *Operating benefits* has β = 0.113 with *p* = 0.073 (greater than 0.05), so it is concluded that there are no indirect effects between those variables.

Table 7. Sum of indirect effects.

To	From	
	TIN	IAV
IMA	0.265 (*p* < 0.001)	
OBE	0.477 (*p* < 0.001)	0.113 (*p* = 0.073)

4.7. Total Effects

Table 8 illustrates the total effects among variables and the associated *p*-value. It is observed that all the total effects are statistically significant, even though one of the indirect effects was not. For example, the relationship between *Technological innovation* and *Operating benefits* has a total effect of β = 0.809 with *p*-value < 0.001, indicating a significant total effect between those variables. We observed that the indirect relationship between *Information availability* and *Operating benefits* was statistically non-significant, but the direct effect and total effects were significant.

Table 8. Total effects.

To	From		
	TIN	IAV	IMA
IAV	0.781 (*p* < 0.001)		
IMA	0.697 (*p* < 0.001)	0.340 (*p* < 0.001)	
OBE	0.809 (*p* < 0.001)	0.428 (*p* < 0.001)	0.331 (*p* < 0.001)

4.8. Sensitivity Analysis

Table 9 illustrates the results of the sensitivity analysis. In this case, the independent variables appear in columns and the dependent variables in rows. The conditional probability value is indicated by the term "If", while the joint or simultaneous probability is marked as "&". Thus, the probability of finding *TIN*+ and *IMA*+ together is only 0.076, but the probability of having *IMA+*, given that *TIN+* has occurred, is 0.545. However, if *TIN+* has occurred, there is only a 0.091 probability of getting *IMA-*, which reassures managers that as long as they invest in innovative ICT, getting poor *Information management* has a low occurrence probability and guarantees ICT implementation. The other relationships between variables are interpreted similarly.

Table 9. Sensitivity analysis.

			TIN		IMA		IAV	
			+	−	+	−	+	−
		Value	0.139	0.165	0.165	0.089	0.165	0.203
IMA	+	0.165	& = 0.076	& = 0.013				
			If = 0.545	If = 0.091				
	−	0.089	& = 0.013	& = 0.063				
			If = 0.091	If = 0.385				
IAV	+	0.165	& = 0.076	& = 0.000	& = 0.089	& = 0.000		
			If = 0.545	If = 0.000	If = 0.538	If = 0.000		
	−	0.203	& = 0.000	& = 0.101	& = 0.013	& = 0.051		
			If = 0.000	If = 0.615	If = 0.063	If = 0.250		
OBE	+	0.177	& = 0.101	& = 0.000	& = 0.089	& = 0.013	& = 0.101	& = 0.000
			If = 0.727	If = 0.000	If = 0.538	If = 0.143	If = 0.615	If = 0.000
	−	0.177	& = 0.000	& = 0.114	& = 0.000	& = 0.076	& = 0.000	& = 0.076
			If = 0.000	If = 0.692	If = 0.000	If = 0.857	If = 0.000	If = 0.375

5. Conclusions

Descriptive analysis in Table 4 indicates that MSME managers are sure about the benefits that ICTs offer in production and management systems. Four of the five items analyzed have medians higher than four, and the most important benefits are the flexibility they can have with customers. That assertion is confirmed by noting that the other items with a median greater than four are in the *Information availability* variable and are associated with customer order follow-up and order management and delivery times. In other words, managers focus on having a close relationship with their customers.

From the sensitivity analysis, the following can be concluded and recommended:

Managers should encourage the implementation of innovative ICT in their production, administrative, and SC systems, as this ensures adequate *Information management* (If = 0.545) and wide *Information availability* (If = 0.545). In addition, *Operating benefits* are guaranteed at high levels (If = 0.727). However, if there are low ICT innovative levels, then it is probable to have low levels of *Information management* (If = 0.385) and *Information availability* (If = 0.615), and also *Operating benefits* are not guaranteed (If = 0.692).

Companies that achieve an adequate and high level of *Information management* can guarantee a higher *Information availability* (If = 0.538) and obtain *Operating benefits* (If = 0.538). However, if they do not achieve adequate *Information management*, in that case, they may have a low *Information availability* (If = 0.250) and, above all, they may not obtain the expected *Operating benefits* (If = 0.857), which represents a high risk.

Finally, companies with higher *Information availability* always obtain higher *Operating benefits* (If = 0.615), and there is no possibility of these being low (If = 0.000). Likewise, if there are low levels of *Information availability*, in that case, high levels of *Operating benefits* (If = 0.000) can never be obtained. On the contrary, there is a possibility that these will be low (If = 0.375).

Regarding the six hypotheses initially proposed, all have been statistically tested, and based on this, the following inferences are made:

The MI as SMEs in Baja California, Mexico should focus their efforts on achieving an adequate level of implementation of innovative ICT in their production and supply chain systems. That effort enables them to facilitate relationships with suppliers, customers, government and society, and departments within the company by managing inventories, tracking orders, and delivery time, as shown in hypothesis H_1. In other words, this means that MIs in Baja California are currently developing a digital environment by incorporating ICT in SC activities and managing information in SC operations through appropriate technology, and greater competitiveness and productivity indexes are obtained. Hence, collaborators in the production processes and the SC members are motivated since they have reliable and timely information to implement activities and control operations, reducing uncertainty.

The integration of innovative ICT in SC activities represents a potential opportunity for MIs; however, the information generated should be made available cautiously to other members, as demonstrated by hypothesis H_2. The practical implication is that MIs are not excluded from integrating concepts and technology of Industry 4.0 because the *Technological innovation* level is directly related to higher sales and online purchases and inventory management. Our findings are also empirical support to Toudert [36], who mentions that despite the digital gap in Mexico, companies show interest in integrating ICTs, which is reflected in more excellent digital communication and having more information related to their production systems.

The information must be available, but there must be adequate management of it, as demonstrated statistically and empirically in hypothesis H_3. These findings are similar to those reported by Durugbo [37], who also indicates that this information must be accurate, complete, transparent, and rapidly disseminated to all concerned parties. Likewise, Avelar-Sosa, García-Alcaraz, Cedillo-Campos and Adarme-Jaimes [29], indicate that *Information availability* minimizes the SC risk level due to good communication among members.

Undoubtedly, implementing ICT in administrative, productive, and SC systems requires a robust economic investment, considered by many companies as a risk. Our research shows that *Technological innovation* is directly related to the *Operating benefits* obtained, as indicated by hypothesis H_4. In other words, ICTs allow companies to generate better flexibility to meet customer needs since many times they must only reschedule production programs, allowing shorter order cycles and having more extensive market coverage. Our findings coincide with García-Alcaraz et al. [38]. They indicate that ICTs implementation speeds up the new product development process. In global terms, Appiah-Otoo and Song [39] associate ICTs with the overall growth of countries.

Having ICTs in production and administrative systems to obtain information is not enough. The justification of these ICTs is only achieved if this information is available and generates benefits. Our study has shown statistically and empirically that these two variables are related, as indicated in hypothesis H_5. Our findings agree with Ali and Kumar [40], who suggest that ICTs have favored farmers' decision-making process in SC, while Ansola et al. [41] point out that ICTs have improved airport operations.

In conclusion, it has been statistically proven that proper *Information management* impacts the *Operating benefits* obtained and that the information systems implemented by the MI favor cost competitiveness and strengthen the relationships with customers and suppliers in a closer way.

6. Limitations and Future Research

This research has reported the ICT control process in MI and the operational benefits, where it is assumed that the planning and implementation processes have already been carried out. However, future research requires an integrative structural equation model in which all stages of the ICT adoption process are analyzed.

Moreover, only the operational benefits have been analyzed in this report; however, the economic benefits obtained should also be explored. Thus, future research will be focused on social and commercial benefits and how these are converted into higher financial incomes.

Author Contributions: Conceptualization, J.E.O.-T. and R.J.P.-L.; methodology, R.J.P.-L. and J.L.G.-A.; software, J.L.G.-A.; validation, M.M.-M. and C.C.-W.; formal analysis, J.L.G.-A. and J.A.L.-B.; investigation, R.J.P.-L.; resources, J.E.O.-T.; data curation, M.M.-M.; writing—original draft preparation, J.L.G.-A. and J.A.L.-B.; writing—review and editing, J.L.G.-A.; visualization, M.M.-M.; supervision, J.E.O.-T.; project administration, J.E.O.-T.; funding acquisition. All authors have read and agreed to the published version of the manuscript.

Funding: This research was funded by the Mexican National Council for Science and Technology (CONACYT) with a scholarship for Rubén Jesús Pérez-López (grant: 542000) and María Mojarro Magaña (grant: 541999) for their doctoral studies at the Autonomous University of Baja California.

Institutional Review Board Statement: Not applicable.

Informed Consent Statement: Not applicable.

Data Availability Statement: The database containing the information analyzed in the structural equation model is available in an Excel sheet at this doi:10.17632/pssdjd5fxx.1.

Acknowledgments: No applicable.

Conflicts of Interest: The authors declare no conflict of interest.

References

1. Hallikas, J.; Korpela, K.; Vilko, J.; Multaharju, S. Assessing Benefits of Information Process Integration in Supply Chains. In Proceedings of the Procedia Manufacturing, Chicago, IL, USA, 9–14 August 2019; pp. 1530–1537.
2. Al-Gahtani, S.S. Empirical investigation of e-learning acceptance and assimilation: A structural equation model. *Appl. Comput. Inform.* **2016**, *12*, 27–50. [CrossRef]
3. Swafford, P.M.; Ghosh, S.; Murthy, N. Achieving supply chain agility through IT integration and flexibility. *Int. J. Prod. Econ.* **2008**, *116*, 288–297. [CrossRef]
4. David, S.-L.; Philip, K.; Edith, S.-L. *Designing and Managing the Supply Chain: Concepts Strategies and Case Studies*; McGraw Hill: New York, NY, USA, 2000.
5. Li, S.; Lin, B. Accessing information sharing and information quality in supply chain management. *Decis. Support. Syst.* **2006**, *42*, 1641–1656. [CrossRef]
6. Llach, J.; Alonso-Almeida, M.D.M. Integrating ICTs and Supply Chain Management: The Case of Micro-Sized Firms. *Hum. Factors Ergon. Manuf. Serv. Ind.* **2015**, *25*, 385–397. [CrossRef]
7. Petrick, I.; Maitland, C.; Pogrebnyakov, N. Unpacking Coordination Benefits in Supply Networks: Findings from Manufacturing SMEs. *J. Small Bus. Manag.* **2016**, *54*, 582–597. [CrossRef]
8. Martínez-Domínguez, M.; Mora-Rivera, J. Internet adoption and usage patterns in rural Mexico. *Technol. Soc.* **2020**, *60*, 101226. [CrossRef]
9. Cuevas-Vargas, H.; Estrada, S.; Larios-Gómez, E. The Effects of ICTs As Innovation Facilitators for a Greater Business Performance. Evidence from Mexico. *Procedia Comput. Sci.* **2016**, *91*, 47–56. [CrossRef]
10. IMMEX. *INDEX Juárez—Monthly Statistic Information (22 January 2021)*; Asociación de Maquiladoras A.C.: Ciudad Juárez, Mexico, 2021; pp. 1–3.
11. IMMEX. *Monthly Statistical Information—Employment*; Maquiladora Association A.C.: Ciudad Juárez, Mexico, 2021; p. 4.
12. Pérez-López, R.J.; Olguín Tiznado, J.E.; Mojarro Magaña, M.; Camargo Wilson, C.; López Barreras, J.A.; García-Alcaraz, J.L. Information Sharing with ICT in Production Systems and Operational Performance. *Sustainability* **2019**, *11*, 3640. [CrossRef]
13. Sandoval Bringas, J.A.; Carreño León, M.A.; Alvarez Rodriguez, F.J. Analysis of the Perception of Students of the Autonomous University of Baja California Sur for the Use of m-Learning. In *Advances in Intelligent Systems and Computing*; AISC: Chicago, IL, USA, 2021; Volume 1192, pp. 102–108.
14. Cruz Estrada, I.; Miranda Zavala, A.M. ICT adoption in restaurants in Puerto Nuevo, Rosarito, Baja California. *Rev. Innovar* **2019**, *29*, 59–75. [CrossRef]
15. Sánchez, C.A.F.; Lagarda, A.M.; Angulo, N.R.; Barceló, J.G.A. The construction of the ICT cluster in the regional development of Baja California. *Interciencia* **2017**, *42*, 132–139.
16. Zhai, X.; An, Y. The relationship between technological innovation and green transformation efficiency in China: An empirical analysis using spatial panel data. *Technol. Soc.* **2021**, *64*, 101498. [CrossRef]
17. Lema, R.; Quadros, R.; Schmitz, H. Reorganising global value chains and building innovation capabilities in Brazil and India. *Res. Policy* **2015**, *44*, 1376–1386. [CrossRef]

18. Orji, I.J.; Kusi-Sarpong, S.; Gupta, H. The critical success factors of using social media for supply chain social sustainability in the freight logistics industry. *Int. J. Prod. Res.* **2020**, *58*, 1522–1539. [CrossRef]
19. Dutta, G.; Kumar, R.; Sindhwani, R.; Singh, R.K. Digital transformation priorities of India's discrete manufacturing SMEs–a conceptual study in perspective of Industry 4.0. *Compet. Rev. Int. Bus. J.* **2020**, *30*, 289–314. [CrossRef]
20. Haj, K.A.; Dhiaf, M.M. Do information and communication technologies affect the performance of a supply chain? Pieces of evidence from the Tunisian food sector. *Yugosl. J. Oper. Res.* **2019**, *29*, 539–552. [CrossRef]
21. Saleem, H.; Li, Y.; Ali, Z.; Ayyoub, M.; Wang, Y.; Mehreen, A. Big data use and its outcomes in supply chain context: The roles of information sharing and technological innovation. *J. Enterp. Inf. Manag.* **2020**, *33*. [CrossRef]
22. Madonsela, N.S. Integration of the Management Information System for Competitive Positioning. *Procedia Manuf.* **2020**, *43*, 375–382. [CrossRef]
23. Van Wijk, R.; Jansen, J.J.; Lyles, M.A. Inter-and intra-organizational knowledge transfer: A meta-analytic review and assessment of its antecedents and consequences. *J. Manag. Stud.* **2008**, *45*, 830–853. [CrossRef]
24. Kumar, R.; Singh, R.K.; Shankar, R. Critical success factors for implementation of supply chain management in Indian small and medium enterprises and their impact on performance. *Iimb Manag. Rev.* **2015**, *27*, 92–104. [CrossRef]
25. Partanen, J.; Kohtamäki, M.; Patel, P.C.; Parida, V. Supply chain ambidexterity and manufacturing SME performance: The moderating roles of network capability and strategic information flow. *Int. J. Prod. Econ.* **2020**, *221*, 107470. [CrossRef]
26. Nisar, T.M.; Prabhakar, G.; Strakova, L. Social media information benefits, knowledge management and smart organizations. *J. Bus. Res.* **2019**, *94*, 264–272. [CrossRef]
27. Garcia-Alcaraz, J.L.; Maldonado-Macias, A.A.; Alor-Hernandez, G.; Sanchez-Ramirez, C. The impact of information and communication technologies (ICT) on agility, operating, and economical performance of supply chain. *Adv. Prod. Eng. Manag.* **2017**, *12*, 29–40. [CrossRef]
28. Al-Tahat, M.D.; Bataineh, K.M. Statistical analyses and modeling of the implementation of agile manufacturing tactics in industrial firms. *Math. Probl. Eng.* **2012**, *2012*. [CrossRef]
29. Avelar-Sosa, L.; García-Alcaraz, J.L.; Cedillo-Campos, M.G.; Adarme-Jaimes, W. Effects of regional infrastructure and offered services in the supply chains performance: Case Ciudad Juarez. *Dyna* **2014**, *81*, 208–217. [CrossRef]
30. García-Alcaraz, J.L.; Prieto-Luevano, D.J.; Maldonado-Macías, A.A.; Blanco-Fernández, J.; Jiménez-Macías, E.; Moreno-Jiménez, J.M. Structural equation modeling to identify the human resource value in the JIT implementation: Case maquiladora sector. *Int. J. Adv. Manuf. Technol.* **2015**, *77*, 1483–1497. [CrossRef]
31. Hoffman, J.I.E. Chapter 9—Outliers and Extreme Values. In *Basic Biostatistics for Medical and Biomedical Practitioners*, 2nd ed.; Hoffman, J.I.E., Ed.; Academic Press: Boston, MA, USA, 2019; pp. 149–155. [CrossRef]
32. Kock, N. *WarpPLS User Manual: Version 6.0*; ScriptWarp Systems: Laredo, TX, USA, 2017; p. 141.
33. Avelar-Sosa, L.; García-Alcaraz, J.L.; Vergara-Villegas, O.O.; Maldonado-Macías, A.A.; Alor-Hernández, G. Impact of traditional and international logistic policies in supply chain performance. *Int. J. Adv. Manuf. Technol.* **2014**, *76*, 913–925. [CrossRef]
34. García-Alcaraz, J.L.; Macías, A.A.M.; Luevano, D.J.P.; Fernández, J.B.; López, A.J.G.; Macías, E.J. Main benefits obtained from a successful JIT implementation. *Int. J. Adv. Manuf. Technol.* **2016**, *86*, 2711–2722. [CrossRef]
35. Ketchen, D.J. *A Primer on Partial Least Squares Structural Equation Modeling*; Sage publications: Thousand Oaks, CA, USA, 2013. [CrossRef]
36. Toudert, D. Brecha digital, uso frecuente y aprovechamiento de Internet en México. *Convergencia* **2019**, *26*. [CrossRef]
37. Durugbo, C. Managing information for collaborative networks. *Ind. Manag. Data Syst.* **2014**, *114*, 1207–1228. [CrossRef]
38. García-Alcaraz, J.L.; Maldonado-Macías, A.A.; Sánchez-Ramírez, C.; Latorre-Biel, J.-I. Role of product, market, and organisational characteristics on NPD benefits. *Int. J. Prod. Dev.* **2018**, *22*, 421–440. [CrossRef]
39. Appiah-Otoo, I.; Song, N. The impact of ICT on economic growth-Comparing rich and poor countries. *Telecommun. Policy* **2021**, *45*, 102082. [CrossRef]
40. Ali, J.; Kumar, S. Information and communication technologies (ICTs) and farmers' decision-making across the agricultural supply chain. *Int. J. Inf. Manag.* **2011**, *31*, 149–159. [CrossRef]
41. Ansola, P.G.; Higuera, A.G.; de las Morenas, J.; García-Escribano, J. Decision Making Platform Supported on ICTs: Application to the Management of Ground Handling Operations at an Airport. *IFAC Proc. Vol.* **2011**, *44*, 13074–13079. [CrossRef]

Article

Marketing Mix Modeling Using PLS-SEM, Bootstrapping the Model Coefficients

Mariano Méndez-Suárez

Department of Market Research and Quantitative Methods, ESIC Business & Marketing School, Pozuelo de Alarcón, 28223 Madrid, Spain; mariano.mendez@esic.edu

Abstract: Partial least squares structural equations modeling (PLS-SEM) uses sampling bootstrapping to calculate the significance of the model parameter estimates (e.g., path coefficients and outer loadings). However, when data are time series, as in marketing mix modeling, sampling bootstrapping shows inconsistencies that arise because the series has an autocorrelation structure and contains seasonal events, such as Christmas or Black Friday, especially in multichannel retailing, making the significance analysis of the PLS-SEM model unreliable. The alternative proposed in this research uses maximum entropy bootstrapping (meboot), a technique specifically designed for time series, which maintains the autocorrelation structure and preserves the occurrence over time of seasonal events or structural changes that occurred in the original series in the bootstrapped series. The results showed that meboot had superior performance than sampling bootstrapping in terms of the coherence of the bootstrapped data and the quality of the significance analysis.

Keywords: partial least squares structural equation modeling (PLS-SEM); PLS-SEM bootstrapping; PLS-SEM with time series; marketing mix modeling; maximum entropy bootstrapping

1. Introduction

Marketing mix models use multiple regression to measure marketing effectiveness and efficiency [1]. In the case of multichannel retailers that sell online and offline and advertise on both offline and Internet media, a common solution to the model marketing mix is chaining multiple regression models (based on conversations with consulting experts), i.e., modeling first the impact of advertising on online sales and then using this information to model offline sales. Recent research [2] proposed using partial least squares structural equation models (PLS-SEM) to measure the simultaneous impact of advertising in multichannel retailer contexts and to measure the effectiveness of the different advertising campaigns on web and store sales [3].

PLS-SEM has some desirable properties for marketing mix modeling because it is a causal modeling approach aimed at maximizing the explained variance of the dependent constructs, and because it is similar to multiple regression analysis, it is appropriate for prediction [4]. Moreover, and very relevant, PLS-SEM avoids the problem of indeterminacy and displays the factor scores [5], allowing the use of latent variable scores measured by one or several indicators in subsequent analyses [6]. Consequently, PLS-SEM is particularly useful for measuring the efficiency of marketing campaigns by attributing sales to each of the advertising channels and calculating marketing ROI [3].

However, because PLS-SEM does not assume normality, lack of extreme values, or symmetry in sample data [7], the parametric significance tests usually employed in linear models cannot be applied to test whether outer loadings and path coefficients are significant. Instead, PLS-SEM relies on a nonparametric sampling bootstrapping procedure [8] to test the significance of estimated coefficients. This bootstrapping methodology involves repeated random sampling with replacement from the original sample to create bootstrap samples. It is a good procedure for estimating sampling distributions under independent and identically distributed (i.i.d.) random variables [9], even in situations in which the i.i.d. setup is slightly violated [10], as with cases in which there might be changes in the mean or variance (i.e., the survey is conducted in different countries or with heterogenous respondents) [11,12].

Although sampling bootstrapping is a proper method to measure the significance of the coefficients in most PLS-SEM applications, it is not recommended for marketing mix time series because the data has internal structure and the sampling bootstrapping method can change the dates of events, such as Black Friday or Christmas, or introduce several additional events or none at all in a given year. It also does not respect the time intervals of the structural changes that the series may have.

As an alternative to sampling bootstrapping, we propose maximum entropy (meboot) bootstrapping [13], which maintains the individual basic shapes of time series and their time dependence structures as the autocorrelation function (ACF) and the partial autocorrelation function (PACF). Additionally, when applying meboot bootstrapping, the results inherit the structure while respecting the dates of special events such as Black Friday as well as the possible structural changes.

Despite its importance, little research has been done in the area of time series significance analysis using PLS-SEM models, especially with regard to marketing mix analysis. Furthermore, current research does not highlight the relevance and importance of the application of consistent bootstrap methodologies for solving these types of problems; this research makes important contributions by filling this void. For these reasons, the overall aim of this paper is to provide a detailed empirical demonstration of the advantages of the suggested meboot bootstrapping procedure in comparison with sampling bootstrapping to calculate the significance of PLS-SEM model parameter estimates in a time series or marketing mix modeling context. To this end, we based our analysis on standardized data from a European consumer electronics multichannel company [2] containing web and store sales and online and offline advertising activities.

Given this aim, the remainder of this paper is structured as follows. First, the theoretical foundations are explained. Then, the data used in this research is analyzed, and next, both bootstrapping methods are applied to finally discuss the results.

2. Theoretical Foundation

2.1. PLS-SEM

PLS-SEM is a technique appropriate for solving marketing mix problems even when very complex relationships exist [14] because the optimization algorithm maximizes the variance explained of the model's endogenous constructs, making it especially appropriate to identify key variables in situations of weak theory [15] or verify whether the hypothesized relationships are empirically acceptable [16], for example, those involving marketing mix model variables. Regarding its statistical properties, PLS-SEM admits single item constructs without identification or convergence problems [17]; moreover, PLS-SEM models can handle extremely non-normal data with asymmetries and very high levels of skewness, for example, those corresponding to marketing events such as Black Friday. PLS-SEM is also appropriate for the typical small sample sizes of marketing mix models, such as in our case of 120 weekly observations corresponding to approximately 2.5 years of weekly data.

Earlier applications of PLS-SEM to solve marketing mix problems focused on better understanding the direct and cross effects of advertising on sales. Early research [18] studied the impact of the interaction of radio and print advertising in the opening of checking and savings accounts at a commercial bank, finding evidence of direct and cross effects between both media. More recent research [19] added Internet advertising variables to measure the impact of print advertising and paid search on a service company, finding a crossover effect on online conversions.

Recently, [2] PLS-SEM applied to marketing mix showed evidence of the amplifying effect of organic search queries on the advertising and, consequently, the sales of a multichannel retailer. Additionally, the PLS-SEM [3] model was used to calculate the ROI of offline and Internet advertising campaigns.

To verify the statistical significance of the PLS-SEM model parameters, the literature proposes using sampling bootstrapping; the next section discusses the reasons.

2.2. Sampling Bootstrapping

The term bootstrapping is inspired by the story of the Baron of Munchausen [20], who explained how he pulled himself and his horse out of a swamp by his own hair, meaning that the Baron saved himself by his own means. In this sense, the homonymous statistical technique developed by Efron [9] is similar because bootstrapping draws conclusions about the characteristics of a population using the sample itself; in other words, given the absence of information about the population, the sample is assumed to be the best estimate of the population [21], making this method very appropriate when, as is the case with PLS-SEM, there is no knowledge about the distribution of the parameters.

To find the empirical sampling distribution of a parameter, bootstrapping generates a number of samples with repetition (recommended: 5000) [4], containing the same amount of data as the original series to be sure that the samples obtained have the same statistical properties as the original sample, i.e., if the data contains 120 observations, as in the present research, 5000 samples with 120 observations are generated; in this way, each resample has the same number of elements as the original sample, and the replacement method transforms the finite sample into an infinite population. For each sample, a PLS-SEM model is calculated, and the data on the coefficients of interest are stored, creating a distribution of 5000 distinct coefficients, one for each of the path coefficients or outer loading models of interest. For example, when analyzing the loadings of the indicator λ, we will obtain 5000 values of the estimate λ^*, these values are then ordered from smallest to largest:

$$\lambda^*_{(1)}, \lambda^*_{(2)}, \ldots, \lambda^*_{(5000)} \qquad (1)$$

Then, the lower and upper bounds of the confidence intervals are identified, i.e., if the desired confidence interval is 95%, the interval goes from the lower bound observation, 5000×0.025, to the upper 5000×0.975 observation, that is, from 125 observations to 4875. The resulting confidence interval (CI) suggests that the population value of λ

$$CI = \left[\lambda^*_{(125)}, \lambda^*_{(4875)}\right] \qquad (2)$$

will be somewhere in between $\lambda^*_{(125)}$ and $\lambda^*_{(4875)}$ with a 95% probability. Once the confidence interval is calculated, if it does not include 0, we may consider that the coefficient is significant at 95%.

However, as stated previously, in many cases, because of the nature of the data, the distribution of the parameters is asymmetric and the percentile method is subject to coverage error as stated by [7], meaning that, for example, a 95% confidence interval may actually be a 90% confidence interval. Hence, it is recommended to construct bias-corrected percentile confidence intervals to make statistical inferences when using PLS-SEM. Using bias-corrected and accelerated (BCa) bootstrap confidence intervals solves this problem by adjusting for biases and skewness in the bootstrap distribution [22]; for a detailed step-by-step explanation of the methodology in a PLS-SEM context, see [23].

In the case of time series data as marketing mix model variables, this methodology has a major drawback because, by definition, resampling does not preserve the order of the data, the autocorrelation structure, or the exact time of marketing-associated events such as Black Friday. To solve these problems, the present research proposes the maximum entropy bootstrapping methodology for analyzing the significance of time series coefficients, which will be explained next.

2.3. Maximum Entropy Bootstrapping

Carlstein [24], aware that time series do not satisfy the i.i.d. hypothesis required by bootstrapping and the problems generated by breaking the internal structure of time series by shuffling the data, proposed a solution convenient for stationary time series consisting of bootstrapping nonoverlapping blocks of observations instead of case-by-case observations; on the basis of this idea, the methodology was improved with the proposal of nonoverlapping moving blocks [25,26]. However, even after these improvements, the methods faced the same problems with respect to violations of the required stationarity property and therefore did not provide any remedy.

As a solution to time series bootstrapping, Vinod and López-de-Lacalle [13] proposed the application of the principle of maximum entropy (ME), explained in depth by [27]. According to Vinod [28], ME is a powerful tool to avoid unnecessary distributional assumptions, such as i.i.d. or stationarity assumptions. ME constructs a population of time series, called ensemble Ω, which can include regime switches, gaps, or jump discontinuities. With $f(x)$ being the density function of x_t, the entropy H (Equation (3)) is defined as:

$$H = E(-\log f(x)), \quad (3)$$

Maximizing the entropy H in a density $f(x)$ function, defined in terms of Shannon information [29], means that we are finding the smoothest possible probability distribution that meets the constraints derived from prior knowledge about the mean and variance of the original series. The meboot algorithm constructs segments of ME density $f(x)$ subject to certain mass- and mean-preserving constraints.

The meboot algorithm [13] is a procedure that generates a large number of replicates, e.g., 5000, of the original series, which can be used for statistical inference; it then applies the "blocking" technique to break the time series into nonoverlapping blocks such that the grand mean of all the simulated samples equals the time average of the original, constructing bootstrap samples, or ensembles, that retain the basic shape and dependence structure of the original data. Figure 1 shows the actual series of web sales used in this research, explained in the next section, as well as two random ensembles generated with the meboot algorithm.

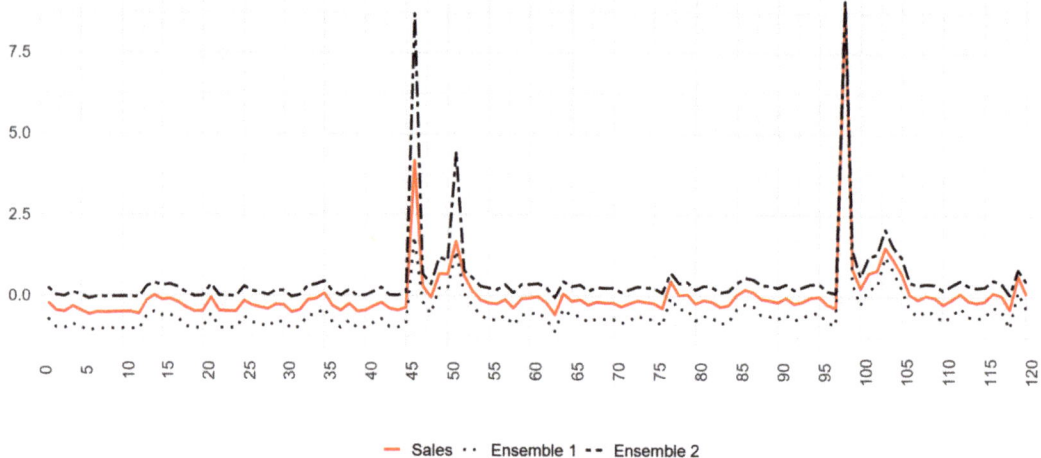

Figure 1. Plot of the standardized EUR series of web sales data used in this research, explained in the next section, and two random ensembles.

Moreover, the approach can be applied in the presence of structural breaks, such as economic crises or recoveries, as well as jumps due to Black Friday sales in which both offline and online sales may "jump" sharply above the mean. For more information on meboot, Vinod [30] provides extensive Monte Carlo evidence that supports the use of the meboot in empirical work and suggests that the meboot confidence intervals are reliable.

3. Materials and Methods

3.1. Data

To conduct the present research, we used data from Méndez-Suárez and Monfort [2], which contains a time series over 120 weeks from a European consumer electronics multichannel retailer, including information on investment in offline, Internet, and paid search advertising, as well as Google queries containing the name of the retailer and the online and offline sales. Table 1 depicts the descriptive statistics of the standardized values of the original data; some variables, such as online Sales, queries, and retargeting, show high levels of skewness and excess kurtosis.

Table 1. Descriptive statistics of the data.

Variables	Median	Min	Max	Skewness	Kurtosis
Online Sales	−0.2	−0.6	9.0	6.8	54.5
Offline Sales	−0.3	−0.7	5.5	3.4	12.4
Queries	−0.3	−0.8	6.7	4.7	26.7
Paid Search	−0.1	−1.8	5.3	1.5	6.3
Store flyer	0.2	−1.1	2.4	0.2	−1.3
TV advertising	0.1	−1.4	4.6	1.1	3.7
Display	0.0	−1.3	3.5	1.3	2.4
Facebook	−0.2	−1.5	3.7	1.0	1.2
Retargeting	0.0	−1.1	7.6	3.7	25.0
Twitter	0.0	−1.2	5.0	1.8	5.8
YouTube	−0.2	−0.9	3.7	1.3	2.0
Christmas	−0.2	−0.2	5.4	5.1	24.6

Note: Data represent standardized EUR with a mean of 0 and standard deviation of 1. Christmas is a dummy binary variable representing Christmas Eve and Epiphany.

3.2. Methods

To compare the results of sampling versus meboot bootstrapping, we used the PLS-SEM model from [2], depicted in Figure 2. The online and offline media in which the multichannel retailer advertised during the period are represented as two reflective latent constructs; the rest of the exogenous variables included in the structural model are single item constructs.

The latent variable online advertising included display, Facebook, Retargeting, Twitter, and YouTube, and the latent variable offline advertising contained store flyers and TV advertising (Equation (4)).

$$Online_t = Display_t \lambda_1 + Facebook_t \lambda_2 + Retargeting_t \lambda_3 + Twitter_t \lambda_4 + Youtube_t \lambda_5 \\ Offline_t = Store\ flyer_t \lambda_6 + TV\ Advertising_t \lambda_7, \quad (4)$$

The structural model contained four endogenous variables (Equation (5)), including queries, explained by online and offline web and store sales, both explained by on and offline advertising, paid search, and Christmas. Paid search was explained by queries.

$$Queries_t = Online_t \beta_1 + Offline_t \beta_2 \\ WebSales_t = Queries_t \beta_3 + Online_t \beta_4 + Offline_t \beta_5 + PaidSearch_t \beta_6 + Christmas_t \beta_7 \\ StoreSales_t = Queries_t \beta_8 + Online_t \beta_9 + Offline_t \beta_{10} + PaidSearch_t \beta_{11} + Christmas_t \beta_{12} , \quad (5) \\ PaidSearch_t = Queries_t \beta_{13}$$

The PLS-SEM model from Figure 2 was used to bootstrap the latent variable outer loadings and the path coefficients using sampling and meboot; the results are presented in the following section.

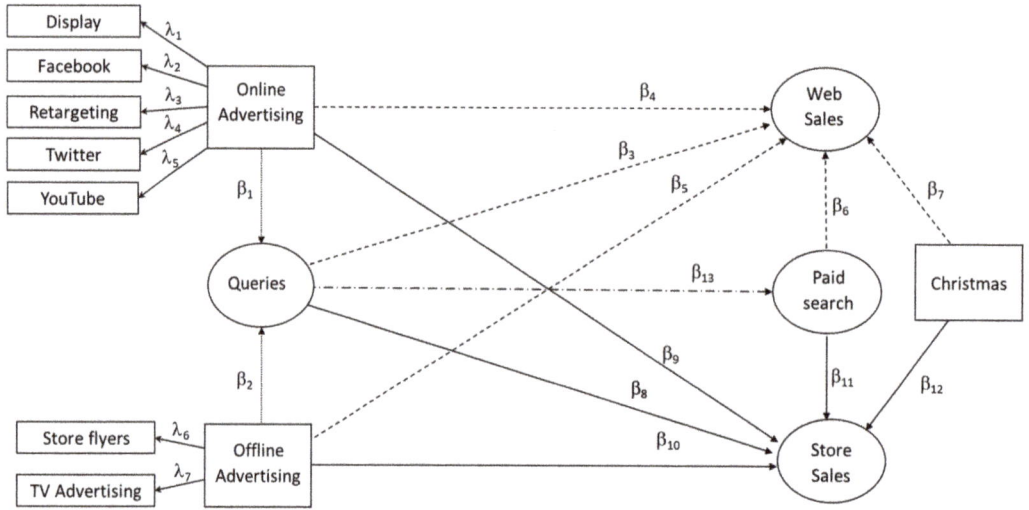

Figure 2. The PLS-SEM model used to illustrate the sample and meboot bootstrapping results comparison. Figure adapted with permission; the article was published in Journal of Business Research, 112, Méndez-Suárez, M.; Monfort, A. The amplifying effect of branded queries on advertising in multichannel retailing, 254–260, Copyright Elsevier (2020).

4. Empirical Results

To compare the results of sampling and meboot, we bootstrapped 5000 subsamples of the PLS-SEM model and calculated the bias-corrected and accelerated (BCa) confidence intervals [7]. Bootstrapping of the structural model employed the R [31] packages, plspm [32], and meboot [13]. The BCa confidence interval calculation in R followed that of Streukens and Leroi-Werelds [23]. The discriminant validity of the model, heterotrait–monotrait (HTMT) ratio of correlations, employed the R semTools package [33].

4.1. Correlations

The correlation of the original series and two random draws of the meboot and sample bootstrap are shown in Table 2a–c. The results showed similar correlations between the original and the bootstrapped variables; there were no significant differences to suggest that one method is better than the other or that one of the methods has major flaws and cannot be used to assess the significance of the results. Next, we analyze the results of the bootstrapped confidence intervals.

4.2. Reliability, Validity, Structural Model, and Fit Assessment

Following [7], to assess the reflective measurement model, we evaluated the composite convergent validity using the average variance explained (AVE), the internal consistency reliability with Cronbach's α, and the discriminant validity using HTMT. The mathematical formulations are represented in Equation (6) (a–d), respectively.

$$(a)\ AVE\xi_j = \frac{1}{K_j}\sum_{k=1}^{K_j}\lambda_{jk}^2; (b)\ Cronbach's\ \alpha = \frac{N\cdot\bar{c}}{1+(N-1)\cdot\bar{c}}; (c)\ Joreskog's\ \rho = \frac{\left(\sum_{i=1}^{N}l_1\right)^2}{\left(\sum_{i=1}^{N}l_1\right)^2 + \sum_{i=1}^{N}var(e_i)}$$

$$(d)\ HTMT_{ij} = \frac{1}{K_iK_j}\sum_{g=1}^{K_i}\sum_{k=1}^{K_j}r_{ig,jh} \div \left(\frac{2}{K_i(K_i-1)}\cdot\sum_{g=1}^{K_i-1}\sum_{k=g+1}^{K_i}r_{ig,ih}\cdot\frac{2}{K_j(K_j-1)}\cdot\sum_{g=1}^{K_j-1}\sum_{k=g+1}^{K_j}r_{jg,jh}\right)^{\frac{1}{2}}$$

(6)

Table 2. (a) Correlation coefficients of the time series, (b) correlation coefficients of one randomly selected series from meboot, and (c) correlation coefficients of one randomly selected series from sampling bootstrap.

	1	2	3	4	5	6	7	8	9	10	11	12
(a) Correlations of Original Series												
1 Online Sales	100											
2 Offline Sales	76	100										
3 Queries	92	75	100									
4 Paid Search	69	68	53	100								
5 Store Flyers	33	30	38	19	100							
6 TV Advertising	23	6	24	6	46	100						
7 Display	32	7	38	12	42	66	100					
8 Facebook	34	20	27	28	28	42	48	100				
9 Retargeting	57	52	44	64	11	8	9	28	100			
10 Twitter	32	3	23	19	17	58	64	51	12	100		
11 YouTube	36	15	26	26	29	47	46	71	34	52	100	
12 Christmas	32	57	29	34	6	2	−1	12	39	−4	12	100
(b) Correlation of one random series, meboot bootstrapping												
	1	2	3	4	5	6	7	8	9	10	11	12
1 Online Sales	100											
2 Offline Sales	81	100										
3 Queries	80	86	100									
4 Paid Search	72	74	60	100								
5 Store Flyers	35	37	45	26	100							
6 TV Advertising	10	11	22	8	41	100						
7 Display	14	28	40	14	38	67	100					
8 Facebook	29	32	28	29	29	42	41	100				
9 Retargeting	46	40	37	67	13	12	12	30	100			
10 Twitter	8	28	20	20	17	55	62	53	12	100		
11 YouTube	25	27	24	26	30	43	38	70	35	53	100	
12 Christmas	56	18	22	31	4	1	−2	13	32	−4	14	100
(c) Correlation of one random series, sampling bootstrapping												
	1	2	3	4	5	6	7	8	9	10	11	12
1 Online Sales	100											
2 Offline Sales	90	100										
3 Queries	93	91	100									
4 Paid Search	59	65	54	100								
5 Store Flyers	39	33	40	18	100							
6 TV Advertising	12	4	20	2	42	100						
7 Display	41	33	46	23	47	63	100					
8 Facebook	31	24	26	31	32	43	53	100				
9 Retargeting	50	55	54	74	22	11	33	39	100			
10 Twitter	44	23	30	33	28	54	61	59	30	100		
11 YouTube	23	11	20	20	32	56	55	71	32	69	100	
12 Christmas	5	36	7	34	−15	−19	−7	−4	24	−15	−11	100

Note: Data values are percentages. Bootstrapped.

The AVE for construct ξ_j is defined as the average of the explained variances λ^2 of each reflective construct. In Cronbach's α, N is the number of low-order components (i = 1, ... , N), and $\bar{c}$ is the average correlation between the lower-order components. In Jöreskog's ρ, l_i is the loading of the lower-order component i on a particular higher-order construct, and var(e_i) is the variance of the measurement error of the lower-order component i. As explained by [34], the HTMT of constructs ξ_i and ξ_j with K_i and K_j indicators, respectively, are the averages of the correlations of indicators across constructs measuring different phenomena relative to the average of the correlations of indicators within the same construct.

Table 3 shows the BCa confidence intervals of the reflective measuring model assessment using both bootstrapping methodologies. For the external loadings of the latent variables (Table 3a), there was agreement between the two methods in terms of the significance of the loadings, but in this case, the width of the intervals is consistently larger when using sampling bootstrapping, which means that there is a much larger level of dispersion of the results when this methodology is used.

Table 3. Assessment of the reflective measurement model latent variables by meboot and sampling bootstrapping. (**a**) Convergent validity of the outer model. (**b**) Reliability of the outer model. (**c**) Discriminant validity.

Indicators	Loadings	95% BCa CI Meboot	CI Amplitude	>0.5?	95% BCa CI Sampling	CI Amplitude	>0.5?
(a) Outer Loading Convergent Validity Bootstrap Results							
Store flyer	0.93	(0.87, 0.93)	0.10	Yes	(0.75, 0.97)	0.22	Yes
TV advertising	0.75	(0.64, 0.83)	0.14	Yes	(0.58, 0.87)	0.30	Yes
Display	0.65	(0.65, 0.75)	0.09	Yes	(0.24, 0.81)	0.57	No
Facebook	0.78	(0.71, 0.80)	0.11	Yes	(0.53, 0.87)	0.34	Yes
Retargeting	0.66	(0.64, 0.79)	0.15	Yes	(0.50, 0.88)	0.38	Yes
Twitter	0.67	(0.65, 0.76)	0.05	Yes	(0.24, 0.86)	0.62	No
YouTube	0.80	(0.67, 0.82)	0.19	Yes	(0.63, 0.88)	0.25	Yes
Latent Variables	AVE	95% BCa CI Meboot	CI Amplitude	>0.5?	95% BCa CI Sampling	CI Amplitude	>0.5?
Online ad	0.51	(0.51, 0.55)	0.04	Yes	(0.35, 0.62)	0.27	No
Offline ad	0.72	(0.66, 0.76)	0.11	Yes	(0.63, 0.8)	0.17	Yes
(b) Latent Variables Internal Consistency Reliability Bootstrap Results							
Latent Variables	Cronbach's Alpha	95% BCa CI Meboot	CI Amplitude	0.60–0.90?	95% BCa CI Sampling	CI Amplitude	0.60–0.90?
Online ad	0.78	(0.77, 0.8)	0.03	Yes	(0.70, 0.84)	0.14	Yes
Offline ad	0.63	(0.6, 0.71)	0.11	Yes	(0.44, 0.77)	0.34	No
Latent Variables	Jöreskog's ρ	95% BCa CI Meboot	CI Amplitude	>0.7?	95% BCa CI Sampling	CI Amplitude	>0.7?
Online ad	0.85	(0.85, 0.87)	0.02	Yes	(0.45, 1)	0.55	No
Offline ad	0.84	(0.82, 0.87)	0.05	Yes	(0.50, 1.39)	0.89	No
(c) Latent Variables Discriminant Validity Bootstrap Results							
Latent Variables	HTMT	95% BCa CI Meboot	CI Amplitude	CI < 1?	95% BCa CI Sampling	CI Amplitude	CI < 1?
Online ad & Offline ad	0.80	(0.73, 0.89)	0.16	Yes	(0.61, 0.99)	0.37	Yes

Note: As per Hair et al. [7], bootstrapped coefficients are corrected and accelerated (BCa).

However, the problems become especially severe when assessing the reflective constructs (Table 3b) because of the width of the sampling bootstrap intervals, which in all cases is three times wider or more compared with the meboot intervals; consequently, the latent variables are not validated in terms of AVE, Cronbach's Alpha, and Jöreskog's ρ; the HTMT is validated but by hundredths of a percent.

The confidence intervals from the regression coefficients (Table 4a) had similar amplitudes and showed similar results with respect to significance in all the paths, except for the offline advertising path to web sales, for which the sampling bootstrap method indicated that offline advertising had a non-significant coefficient on web sales; in other words, offline advertising does not impact the sales of the web store.

Table 4. Evaluation of the structural model. (**a**) The model's regression coefficients and their significance based on meboot and sampling bootstrapping. (**b**) The model's predictive accuracy based on meboot and sampling bootstrapping.

(a) Regression Coefficients Bootstrap Results								
Endogenous Variables	Exogenous Variables	Path Coefficient	95% BCa CI Meboot	CI Amplitude	Significance ($p < 0.05$)?	95% BCa CI Sampling	CI Amplitude	Significance ($p < 0.05$)?
Web sales	Online ad	0.14	(0.07, 0.32)	0.25	Yes	(0.05, 0.24)	0.19	Yes
Web sales	Offline ad	−0.05	(−0.14, −0.002)	0.14	Yes	(−0.12, 0.02)	0.14	No
Web sales	Queries	0.75	(0.58, 0.89)	0.31	Yes	(0.62, 0.85)	0.23	Yes
Web sales	Paid Search	0.24	(0.06, 0.32)	0.26	Yes	(0.15, 0.39)	0.23	Yes
Web sales	Christmas	−0.15	(−0.15, 0.13)	0.29	No	(−0.14, 0.12)	0.26	No
Store sales	Online ad	−0.18	(−0.24, −0.04)	0.19	Yes	(−0.36, −0.01)	0.35	Yes
Store sales	Offline ad	0.05	(−0.02, 0.12)	0.14	No	(−0.14, 0.16)	0.30	No
Store sales	Queries	0.52	(0.27, 0.65)	0.38	Yes	(0.42, 0.75)	0.34	Yes
Store sales	Paid Search	0.39	(0.31, 0.52)	0.21	Yes	(0.12, 0.59)	0.46	Yes
Store sales	Christmas	0.32	(0.23, 0.44)	0.21	Yes	(0.08, 0.53)	0.45	Yes
Queries	Online ad	0.38	(0.27, 0.47)	0.20	Yes	(0.12, 0.59)	0.47	Yes
Queries	Offline ad	0.20	(0.13, 0.3)	0.17	Yes	(0.08, 0.35)	0.27	Yes
Paid Search	Queries	0.54	(0.36, 0.62)	0.26	Yes	(0.36, 0.66)	0.30	Yes

(b) Predictive accuracy of the structural model evaluated with the magnitude of the explained variance, R^2					
Endogenous Variables	R^2	95% BCa CI Meboot	CI Amplitude	95% BCa CI Sampling	CI Amplitude
Queries	0.26	(0.17, 0.3)	0.13	(0.06, 0.36)	0.30
Paid Search	0.29	(0.13, 0.39)	0.25	(0.11, 0.42)	0.31
Store sales	0.79	(0.77, 0.94)	0.17	(0.68, 0.93)	0.25
Web sales	0.92	(0.91, 0.95)	0.04	(0.92, 0.97)	0.04

Note: As per Hair et al. [7], bootstrapped coefficients are corrected and accelerated (BCa).

As [35] stated, the different meaning of the term fit does not depend on whether covariance-based SEM or variance-based SEM is used but on whether confirmatory or explanatory research is performed (see [36]). Since in explanatory research, as in this case, we would like to explain as much variation as possible in a dependent variable, the R^2 is the natural measure of fit; however, as occurred in the assessment of the reflective construct outer loadings, the confidence intervals of the R^2 (Table 4b) of the sampling bootstrapped values were widespread and invalidated the model, contrary to the meboot values, which showed high levels of fit in line with the results of the model application shown in Figure 3.

To understand what really explains the differences between the bootstrapping methodologies, we need to visually inspect the entire time series. Figure 3 shows the original series, and two random paths of the sampling and meboot series both for online and offline sales. The sampling bootstrapped series added jumps to sales corresponding to events such as Christmas and Black Friday but at very different times from those occurring in the original series, and, for example, in the case of offline sales (Figure 3a), it included up to 10 jumps, only one of which corresponded to the date on which it occurred; however, at the times these events occur, the sampling bootstrapped series did not reflect them. On the other hand, in the meboot series, the jumps occurred at the same times as in the original series; however, as expected for the maximum entropy modeling, some replicas of the original series were more pronounced than others.

20. Raspe, R.E. *The Surprising Adventures of Baron Munchausen*; Standard Ebooks: Nevada County, CA, USA, 1781.
21. Abdi, H.; Chin, W.W.; Vinzi, V.E.; Russolillo, G.; Trinchera, L. New Perspectives in Partial Least Squares and Related Methods. In *Springer Proceedings in Mathematics and Statistics*; Abdi, H., Chin, W.W., Esposito Vinzi, V., Russolillo, G., Trinchera, L., Eds.; Springer: New York, NY, USA, 2013; Volume 56, pp. 201–208, ISBN 978-1-4614-8282-6.
22. Efron, B. Better bootstrap confidence intervals. *J. Am. Stat. Assoc.* **1987**, *82*, 171–185. [CrossRef]
23. Streukens, S.; Leroi-Werelds, S. Bootstrapping and PLS-SEM: A step-by-step guide to get more out of your bootstrap results. *Eur. Manag. J.* **2016**, *34*, 618–632. [CrossRef]
24. Carlstein, E. The Use of Subseries Values for Estimating the Variance of a General Statistic from a Stationary Sequence. *Ann. Stat.* **1986**, *14*, 1171–1179. [CrossRef]
25. Kunsch, H.R. The Jackknife and the Bootstrap for General Stationary Observations. *Ann. Stat.* **1989**, *17*, 1217–1241. [CrossRef]
26. Liu, R.Y.; Singh, K. Moving blocks jackknife and bootstrap capture weak dependence. *Explor. Limits Bootstrap* **1992**, *225*, 248.
27. Baldwin, R.A. Use of Maximum Entropy Modeling in Wildlife Research. *Entropy* **2009**, *11*, 854–866. [CrossRef]
28. Vinod, H.D. *Maximum Entropy Bootstrap Algorithm Enhancements*; Discussion Paper Series; Fordham University: Bronx, NY, USA, 2013; Volume 2013-04.
29. Shannon, C.E. A Mathematical Theory of Communication. *Bell Syst. Tech. J.* **1948**, *27*, 379–423. [CrossRef]
30. Vinod, H. New bootstrap inference for spurious regression problems. *J. Appl. Stat.* **2016**, *43*, 317–335. [CrossRef]
31. R Core Team R. *A Language and Environment for Statistical Computing*; R Foundation for Statistical Computing: Vienna, Austria, 2021.
32. Sanchez, G. *PLS Path Modeling with R*; Trowchez Editions: Berkeley, CA, USA, 2013.
33. Jorgensen, T.D.; Pornprasertmanit, S.; Schoemann, A.M.; Rosseel, Y. semTools: Useful Tools for Structural Equation Modeling, Version 0.5-5; R Packag. 2021. Available online: https://cran.r-project.org/web/packages/semTools/semTools.pdf (accessed on 1 August 2021).
34. Henseler, J.; Ringle, C.M.; Sarstedt, M. A new criterion for assessing discriminant validity in variance-based structural equation modeling. *J. Acad. Mark. Sci.* **2015**, *43*, 115–135. [CrossRef]
35. Henseler, J.; Hubona, G.; Ray, P.A. Using PLS path modeling in new technology research: Updated guidelines. *Ind. Manag. Data Syst.* **2016**, *116*, 2–20. [CrossRef]
36. Henseler, J. Partial least squares path modeling: Quo vadis? *Qual. Quant.* **2018**, *52*, 1–8. [CrossRef]
37. Hair, J.F.; Sarstedt, M.; Hopkins, L.; Kuppelwieser, V.G. Partial least squares structural equation modeling (PLS-SEM): An emerging tool in business research. *Eur. Bus. Rev.* **2014**, *26*, 106–121. [CrossRef]

Article

Moderating Effect of Proactivity on Firm Absorptive Capacity and Performance: Empirical Evidence from Spanish Firms

Rafael Sancho-Zamora [1,*], Isidro Peña-García [1], Santiago Gutiérrez-Broncano [2] and Felipe Hernández-Perlines [3]

[1] Department of Business Administration, Faculty of Law and Social Science, University of Castilla-La Mancha, Calle de las Cuadras, 2, 13003 Ciudad Real, Spain; Isidro.Pena@uclm.es
[2] Department of Business Administration, Faculty Social Science, University of Castilla-La Mancha, Avda. Real Fábrica de Seda, s/n, 45600 Talavera de la Reina, Spain; Santiago.Gutierrez@uclm.es
[3] Department of Business Administration, Faculty of Law and Social Science, University of Castilla-La Mancha, Cobertizo de San Pedro Mártir, s/n, 45071 Toledo, Spain; Felipe.HPerlines@uclm.es
* Correspondence: Rafael.Sancho@uclm.es

Abstract: The purpose of this study was to understand how proactivity can affect the relationship between absorptive capacity and organisational performance. Most previous studies have ignored the role of proactivity in this relationship and have not considered the multidimensional nature of absorptive capacity. A questionnaire was sent to 800 CEOs of Spanish companies from different sectors, procuring a response rate of 38.25%. A structural equation model was applied to test the hypothesis. This study confirms the positive effect that absorptive capacity has on business performance and the moderating role of proactivity in this relationship. Companies that develop their capacity to absorb information from the environment achieve better results. Furthermore, if they engage in proactive behaviour within their company, this relationship is stronger. Future research should include more capacities that are related to knowledge and business performance (i.e., learning capability, innovation capacity, etc.). This study contributes to the understanding of how to manage a company's knowledge in an appropriate way. It sheds new light on how knowledge management should be conducted, emphasising not only the gathering of information but also the promotion of a proactive attitude on the part of employees to achieve the goal of better performance.

Keywords: proactivity; absorptive capacity; potential absorptive capacity; realised absorptive capacity; structural equation modelling

Citation: Sancho-Zamora, R.; Peña-García, I.; Gutiérrez-Broncano, S.; Hernández-Perlines, F. Moderating Effect of Proactivity on Firm Absorptive Capacity and Performance: Empirical Evidence from Spanish Firms. *Mathematics* **2021**, *9*, 2099. https://doi.org/10.3390/math9172099

Academic Editors: María del Carmen Valls Martínez and Pedro Antonio Martín Cervantes

Received: 26 July 2021
Accepted: 24 August 2021
Published: 30 August 2021

Publisher's Note: MDPI stays neutral with regard to jurisdictional claims in published maps and institutional affiliations.

Copyright: © 2021 by the authors. Licensee MDPI, Basel, Switzerland. This article is an open access article distributed under the terms and conditions of the Creative Commons Attribution (CC BY) license (https://creativecommons.org/licenses/by/4.0/).

1. Introduction

Business success is currently determined by the ability to innovate and adapt to changes in the environment. In an environment in which organisations must confront new challenges, and against a backdrop of volatility and uncertainty, the ability to adapt and display a disruptive attitude on the part of each and every member of the organisation are what makes the difference [1].

Knowledge is a key aspect in this context [2,3], and it has consequently become necessary for management to use external sources of information in order to provide a more appropriate response to the complexities of a rapidly changing dynamic environment [4,5]. Having external knowledge is paramount if a company's performance is to be ensured [6,7]. Moreover, merely acquiring knowledge is not sufficient; it is necessary to use that knowledge correctly, and companies must, therefore, invest in this process [8].

This path from knowledge acquisition to its productive use is far from easy, and many companies have difficulties with the process [9,10]. In order to achieve this, companies must develop their absorptive capacity (ACAP) [11], a notion that is increasingly recognised by researchers as a source of competitive advantage [12–17].

ACAP is an ability that companies must develop in order to maintain a competitive advantage, as it favours adaptation to the changes occurring in their increasingly competitive and complex environments [18]. It refers to a company's ability to recognise the value of new information, assimilate it and apply it for business purposes [19]. According to Kale et al. [20], it enables enterprises to creatively use all the external information it gathers, which serves to provide a better performance in terms of new product development, market share and profitability.

Numerous studies on business performance have been conducted in the field of business management, strategy, knowledge management and other related fields since the authors of [19] first introduced the concept of ACAP in 1990 [21]. However, despite the existence of a large number of studies addressing them, the literature on this topic is confronted with an "ambiguity problem" mainly about conceptualisation as well as measurement. This is because researchers have used ACAP as a general-purpose construct, overlooking previous studies on the topic that have presented, for example, its multidimensional nature, among other relevant theories, and because of the lack of understanding of the nature of the construct itself [18,21]. Furthermore, few studies examine ACAP in depth [20], despite that, of all the dynamic capabilities, ACAP is among the so-called high-level capabilities, as it serves to explain both the success and failure of a company, in addition to favouring the creation of wealth and participation in the creation of a competitive advantage over time [22].

Studies have found that ACAP is related to competitive advantage, innovation [23], value creation [24], customer loyalty and satisfaction [8], etc. However, few studies conduct in-depth research into the functioning of ACAP in regard to unravelling the complexities of organisational and knowledge management capabilities [22]. Despite the fact that many scientific papers have developed the theoretical foundations of ACAP since the 1990s, there is still clearly a lack of empirical research [25–27].

Moreover, in order to achieve high business performance in this complex, dynamic and volatile business environment faced by modern organisations, it forces management and employees not only to adapt and adjust to the major demands of the environment but also to act proactively to prevent conflict from arising. Now more than ever, the knowledge economy, with its inherent ambiguity, novelty and complexity, has dictated that organisations and their most successful members must embrace a less structured organisational environment in which autonomy, self-governance, opportunity recognition, personal initiative and capitalisation, collaboration and adaptation are in greater demand [28–30].

Due to the aforementioned, it has become necessary to acquire and use knowledge as well as to have a modern and proactive organisation. Organisations that are able to acquire and use relevant external knowledge and, in addition, have proactive behaviour that anticipates changes in the environment and promotes internal changes, will contribute to organisational effectiveness [31,32] and the achievement of better results. These proactive companies are the first to notice the early signs of change in their environment, able to see the opportunities for competitive advantage arising from these changes and can turn them into reality. The proactive strategic posture supports communication and knowledge sharing within the company and positively influences knowledge acquisition and utilisation [33], which supports engagement and employment initiatives [34].

Organisations that have this external information but lack employees who can proactively and collaboratively address complex and unexpected issues will, conversely, have greater difficulty in achieving the expected business performance [35].

For this study, our objective was to analyse the effect of ACAP on business performance from a multidimensional point of view. In addition, we endeavoured to discover the role played by proactive behaviour within this relationship, differentiating companies that are proactive from those that are not. In order to achieve this, we considered the different dimensions of ACAP and their relationships with business performance. We also analysed the moderating effect that proactivity has on this relationship. Finally, conclusions and practical recommendations for companies were established, along with the limitations and future lines of research that are included at the end of this work, together with the references used.

2. Absorptive Capacity and Performance

As mentioned above, ACAP refers to the ability to locate new ideas and incorporate them into an organisation's processes, mainly in regard to those aspects that are considered most important for organisational competitiveness [17,19].

ACAP was initially defined as the ability to recognise the value of new information, assimilate it and apply it for business purposes [19]. Absorptive capacity, when understood in this manner, incorporates both the need to value and acquire knowledge from the external environment and the internal processes of learning from past experience and current actions [36].

However, Sakhdari [37] highlighted the great scarcity of theories analysing how organisational mechanisms can affect different aspects of ACAP. The focus of this study was on what ACAP predicts, and it is related to fields such as dynamic capability [38,39], organisational learning [40,41] and knowledge management [42,43].

ACAP plays a central role in research, following the company's knowledge-based view [44]. One part of a company's knowledge is developed internally, and another part is acquired from different external sources [45,46].

Taking the resource-based view as a basis, ACAP can be seen as a strategically valuable capability [47,48]. This capability is a socially complex routine by which companies acquire, assimilate, transform and exploit knowledge with the intention of creating value and gaining a competitive advantage [49].

ACAP can, therefore, improve a company's performance by exploiting firm-specific internal and external competencies to cope with changes in the environment [38]. ACAP allows companies to develop skills with which to detect knowledge and information from outside the organisation that may be useful to them. They can then internalise and adapt that information to their specific needs and exploit it for their business objectives, converting it into business results. Organisations that have a substantial knowledge base in a particular field, therefore, tend to have a high ACAP and will be able to evaluate and act on new information or ideas that develop within that field of knowledge [17,50]. ACAP is undoubtedly a key intangible asset for companies' competitiveness [19,50,51] and, consequently, it improves business performance.

The existing literature presents a multitude of studies that have analysed the positive effect ACAP has on different business variables. Companies that are able to assimilate new knowledge intensify their learning and become more efficient, have an easier time creating new products [52], better exploit technological advances that are closely linked to innovative behaviour [53] and improve their financial performance [17]. After conducting a meta-analysis, Song et al. [54] found that ACAP has a significant positive effect on companies' performance. However, there were also studies [55,56] that found an inverted U-shaped relationship between these variables, qualifying that ACAP increases a company's financial performance but only to a certain extent, and there is a point at which a higher ACAP is associated with lower levels of new product development.

Nevertheless, this higher capacity to acquire and exploit new knowledge is expected to influence the company's innovation and, hence, superior performance. The reason offered to explain this direct and positive relationship is that companies must continually strive to develop their knowledge bases if they are to thrive and remain competitive [57].

Although this relationship has been examined in previous research [26,58,59], few studies have considered its multidimensional character and thus have analysed each particular dimension of ACAP [60]. Zahra and George [17] distinguish between the potential ACAP and the realised ACAP discussed in Lichtenthaler [56], stating that the potential ACAP is related to realised absorptive capacity, which is in turn related to organisational performance [61].

As stated above, Zahra and George [17] established that ACAP has two subsets and four dimensions: potential ACAP (comprising knowledge acquisition and assimilation capabilities) (PACAP) and realised ACAP (focusing on knowledge transformation and exploitation capabilities) (RACAP), which may have different influences on the creation and maintenance of a company's competitive advantage. Although the authors of one study expressed a contrary argument to that of Zahra and George, stating certain ambiguities and omissions regarding some important research contributions [62], we understand that these are small contributions that do not add excessive value.

However, this distinction has been used in subsequent research [14,63–68] and clarifies the different dimensions included in absorptive capacity. Acquisition refers to a company's ability to identify and acquire external knowledge about itself from surrounding information [67]. Assimilation also refers to a company's ability to develop useful processes and routines with which to analyse, interpret and understand externally acquired knowledge [66]. Transformation means developing and refining these routines in order to combine existing knowledge with acquired and assimilated knowledge for future use [17].

Companies that wish to improve their performance must develop and manage all these dimensions simultaneously [17]. That is, companies that invest only in the capacity to acquire and assimilate external knowledge (i.e., PACAP) will succeed in generating a new and expanded knowledge base, but this alone will not necessarily lead to superior performance unless the new knowledge is exploited in new products and processes [69–71]. Otherwise, the costs of acquiring this new information may easily outweigh the resulting benefits [19]. Conversely, companies that limit themselves solely and exclusively to transforming and exploiting new knowledge (i.e., RACAP) will achieve short-term benefits but will not develop a new and innovative knowledge base [72], thus harming the company's competitiveness in the future. A multidimensional approach is important for a detailed understanding of the effects of ACAP on companies' performance [73].

We therefore argue that having external information alone may not have an impact on a company's performance, but it is a necessary first step for an organisation to have ACAP. This is a dimension (PACAP) that is of the utmost importance for the realisation of ACAP (RACAP). Acquiring information and applying the use dimension, which includes the assimilation, transformation and exploitation of the information, was found to affect a company's performance [20]; thus, we proposed the following two hypotheses:

Hypothesis 1a (H1a). *Potential ACAP is positively related to business performance.*

Hypothesis 1b (H1b). *Realised ACAP is positively related to business performance.*

3. Proactivity as a Moderating Factor: Proactive Firms

Venkataraman [74] defines proactivity as the processes with the objective of anticipating and operating on future needs by "seeking new opportunities that may or may not be related to the usual course of action, introducing new products and brands ahead of the competition and eliminating operations that are strategically in the maturity or decline stages of the life cycle".

Proactivity has emerged as an extremely important behaviour in organisations and was shown to correlate with positive organisational and individual outcomes [35]. In organisational literature, proactivity, or the proactive behaviour companies engage in, involves actions related to anticipatory or change-oriented behaviour [75]. This type of proactive behaviour means that individuals can promote, change and contribute to organisational effectiveness in an anticipatory manner [31]. Proactive behaviour involves acting in advance of a future situation, rather than merely reacting or adapting when problems arise. It means making things happen rather than just watching things happen or waiting for something to happen [35].

Proactive companies have a vision of the future and aim to anticipate and act on future needs to shape their environment rather than reactively adapting to changes in the environments in which they operate [76]. Proactive companies are characterised by anticipating social changes and adjusting their internal structures in order to achieve congruence with future needs [77]. These companies, in addition to being proactive, are also more likely to enjoy learning curve effects than non-proactive companies are, signifying that, although a company does not need to be proactive to be innovative and competitive, proactivity may foster the relationship between ACAP and the company's performance.

This can be understood as a future perspective from which companies seek opportunities to develop and introduce new products in order to obtain the advantages of being the pioneers capitalising on emerging opportunities and, thus, of being able to shape the direction of their environment.

Delmas, Hoffman and Kuss [65] found that a company's ability to generate competitive advantage is directly related to its competitors' ability to imitate its strategy. As ACAP and proactivity are difficult to imitate (because they depend on complex and often tacit processes), these elements must, therefore, have an impact on business performance. In a study of 246 Spanish technology companies, García-Morales et al. [78] analysed how technological ACAP and technological proactivity influence organisational learning and innovation and demonstrated how these dynamic capabilities affect organisational performance.

Proactivity, as mentioned above, refers to a company's ability to anticipate future needs, seek new opportunities and take initiative [79]. When these companies also have relevant external information because they have a potential absorptive capacity, they are more likely to become leaders rather than followers and thus are more likely to make substantial changes to their environment by introducing new products, technologies or management techniques [76]. Proactive companies that are also the first to notice the early signs of change in their business environment are pioneers in discovering the potential opportunities for competitive advantage that arise from these changes.

Furthermore, two principal views of dynamic capabilities [38,80] are similar but have a notable discrepancy in regard to the question of whether dynamic capabilities have the potential to explain sustainable competitive advantage in rapidly changing business environments [81]. Nevertheless, Peteraf et al. [81] were able to reconcile both perspectives by considering contingent relevant factors [82]. Holding a proactive strategic position, therefore, supports communication and knowledge sharing within the company and positively influences the relationship between ACAP and performance [33]. Therefore, we proposed the following hypotheses (see Figure 1):

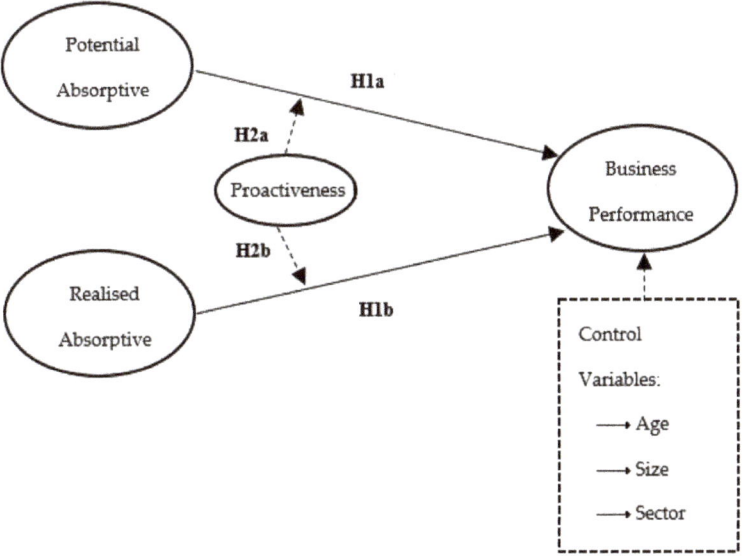

Figure 1. Theoretical Model.

Hypothesis 2a (H2a). *The influence of potential ACAP on business performance will be greater for proactive firms.*

Hypothesis 2b (H2b). *The influence of realised ACAP on business performance will be greater for proactive firms.*

4. Methodology

4.1. Data Collection

The data were obtained from a questionnaire mailed to 800 randomly selected small- and medium-sized enterprises in the Spanish autonomous community of Castilla–La Mancha. The information was collected directly from the company's managing director. The contacts were obtained from the SABI database, and active companies belonging to different sectors of activity in both the industrial and service sectors were selected. A total of 315 questionnaires were obtained, of which 9 were rejected as incomplete (see Table 1).

Table 1. Technical data employed in this research.

Sample size	15,853 companies
	800 companies randomly selected
Unit of analysis	Company
Scope	Castilla–La Mancha (Spain)
Valid responses/response rate	306/38.25%
Confidence level	95%
Error rate	5.55%
Informant	CEO
Data	October–December 2019

Table 2 shows the sectors and the activity to which the companies that participated in this study belong.

Table 2. Sector and activity of the companies analysed.

Sectors (CNAE)	Code	Activity	Number	Percentage
62, 69, 70, 71, 73	1	Specialised consulting services	75	24.50%
41, 43	3	Construction	65	21.24%
55, 56, 46, 47, 68	2	Retail and accommodation services	96	31.37%
10, 11, 14, 18, 21, 23,25, 26, 27, 28, 31	4	Manufacturing	70	22.87%

The statistical power of the sample used in this study was 0.998 and was calculated using Cohen's retrospective test [83], which can be obtained using the G* Power 3.1.9.2 programme [84]. The value obtained affirms that the sample used in this study had an adequate statistical power, as it was above the threshold of 0.80 established by Cohen [83].

4.2. Measurement of Variables

All the variables were measured using a 7-point Likert scale ranging from 1 (strongly disagree) to 7 (strongly agree). The following specific variables were used in this study:

(a) Measurement of potential absorptive capacity. PACAP was operationalised as a second-order A-type composite based on the acquisition capacity (3 items) and the assimilation capacity (4 items). The measurements were performed using the scales proposed by Cohen and Levinthal [19] and Lane et al. [25]. This scale was validated by Flaten et al. [66] and by Hernández-Perlines et al. [85]. ("Management expects employees to have information beyond/outside our industry/sector.")

(b) Measurement of realised absorptive capacity. RACAP was operationalised as a second-order A-type composite on the basis of the transformation capacity (4 items) and the exploitation capacity (3 items). The measurements were conducted using the scales proposed by Cohen and Levinthal [19] and Lane et al. [25]. This scale was validated by Flaten et al. [66] and by Hernández-Perlines et al. [85]. ("Our employees are able to apply the new knowledge in their workplace.")

(c) Performance measurement. Performance was measured by employing an overall measure of a company's performance that assesses the perception of that company's performance in relation to that of its competitors [86]. The use of perception or satisfaction measures as determinants of a company's performance is increasingly common in research [87]. Performance was operationalised as a first-order A-type composite. The 4 items used in this research were sales growth, profit growth, market share growth and return on equity growth, all of which were extracted from a combination of the scales proposed by [88–92]. This scale was validated by Hernández-Perlines et al. [93].

(d) Control variables. Size (number of employees), sector and age (number of years in operation), as proposed by Chrisman et al. [94] and validated by Ibarra-Cisneros and Hernández-Perlines [95], were used as control variables in this research. All the control variables were operationalised as first-order A-type composites.

4.3. Methodology

To analyse the results and test both the direct and moderating hypotheses proposed in this paper, the multivariate partial least squares (PLS) quantitative structural equation technique was employed. The choice of this data analysis method is justified for the following reasons:

(a) It is an appropriate analysis method when research is in the early stages of developing new theoretical constructs [96,97];
(b) It is an analysis method that is characterised by its predictive nature, thus allowing it to address the research questions posed [98,99];
(c) This analysis method makes it possible to observe the different causal relationships between the variables analysed [100,101];
(d) It is a suitable data analysis method when the sample is not large [102,103];
(e) It is a method that allows for the analysis of complex relationships among models [104], and PLS-SEM can handle non-normal data.

5. Results

The software employed for data analysis using PLS-SEM was SmartPLS v.3.3.3 [97]. The results were analysed following the recommendations of Barclay, Higgins and Thompson [105] and Hair Jr., Sarstedt, Ringle and Gudergan [106], who advised to first evaluate the measurement model and then evaluate the structural model.

To follow the evaluation process of both the measurement model and the structural model, the variables of this study were modelled following the method described by Sarstedt [99] to analyse them with PLS:

(a) The PACAP was operationalised as a second-order A-type compound;
(b) The RACAP was operationalised as a second-order A-type compound;
(c) Performance was operationalised as a first-order A-type composite;
(d) The three control variables (age, sector and size) were operationalised as a first-order A-type composite.

To evaluate the measurement model, the variables were checked for reliability and adequate levels of convergent and discriminant validity, following the recommendations of [106]. The following indicators were used for this purpose [105–107]:

(a) Composite reliability: composite reliability should, according to [108], have values above 0.7, with appropriate values between 0.7 and 0.9 [98]. All the model indi-

cators had acceptable composite reliability values (see Table 4). Furthermore, the composite reliability had no redundancy problems because no value was higher than 0.95 [109,110];
(b) Cronbach's alpha: Cronbach's alpha values above 0.7 [108]. In our case, Cronbach's alpha was higher than this value for all the variables (see Table 4);
(c) Rho A: the Rho A must be greater than 0.7 [111] and must lie between the values of composite reliability and Cronbach's alpha [98]. This condition was met for all the variables (see Table 4);
(d) Average variance extracted (AVE) can be used to assess the convergent validity of each composite. [108] recommend a value higher than 0.5 for the AVE. This condition was valid for our data (see Table 4);
(e) Heterotrait–monotrait ratio (HTMT): this ratio enables the measurement of discriminant validity, and it is necessary to check that the correlation between each pair of constructs is not greater than the square root value of the AVE of each construct. For discriminant validity to hold, HTMT values must be less than 0.85 [103]. Discriminant validity is confirmed when the indicated values are met (see Table 4).
(f) Cronbach's alpha, composite reliability and average variance extracted (AVE) of the first order composites are listed (see Table 3).

Table 3. Cronbach's alpha, composite reliability and average variance extracted (AVE) of the first order composites.

	Path Coefficient	Cronbach's Alpha	Composite Reliability	AVE
Acquisition capacity	0.942	0.871	0.921	0.795
Assimilation capacity	0.893	0.917	0.915	0.800
Transformation capacity	0.911	0.844	0.895	0.683
Exploitation capacity	0.905	0.891	0.930	0.820

Table 4. Correlation matrix, composite reliability, convergent and discriminant validity, heterotrait–monotrait ratio (HTMT) and descriptive statistics.

Construct	AVE	Composite Reliability	PACAP	RACAP	PROAC	PERF
1. Potential ACAP (PACAP)	0.874	0.935	0.934 *			G
2. Realised ACAP (RACAP)	0.891	0.942	0.622	0.9 *		
3. Proactivity (PROAC)	0.764	0.785	0.382	0.342	0.860 *	
4. Performance (PERF)	0.723	0.913	0.270	0.194	0.206	0.850 *
Heterotrait–Monotrait ratio (HTMT)						
1. Potential ACAP (PACAP)						
2. Realised ACAP(RACAP)			0.223			
3. Proactivity (PROAC)			0.703	0.627		
4. Performance (PERF)			0.106	0.192	0.211	
Cronbach's alpha			0.880	0.888	0.795	0.875
Rho A			0.877	0.890	0.789	0.845
Mean			4.16	4.20	4.10	3.96
SD			1.24	1.31	1.05	0.99

Note: the mean and standard deviation values of each of the second-order composites have been calculated from the mean values of the different first-order composites of which they are composed. (*) The values of the diagonal have been obtained from the square root of the AVE of each compound.

To complete the verification of the discriminant validity, we also computed the HTMT inference from the bootstrapping option (5000 subsamples). When the resulting interval contains values of less than 1, discriminant validity exists, and our data meet this requirement (see Table 5).

Table 5. HTMT inference.

	Original Sample (O)	Sample Mean (M)	5.0%	95.0%	Sample Mean (M)	Bias	5.0%	95.0%
PACAP -> PERF	0.228	0.394	0.118	0.542	0.209	0.009	0.085	0.577
RACAP -> PERF	0.344	0.752	0.050	0.624	0.347	0.003	0.026	0.606
PROAC -> PERF	0.201	0.421	0.009	0.340	0.201	0.000	0.177	0.314

Once the convergent and discriminant validity of the measurement model had been ensured, we proceeded to check the relationships between the different variables in order to conduct an analysis of the structural model.

The analysis of the structural model confirmed that PACAP has a positive impact on performance since the path coefficient is 0.228 (higher than 0.2, which [112] established as the minimum limit). Moreover, this effect is significant (the t value is 3.976 when using the one-tailed t (4.999) and $p < 0.001$ as a basis) (see Table 6 and Figure 2). The first hypothesis is, therefore, confirmed.

Table 6. Structural model.

Model	R^2	ß	t Malue	Hypothesis
Direct model:				
PACAP > PERF	0.245	0.228	3.979	H_{1a}: Supported
RACAP > PERF		0.381	4.527	H_{1b}: Supported
Moderation model:				
Moderation of PROAC in PACAP > PERF	0.336	0.216	4.041	H_{2a}: Supported
Moderation of PROAC in RACAP > PERF	0.363	0.350	4.598	H_{2b}: Supported

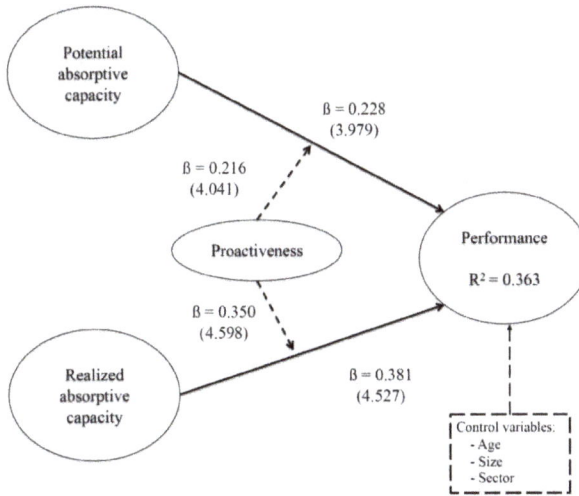

Figure 2. Structural Equation Model. t values are in parentheses.

Furthermore, RACAP positively and significantly influences performance (the path coefficient is 0.381 and the t value is 4.527 when using the one-tailed t at t (4.999) and $p < 0.01$ as a basis) (see Table 6 and Figure 2). The second hypothesis is, therefore, confirmed.

PACAP and RACAP have a positive effect on performance and explain 24.5% of the variance.

The moderating effect of proactivity was determined by first calculating its influence on PACAP and performance. In this case, the moderating effect is positive and significant, as the path coefficient is 0.216 and the t value is 4.041. The moderating effect of proactivity increases the influence of PACAP on performance to explain 33.6% of its variance (see

Table 5 and Figure 2). Finally, the moderating effect of proactivity on the influence of PACAP on performance is large, with an f2 value of 0.44 [112].

We subsequently calculated the moderating effect of proactivity on the influence of RACAP and performance. In this case, the moderating effect is positive and significant, as the path coefficient is 0.350 and the t value is 4.598. The moderating effect of proactivity increases the influence of PACAP on performance to explain 36.3% of its variance (see Table 6 and Figure 2). Henseler, Fassott, Dijkstra and Wilson [113] recommend estimating the effect size (f2) to determine the strength of the moderating effect. According to these authors, values of 0.02, 0.15 and 0.35 correspond to weak, moderate and strong moderating effects, respectively. In our case, the moderating effect of proactivity on the ratio of potential absorptive capacity to performance is 0.25 for f2 (moderate effect) while the moderating effect of proactivity on the ratio of realised absorptive capacity is 0.33 for f2 (moderate effect, almost strong). The explanation for this higher intensity of the moderating effect of proactivity on realised capacity may be that, in this case, there is greater tangibility and, therefore, it is more perceptible than in the first case.

None of the control variables have an influence that can be considered relevant (the path coefficients are less than 0.2), and they are not significant (their value is less than the recommended value, $p < 0.001$) (see Table 7).

Table 7. Control variables.

Variable	ß	t Value
Age	−0.046	0.670
Sector	−0.082	0.423
Size	−0.029	0.547

In order to complete the analysis of the structural model, the goodness of fit of the model was calculated by employing the standardised root mean square residual (SRMR) proposed by Hu and Bentler [114] and Henseler et al. [103]. In our case, the SRMR value was 0.063 (lower than 0.08 is adequate, as recommended by Henseler et al. [103]).

6. Discussion

This study raises awareness about the importance of ACAP in business performance. Using the most recent literature on dynamic capabilities and performance as a basis, we conducted a study to empirically demonstrate the importance of absorption capacity. The results of this study provide important evidence supporting the interplay between ACAP and proactivity and their contribution to business performance improvements. The results obtained indicate that companies that have managed to accompany their effort to engage with external knowledge with a proactive attitude should expect a better performance.

The theoretical framework employed in this study is focused on the theory of absorptive capacity, knowledge management and a resource and capabilities-based view. The relevant literature indicates that the relationship between performance and ACAP is positive, and improving ACAP will, therefore, increase business performance [3]. According to previous studies and the data analysed in this paper, we can state that companies with greater absorption capacity make much more effective use of all the information captured from exterior sources and improve their performance. In highly changing environments, this circumstance is fundamental for the improvement of their processes and products to improve their competitive position. The literature on ACAP postulates that greater investment in knowledge creation increases absorptive capacity, which ultimately helps the company to achieve higher innovative output and financial performance.

Furthermore, proactive sharing by companies will allow them to take advantage of market opportunities [115], thus anticipating future problems. Proactivity is ultimately the company's ability to engage resources by introducing new products and services ahead of competitors on the basis of predictions of future demand [116–118].

The main contribution of this paper to the existing literature is the discovery that absorption capacity becomes business performance mainly when proactivity is involved in this process. The role of proactivity in the relationship between ACAP and performance is a current topic in the literature. The results of this research on the moderating role played by proactivity in the effect of ACAP on firm performance contribute to the understanding of the effects of the use of strategically and rapidly absorbed and renewed information.

Although this work has made it possible to test the working hypotheses selected, it has some limitations that could provide opportunities for further research. Business performance, however, involves more capacities that could be analysed in the future. Another limitation lies in the regional nature of this work. Given the global nature of the economy, it is necessary to test its applicability in broader and more diverse contexts. Finally, as with all cross-sectional research, the testing of the hypotheses took place at a specific point in time. Although it is likely that the conditions under which the data were collected will not change substantially, there is no guarantee that this is definitively the case and researchers must, therefore, carefully interpret causality between constructs.

Author Contributions: Conceptualisation, R.S.-Z. and F.H.-P.; methodology, R.S.-Z. and F.H.-P.; software, I.P.-G., S.G.-B. and F.H.-P.; validation, R.S.-Z., F.H.-P. and I.P.-G.; formal analysis, R.S.-Z., I.P.-G. and S.G.-B.; investigation, I.P.-G. and S.G.-B.; resources, R.S.-Z. and S.G.-B.; data curation, R.S.-Z., F.H.-P., I.P.-G. and S.G.-B.; writing—original draft preparation, R.S.-Z., F.H.-P., I.P.-G. and S.G.-B.; writing—review and editing, R.S.-Z., F.H.-P., I.P.-G. and S.G.-B.; visualisation, R.S.-Z., F.H.-P., I.P.-G. and S.G.-B.; supervision, R.S.-Z.; project administration, R.S.-Z. All authors have read and agreed to the published version of the manuscript.

Funding: This research was funded by the Faculty of Law and Social Sciences of the University of Castilla–La Mancha.

Data Availability Statement: The data presented in this study are available from the corresponding author upon request. The data are not publicly available due to the requirements of ethical approval.

Acknowledgments: The research team is grateful for the invaluable collaboration of the Business Administration Department of the University of Castilla–La Mancha.

Conflicts of Interest: The authors declare no conflict of interest.

References

1. Randstad. La Gestión del Talento en Entornos VUCA. 2021. Available online: https://www.randstad.es/tendencias360/la-gestion-del-talento-en-entornos-vuca/ (accessed on 10 June 2021).
2. Abecassis-Moedas, C.; Mahmoud-Jouini, S.B. Absorptive capacity and source-recipient complementarity in designing new products: An empirically derived framework. *J. Prod. Inn. Manag.* **2008**, *25*, 473–490. [CrossRef]
3. Lane, P.J.; Salk, J.E.; Lyles, M.A. Absorptive capacity, learning, and performance in international joint ventures. *Strateg. Manag. J.* **2001**, *22*, 1139–1161. [CrossRef]
4. Lane, P.J.; Lubatkin, M. Relative absorptive capacity and interorganizational learning. *Strateg. Manag. J.* **1998**, *19*, 461–477. [CrossRef]
5. O'Connor, G.C. Major Innovation as a Dynamic Capability: A Systems Approach. *J. Prod. Innov. Manag.* **2008**, *25*, 313–330. [CrossRef]
6. Shaw, G.; Williams, A. Knowledge transfer and management in tourism organisations: An emerging research agenda. *Tour. Manag.* **2009**, *30*, 325–335. [CrossRef]
7. King, B.E.; Breen, J.; Whitelaw, P.A. Hungry for Growth? Small and Medium-sized Tourism Enterprise (SMTE) Business Ambitions, Knowledge Acquisition and Industry Engagement. *Int. J. Tour. Res.* **2012**, *16*, 272–281. [CrossRef]
8. Tzokas, N.; Kim, Y.-A.; Akbar, H.; Al-Dajani, H. Absorptive capacity and performance: The role of customer relationship and technological capabilities in high-tech SMEs. *Ind. Mark. Manag.* **2015**, *47*, 134–142. [CrossRef]
9. Hull, C.E.; Covin, J.G. Learning capability, technological parity, and innovation mode use. *J. Prod. Innov. Manag.* **2010**, *27*, 97–114. [CrossRef]
10. McGrath, R.G. Exploratory learning, innovative capacity, and managerial oversight. *Acad. Manag. J.* **2001**, *44*, 118–131.
11. Cohen, W.M.; Levinthal, D. Innovation and Learning: The Two Faces of R & D. *Econ. J.* **1989**, *99*, 569. [CrossRef]
12. Adams, G.L.; Lamont, B.T. Knowledge management systems and developing sustainable competitive advantage. *J. Knowl. Manag.* **2003**, *7*, 142–154. [CrossRef]
13. Darroch, J. Knowledge management, innovation and firm performance. *J. Knowl. Manag.* **2005**, *9*, 101–115. [CrossRef]

14. Jansen, J.J.P.; Bosch, F.A.J.V.D.; Volberda, H. Managing Potential and Realized Absorptive Capacity: How do Organizational Antecedents Matter? *Acad. Manag. J.* **2005**, *48*, 999–1015. [CrossRef]
15. Marqués, D.; Simón, P.F.J.G. The effect of knowledge management practices on firm performance. *J. Knowl. Manag.* **2006**, *10*, 143–156. [CrossRef]
16. Tu, Q.; Vonderembse, M.A.; Ragu-Nathan, T.; Sharkey, T.W. Absorptive capacity: Enhancing the assimilation of time-based manufacturing practices. *J. Oper. Manag.* **2005**, *24*, 692–710. [CrossRef]
17. Zahra, S.A.; George, G. Absorptive capacity: A review, reconceptualization, and extension. *Acad. Manag. Rev.* **2002**, *27*, 185–203. [CrossRef]
18. Jiménez-Barrionuevo, M.M.; García-Morales, V.J.; Molina, L.M. Validation of an instrument to measure absorptive capacity. *Technovation* **2011**, *31*, 190–202. [CrossRef]
19. Cohen, W.M.; Levinthal, D. Absorptive Capacity: A New Perspective on Learning and Innovation. *Adm. Sci. Q.* **1990**, *35*, 128. [CrossRef]
20. Kale, E.; Aknar, A.; Başar, Ö. Absorptive capacity and firm performance: The mediating role of strategic agility. *Int. J. Hosp. Manag.* **2018**, *78*, 276–283. [CrossRef]
21. Cordero, L.; Ferreira, J.J. Absorptive capacity and organizational mechanisms. *Rev. Int. Bus. Strateg.* **2019**, *29*, 61–82. [CrossRef]
22. Garay, L.; Font, X.; Pereira-Moliner, J. Understanding sustainability behaviour: The relationship between information acqui-sition, proactivity and performance. *Tour. Manag.* **2017**, *60*, 418–429. [CrossRef]
23. Thomas, R.; Wood, E. Innovation in tourism: Re-conceptualising and measuring the absorptive capacity of the hotel sector. *Tour. Manag.* **2014**, *45*, 39–48. [CrossRef]
24. Valentina, N.; Passiante, G. Impacts of Absorptive Capacity on Value Creation. *Anatolia* **2009**, *20*, 269–287. [CrossRef]
25. Lane, P.J.; Koka, B.R.; Pathak, S. The Reification of Absorptive Capacity: A Critical Review and Rejuvenation of the Construct. *Acad. Manag. Rev.* **2006**, *31*, 833–863. [CrossRef]
26. Lichtenthaler, U. Absorptive capacity, environmental turbulence, and the complementarity of organizational learning processes. *Acad. Manag. J.* **2009**, *52*, 822–846. [CrossRef]
27. Murovec, N.; Prodan, I. Absorptive capacity, its determinants, and influence on innovation output: Cross-cultural validation of the structural model. *Technovation* **2009**, *29*, 859–872. [CrossRef]
28. Parker, S.K.; Wang, Y. Helping people to make things happen: A framework for proactivity at work. *Int. Coach. Psychol. Rev.* **2015**, *10*, 62–75.
29. Strauss, K.; Griffin, M.A.; Parker, S.K.; Mason, C.M. Building and Sustaining Proactive Behaviors: The Role of Adaptivity and Job Satisfaction. *J. Bus. Psychol.* **2013**, *30*, 63–72. [CrossRef]
30. Wihler, A.; Blickle, G.; Ellen, I.B.P.; Hochwarter, W.A.; Ferris, G.R. Personal Initiative and Job Performance Evaluations: Role of Political Skill in Opportunity Recognition and Capitalization. *J. Manag.* **2014**, *43*, 1388–1420. [CrossRef]
31. Griffin, M.; Neal, A.; Parker, S. A New Model of Work Role Performance: Positive Behavior in Uncertain and Interdependent Contexts. *Acad. Manag. J.* **2007**, *50*, 327–347. [CrossRef]
32. Min, B.H.; Oh, Y. How Do Performance Gaps Affect Improvement in Organizational Performance? Exploring the Mediating Roles of Proactive Activities. *Public Perform. Manag. Rev.* **2020**, *43*, 766–789. [CrossRef]
33. Bierly, P.E.; Damanpour, F.; Santoro, M.D. The Application of External Knowledge: Organizational Conditions for Exploration and Exploitation. *J. Manag. Stud.* **2009**, *46*, 481–509. [CrossRef]
34. Tuppura, A.; Toppinen, A.; Jantunen, A. Proactiveness and corporate social performance in the global forest industry. *Int. For. Rev.* **2013**, *15*, 112–121. [CrossRef]
35. Joo, B.K.B.; Bennett, R.H., III. The influence of proactivity on creative behavior, organizational commitment, and job perfor-mance: Evidence from a Korean multinational. *J. Int. Interdisc. Bus. Res.* **2018**, *5*, 1–20.
36. Easterby-Smith, M.; Graça, M.; Antonacopoulou, E.; Ferdinand, J. Absorptive Capacity: A Process Perspective. *Manag. Learn.* **2008**, *39*, 483–501. [CrossRef]
37. Sakhdari, K. Absorptive capacity: Review and research agenda. *J. Org. Stud. Inn.* **2016**, *3*, 34–50.
38. Teece, D.J.; Pisano, G.; Shuen, A. Dynamic Capabilities and Strategic Management. *Strateg. Manag. J.* **1997**, *18*, 509–533. [CrossRef]
39. Zollo, M.; Winter, S.G. Deliberate Learning and the Evolution of Dynamic Capabilities. *Organ. Sci.* **2002**, *13*, 339–351. [CrossRef]
40. Akgün, A.E.; Lynn, G.S.; Byrne, J.C. Organizational Learning: A Socio-Cognitive Framework. *Hum. Relat.* **2003**, *56*, 839–868. [CrossRef]
41. Easterby-Smith, M. Disciplines of Organizational Learning: Contributions and Critiques. *Hum. Rel.* **1997**, *50*, 1085–1113. [CrossRef]
42. Chiva, R.; Allegre, J. Organisational Learning and Organizational Knowledge: The Integration of Two Approache. *Manag. Learn.* **2005**, *36*, 49–68. [CrossRef]
43. Oshri, I.; Pan, S.L.; Newell, S. Managing Trade-offs and Tensions between Knowledge Management Initiatives and Expertise Development Practices. *Manag. Learn.* **2006**, *37*, 63–82. [CrossRef]
44. Grant, R.M. Toward a knowledge-based theory of the firm. *Strateg. Manag. J.* **1996**, *17*, 109–122. [CrossRef]
45. Fleck, J.; Howells, J. Technology, the technology complex and the paradox of technological determinism. *Technol. Anal. Strateg. Manag.* **2001**, *13*, 523–531. [CrossRef]

46. Flor, M.L.; Oltra, M.J. An exploratory analysis of the relationship between absorptive capacity and business strategy. *Technol. Anal. Strateg. Manag.* **2013**, *25*, 1103–1117. [CrossRef]
47. Collis, D.J. Research Note: How Valuable are Organizational Capabilities? *Strateg. Manag. J.* **1994**, *15*, 143–152. [CrossRef]
48. Hall, R. The Strategic Analysis of Intangible Resources. *Strateg. Manag. J.* **1992**, 135–144. [CrossRef]
49. George, G.; Zahra, S.A.; Wheatley, K.; Khan, R. The effects of alliance portfolio characteristics and absorptive capacity on per-formance: A study of biotechnology firms. *J. High Technol. Manag. Res.* **2001**, *12*, 205–227. [CrossRef]
50. Dyer, J.; Singh, H. The relational view: Cooperative strategy and sources of interorganizational competitive advantage. *Acad. Manag. Rev.* **1998**, *23*, 660–679. [CrossRef]
51. Haro-Domínguez, M.C.; Arias-Aranda, D.; Lloréns-Montes, F.J.; Moreno, A.R. The Impact of absorptive capacity on techno-logical acquisitions engineering consulting companies. *Technovation* **2007**, *27*, 417–425. [CrossRef]
52. Zander, U.; Kogut, B. Knowledge and the Speed of the Transfer and Imitation of Organizational Capabilities: An Empirical Test. *Organ. Sci.* **1995**, *6*, 76–92. [CrossRef]
53. Van den Bosch, F.; Volberda, H.; de Boer, M. Coevolution of firm absorptive capacity and knowledge environment: Organiza-tional forms and combinative capabilities. *Org. Sci.* **1999**, *10*, 551–568. [CrossRef]
54. Song, Y.; Gnyawali, D.R.; Srivastava, M.K.; Asgari, E. In Search of Precision in Absorptive Capacity Research: A Synthesis of the Literature and Consolidation of Findings. *J. Manag.* **2018**, *44*, 2343–2374. [CrossRef]
55. Stock, G.N.; Greis, N.P.; Fischer, W.A. Absorptive capacity and new product development. *J. High Technol. Manag. Res.* **2001**, *12*, 77–91. [CrossRef]
56. Lichtenthaler, U. Absorptive capacity and firm performance: An integrative framework of benefits and downsides. *Technol. Anal. Strateg. Manag.* **2015**, *28*, 664–676. [CrossRef]
57. Griffiths-Hemans, J.; Grover, R. Setting the stage for creative new products: Investigating the idea fruition process. *J. Acad. Mark. Sci.* **2006**, *34*, 27–39. [CrossRef]
58. Bergh, D.D.; Lim, E.N.-K. Learning how to restructure: Absorptive capacity and improvisational views of restructuring actions and performance. *Strateg. Manag. J.* **2008**, *29*, 593–616. [CrossRef]
59. Yeoh, P.-L. Realized and Potential Absorptive Capacity: Understanding Their Antecedents and Performance in the Sourcing Context. *J. Mark. Theory Pract.* **2009**, *17*, 21–36. [CrossRef]
60. Volberda, H.W.; Foss, N.J.; Lyles, M.A. PERSPECTIVE—Absorbing the Concept of Absorptive Capacity: How to Realize Its Potential in the Organization Field. *Organ. Sci.* **2010**, *21*, 931–951. [CrossRef]
61. Bolívar-Ramos, M.T.; García-Morales, V.J.; Martín-Rojas, R. The effects of information technology on absorptive capacity and organizational performance. *Technol. Anal. Strateg. Manag.* **2013**, *25*, 905–922. [CrossRef]
62. Todorova, G.; Durisin, B. Absorptive capacity: Valuing a reconceptualization. *Acad. Manag. Rev.* **2007**, *32*, 774–786. [CrossRef]
63. Ali, M.; Kan, K.A.S.; Sarstedt, M. Direct and configurational paths of absorptive capacity and organizational innovation to successful organizational performance. *J. Bus. Res.* **2016**, *69*, 5317–5323. [CrossRef]
64. Camisón, C.; Forés, B. Knowledge absorptive capacity: New insights for its conceptualization and measurement. *J. Bus. Res.* **2010**, *63*, 707–715. [CrossRef]
65. Delmas, M.; Hoffmann, V.H.; Kuss, M. Under the Tip of the Iceberg: Absorptive Capacity, Environmental Strategy, and Competitive Advantage. *Bus. Soc.* **2011**, *50*, 116–154. [CrossRef]
66. Flatten, T.; Greve, G.I.; Brettel, M. Absorptive Capacity and Firm Performance in SMEs: The Mediating Influence of Strategic Alliances. *Eur. Manag. Rev.* **2011**, *8*, 137–152. [CrossRef]
67. Fosfuri, A.; Tribó, J.A. Exploring the antecedents of potential absorptive capacity and its impact on innovation performance. *Omega* **2008**, *36*, 173–187. [CrossRef]
68. Thérin, F. Absorptive capacity: An empirical test of Zahra and George's contribution in small business settings. *Gestion 2000* **2007**, *24*, 17–30.
69. Baker, T.; Miner, A.S.; Eesley, D.T. Improvising firms: Bricolage, account giving and improvisational competencies in the founding process. *Res. Policy* **2003**, *32*, 255–276. [CrossRef]
70. Morgan, R.E.; Turnell, C.R. Market-based Organizational Learning and Market Performance Gains. *Br. J. Manag.* **2003**, *14*, 255–274. [CrossRef]
71. Murray, S.; Peyrefitte, J. Knowledge Type and Communication Media Choice in the Knowledge Transfer Process. *J. Man. Issues* **2007**, *19*, 111–133.
72. Ahuja, G.; Lampert, C. Entrepreneurship in the large corporation: A longitudinal study of how established firms create breakthrough inventions. *Strateg. Manag. J.* **2001**, *22*, 521–543. [CrossRef]
73. Brettel, M.; Greve, G.I.; Flatten, T.C. Giving up linearity: Absorptive capacity and performance. *J. Manag. Issues* **2011**, *23*, 164–189.
74. Venkatraman, N. Strategic Orientation of Business Enterprises: The Construct, Dimensionality, and Measurement. *Manag. Sci.* **1989**, *35*, 942–962. [CrossRef]
75. Grant, A.M.; Ashford, S.J. The dynamics of proactivity at work. *Res. Organ. Behav.* **2008**, *28*, 3–34. [CrossRef]
76. Miller, D.; Friesen, P.H. Archetypes of Strategy Formulation. *Manag. Sci.* **1978**, *24*, 921–933. [CrossRef]
77. Sethi, S.P. A conceptual framework for environmental analysis of social issues and evaluation of business response patterns. *Acad. Manag. Rev.* **1979**, *4*, 63–74. [CrossRef]

78. García-Morales, V.J.; Ruiz-Moreno, A.; Llorens-Montes, F.J. Effects of technology absorptive capacity and technology pro-activity on organizational learning, innovation and performance: An empirical examination. *Tech. Anal. Strateg. Manag.* **2007**, *19*, 527–558. [CrossRef]
79. Lumpkin, G.T.; Dess, G.G. Clarifying the entrepreneurial orientation construct and linking it to performance. *Acad. Manag. Rev.* **1996**, *21*, 135–172. [CrossRef]
80. Eisenhardt, K.M.; Martin, J.A. Dynamic capabilities: What are they? *Strateg. Manag. J.* **2000**, *21*, 1105–1121. [CrossRef]
81. Di Stefano, G.; Peteraf, M. The elephant in the room of Dynamic Capabilities: Bringing two diverging conversations together. *Acad. Manag. Proc.* **2012**, *34*, 1389–1410. [CrossRef]
82. Rodríguez, M.A.; Martín, E. Born-Global SMEs, Performance, and Dynamic Absorptive Capacity: Evidence from Spanish Firms. *J. Small Bus. Manag.* **2019**, *57*, 298–326. [CrossRef]
83. Cohen, J.A. Power primer. *Psych. Bull.* **1992**, *112*, 155. [CrossRef]
84. Faul, F.; Erdfelder, E.; Buchner, A.; Lang, A.-G. Statistical power analyses using G*Power 3.1: Tests for correlation and regression analyses. *Behav. Res. Methods* **2009**, *41*, 1149–1160. [CrossRef]
85. Hernández-Perlines, F.; Moreno-García, J.; Yáñez-Araque, B. Training and business performance: The mediating role of absorptive capacities. *SpringerPlus* **2016**, *5*, 1–16. [CrossRef]
86. Olson, E.M.; Slater, S.F.; Hult, G.T.M. The Performance Implications of Fit among Business Strategy, Marketing Organization Structure, and Strategic Behavior. *J. Mark.* **2005**, *69*, 49–65. [CrossRef]
87. Manzano-García, G.; Ayala-Calvo, J.-C. Entrepreneurial Orientation: Its Relationship with the Entrepreneur's Subjective Success in SMEs. *Sustainability* **2020**, *12*, 4547. [CrossRef]
88. Chirico, F.; Sirmon, D.G.; Sciascia, S.; Mazzola, P. Resource orchestration in family firms: Investigating how entrepreneurial orientation, generational involvement, and participative strategy affect performance. *Strateg. Entrep. J.* **2011**, *5*, 307–326. [CrossRef]
89. Kellermanns, F.W.; Eddleston, K.A.; Zellweger, T.M. Article Commentary: Extending the Socioemotional Wealth Perspective: A Look at the Dark Side. *Entrep. Theory Pract.* **2012**, *36*, 1175–1182. [CrossRef]
90. Krauss, S.I.; Frese, M.; Friedrich, C.; Unger, J.M. Entrepreneurial orientation: A psychological model of success among southern African small business owners. *Eur. J. Work. Organ. Psychol.* **2005**, *14*, 315–344. [CrossRef]
91. Naldi, L.; Nordqvist, M.; Sjöberg, K.; Wiklund, J. Entrepreneurial orientation, risk taking, and performance in family firms. *Fam. Bus. Rev.* **2007**, *20*, 33–47. [CrossRef]
92. Wiklund, J.; Shepherd, D. Knowledge-based resources, entrepreneurial orientation, and the performance of small and medi-umsized businesses. *Strateg. Manag. J.* **2003**, *24*, 1307–1314. [CrossRef]
93. Hernández-Perlines, F.; Covin, J.G.; Ribeiro-Soriano, D.E. Entrepreneurial orientation, concern for socioemotional wealth preservation, and family firm performance. *J. Bus. Res.* **2021**, *126*, 197–208. [CrossRef]
94. Chrisman, J.J.; Chua, J.H.; Sharma, P. Trends and Directions in the Development of a Strategic Management Theory of the Family Firm. *Entrep. Theory Pract.* **2005**, *29*, 555–575. [CrossRef]
95. Ibarra-Cisneros, M.-A.; Hernandez-Perlines, F. Entrepreneurial orientation, absorptive capacity and business performance in SMEs. *Meas. Bus. Exce.* **2019**, *24*, 417–429. [CrossRef]
96. Gefen, D.; Rigdon, E.E.; Straub, D. Editor's comments: An update and extension to SEM guidelines for administrative and social science research. *Mis Q.* **2011**, *35*, 3–14. [CrossRef]
97. Ringle, C.M.; Wende, S.; Becker, J.M. Smart PLS Boenningstedt: SmartPLS GmbH. 2015. Available online: https://www.smartpls.com (accessed on 5 March 2021).
98. Hair, J.F.; Risher, J.; Sarstedt, M.; Ringle, C.M. When to use and how to report the results of PLS-SEM. *Eur. Bus. Rev.* **2019**, *31*, 2–24. [CrossRef]
99. Sarstedt, M.; Mooi, E.A. *A Concise Guide to Market Research: The Process, Data, and Methods Using IBM SPSS Statistics*; Springer: Berlin, Germany, 2019.
100. Jöreskog, K.G.; Wold, H.O. *Systems Under Indirect Observation: Causality, Structure, Prediction*; North Holland: Amsterdam, The Netherlands, 1982.
101. Astrachan, J.H.; Jaskiewicz, P. Emotional Returns and Emotional Costs in Privately Held Family Businesses: Advancing Traditional Business Valuation. *Fam. Bus. Rev.* **2008**, *21*, 139–149. [CrossRef]
102. Reinartz, W.; Haenlein, M.; Henseler, J. An empirical comparison of the efficacy of covariance-based and variance-based SEM. *Int. J. Res. Mark.* **2009**, *26*, 332–344. [CrossRef]
103. Henseler, J.; Ringle, C.M.; Sarstedt, M. A new criterion for assessing discriminant validity in variance-based structural equation modeling. *J. Acad. Mark. Sci.* **2014**, *43*, 115–135. [CrossRef]
104. Hair, J.F.; Sarstedt, M.; Ringle, C.M. Rethinking some of the rethinking of partial least squares. *Eur. J. Mark.* **2019**, *53*, 566–584. [CrossRef]
105. Barclay, D.; Higgins, C.; Thompson, R. The partial least squares (PLS) approach to causal modeling: Personal computer adoption and use as an illustration. *Tech. Stud.* **1995**, *2*, 285–309.
106. Hair, J.F., Jr.; Sarstedt, M.; Ringle, C.M.; Gudergan, S.P. *Advanced Issues in Partial Least Squares Structural Equation Modeling*; Sage Publications: Thousand Oaks, CA, USA, 2017.

107. Roldan, J.L.; Sanchez-Franco, M.J. Variance-based structural equation modeling: Guidelines for using partial least squares in information systems research. In *Research Methodologies, Innovations and Philosophies in Software Systems Engineering and Information systems*; Mora, M., Gelman, O., Steenkamp, A., Raisinghani, M., Eds.; IGI Global: Hershey, PA, USA, 2012; pp. 193–221.
108. Fornell, C.; Larcker, D.F. Evaluating structural equation models with unobservable variables and measurement error. *J. Mark. Res.* **1981**, *18*, 39–50. [CrossRef]
109. Diamantopoulos, A.; Sarstedt, M.; Fuchs, C.; Wilczynski, P.; Kaiser, S. Guidelines for choosing between multi-item and single-item scales for construct measurement: A predictive validity perspective. *J. Acad. Mark. Sci.* **2012**, *40*, 434–449. [CrossRef]
110. Drolet, A.L.; Morrison, D.G. Do We Really Need Multiple-Item Measures in Service Research? *J. Serv. Res.* **2001**, *3*, 196–204. [CrossRef]
111. Dijkstra, T.K.; Henseler, J. Consistent Partial Least Squares Path Modeling. *MIS Q.* **2015**, *39*, 297–316. [CrossRef]
112. Chin, W.W. *Commentary: Issues and Opinion on Structural Equation Modeling*; Management Information Systems Research Center, University of Minnesota: Minneapolis, MN, USA, 1998.
113. Chin, W.W. *How to Write Up and Report PLS Analyses*; Springer: Berlin/Heidelberg, Germany, 2010; pp. 655–690.
114. Hu, L.T.; Bentler, P.M. Fit indices in covariance structure modeling: Sensitivity to underparameterized model misspecification. *Psychol. Methods* **1998**, *3*, 424. [CrossRef]
115. Kellermanns, F.W.; Eddleston, K.A.; Barnett, T.; Pearson, A. An Exploratory Study of Family Member Characteristics and Involvement: Effects on Entrepreneurial Behavior in the Family Firm. *Fam. Bus. Rev.* **2008**, *21*, 1–14. [CrossRef]
116. Covin, J.G.; Slevin, D.P. Strategic management of small firms in hostile and benign environments. *Strateg. Manag. J.* **1989**, *10*, 75–87. [CrossRef]
117. Rauch, A.; Wiklund, J.; Freese, M.; Lumpkin, G.T. Entrepreneurial orientation and business performance: Cumulative empirical evidence. Presented at the 23rd Babson College Entrepreneurship Research Conference, Glasgow, UK, 1 January 2004.
118. Ma, C.; Lin, X.; Chen, Z.X.; Wei, W. Linking perceived overqualification with task performance and proactivity? An examination from self-concept-based perspective. *J. Bus. Res.* **2020**, *118*, 199–209. [CrossRef]

Article

Learning Mathematics of Financial Operations during the COVID-19 Era: An Assessment with Partial Least Squares Structural Equation Modeling

María del Carmen Valls Martínez *, Pedro Antonio Martín-Cervantes, Ana María Sánchez Pérez and María del Carmen Martínez Victoria

Department of Economics and Business, University of Almería, 04120 Almería, Spain; pmc552@ual.es (P.A.M.-C.); asp884@ual.es (A.M.S.P.); mcmvic@ual.es (M.d.C.M.V.)
* Correspondence: mcvalls@ual.es

Citation: Valls Martínez, M.d.C.; Martín-Cervantes, P.A.; Sánchez Pérez, A.M.; Martínez Victoria, M.d.C. Learning Mathematics of Financial Operations during the COVID-19 Era: An Assessment with Partial Least Squares Structural Equation Modeling. *Mathematics* **2021**, *9*, 2120. https://doi.org/10.3390/math9172120

Academic Editor: Maria C. Mariani

Received: 5 August 2021
Accepted: 26 August 2021
Published: 1 September 2021

Publisher's Note: MDPI stays neutral with regard to jurisdictional claims in published maps and institutional affiliations.

Copyright: © 2021 by the authors. Licensee MDPI, Basel, Switzerland. This article is an open access article distributed under the terms and conditions of the Creative Commons Attribution (CC BY) license (https://creativecommons.org/licenses/by/4.0/).

Abstract: The COVID-19 pandemic has affected all walks of life, including education. Universities have been forced to teach in a blended or online environment, which has led professors to adapt their traditional teaching–learning methodologies. The professors of Mathematics of Financial Operations at the University of Almeria (Spain) have created video tutorials so that students can autonomously prepare the theoretical part of the subject, leaving the face-to-face classes for practical exercises. This article aims to analyze the effectiveness of video tutorials and the autonomy finally achieved by students in their learning. For this purpose, a questionnaire was carried out in which, through 21 questions, the constructs Autonomy, Effectiveness, Depth, Format, Challenge, and Use were assessed. Based on these six latent variables, the proposed model using the Partial Least Squares Structural Equation Modeling (PLS-SEM) methodology revealed that students considered the Format and Depth of the video tutorials crucial for genuinely effective performance learning and promoting their autonomy. On the other hand, the variables Challenge and Use were poorly rated. This article presents an original valuation model, which has the virtue of achieving a prediction of 78.6% and, in addition, has high predictive power.

Keywords: video tutorials; blended learning; online learning; financial mathematics; COVID-19; autonomy; effectiveness; PLS-SEM

1. Introduction

At this point, practically everything may have been said about the global expansion of COVID-19 [1], although sadly, there is still a long way to its definitive eradication [2]. As an irony of the destiny, the global expansion of this pandemic followed an expansive pattern quite similar to the black plague of the 14th century, with which it presents a series of resemblances [3]. In a certain sense, it is also comparable with the effects that the Spanish flu brought to world economies and societies at the beginning of the 20th century [4], slowing down its growth and drastically paralyzing any human activity. Even though the World Health Organization decreed on 11 March 2020 that COVID-19 was a global pandemic and previously warned of the seriousness of its potential menace [5], the early-warning mechanisms inexplicably failed during the first two weeks of March 2020 in all countries, contributing to the spreading of the pandemic around the world.

Regardless of the complexities of this problem, essential human activities such as education could under no circumstances cease; hence, public authorities adopted different measures in order to combat the outbreak while trying to ensure the continuation, as far as possible, of normal daily life prior to the pandemic. In this context, the Spanish government opted for declaring a State of Emergency [6], guaranteeing the non-interruption of education through online instruction.

One of the key subjects in the holistic development of students is mathematics [7], as outlined by Inglis and Foster [8] in their semi-centennial perspective. Nevertheless, the entry of the COVID-19 on the scene has meant a redefinition of the student–professor relationship compared to before the pandemic [9]. Hence, the teaching of this field of knowledge given the current scenario has been denominated as a "Historical Disadvantage" by Chirinda et al. [10]. In fact, compared to very recent times in which the evaluation of mathematics learning support could only take the face-to-face mode into account in daily teaching [11], the difficulties caused by the pandemic have represented a disruptive phenomenon with respect to traditional teaching, and, even more so in the teaching of mathematics, since it has resulted not only in the quasi-obligatory decision to opt for online teaching but also in the implementation of practically any element of support coming from the new technologies [12]. In this sense, online teaching of mathematics has been successfully introduced in all educational stages or cycles such as primary school [13], secondary education [10,12,14–16], or at the university level, where these types of initiatives can be found around the world at university systems of countries that are geographically distant and sociocultural heterogeneous, such as the United States [17], Indonesia [18], South Africa [19], or Slovakia [20]. It follows from these facts that the teaching of mathematics from the advent of the COVID-19 onwards is likely to be restricted to automated teaching such that humans will have to limit themselves to interacting with all kinds of technological means [21].

The Spanish universities precisely took this path after the enactment of the State of Emergency [6], under which the faculty as a whole had to take charge of the urgent situation, like their colleagues in other latitudes, by carrying out their teaching via online and applying as teaching resources all technological means available at their disposal. In this context was subsumed the Department of Business and Economics of the University of Almería, Spain, when it was necessary not to cease the educative process; hence it was decided to elaborate video tutorials to support the implementation of the subject Mathematics of Financial Operations, part of which was taught online.

Thus, at the end of the 2020–2021 academic year, a strictly anonymous online questionnaire was distributed among the entire student body. The professors sought to analyze the impact of video tutorials elaborated ad hoc on the students' final performance of this academic discipline. This survey was designed by considering the three elementary dimensions schematized by Glass and Sue [22], emphasizing the student's preferences, satisfaction, and degree of self-learning. All surveys were tabulated and modeled according to the Partial Least Squares Structural Equation Modelling methodology (also known as PLS-SEM) [23–25], a set of procedures that derive from three prior lines of research: the iterative models aimed at obtaining solutions for single-component, multi-component models, and canonical correlations [26], the nonlinear iterative partial least squares (NIPALS) algorithm [27] and from other pre-existing models that opted for the inclusion of latent variables [28].

Although the initial scope of application of this methodology was originally focused on marketing [28], today it is applied to multiple areas of business and economics [29], such as finance [30] or management accounting [25] and other domains of knowledge such as the information sciences [31], based on its versatility and flexibility. It has been designed, according to Hair et al. [32], as a "silver bullet" of quantitative research that can be used both for exploratory and predictive purposes, being especially reliable when it comes to validating models or verifying hypotheses [31]. Most of the works that have previously analyzed the incidence of explanatory videos in the learning process of mathematics start from a purely exploratory perspective [18]. However, our work also applies a predictive dimension, since the model obtained has been optimally adjusted to the employed sample through the PLS-SEM methodology. To the best of our knowledge, this approach has been used for the first time in the analysis of a questionnaire related to the performance of mathematics teaching during the COVID-19 era, highlighting eminently positive feedback from students who consider that video tutorials have been helpful to shortening their study hours, giving priority to factors such as the visual aspect and the format of videos produced by the professors in charge of this unforgettable pilot experience.

This research is structured as follows: First, Section 2 contextualizes the repercussions and effects of the employment of video tutorials in the contemporary teaching–learning process, establishing the hypotheses involved in this research. Next, Section 3 defines the dataset and scale used and the constructs and indicators elaborated, defining the existing interrelationships between the different variables. Subsequently, Section 4 gathers the results obtained from applying the PLS-SEM methodology on the starting dataset, ending with Section 5, which detail the main discussions and conclusions reached in light of this manuscript.

2. Theoretical Background

Technological progress is a reality in the higher education field, not least since the COVID-19 pandemic. Most universities moved to online learning environments to replace face-to-face classes [33]. The concept of e-learning includes the use of new technologies and the Internet to improve the quality of learning by facilitating access to resources as well as remote exchanges and collaboration [34].

Today, technology is mainly used in virtual and asynchronous environments, while its inclusion with face-to-face classes is known as blended learning. Focusing on the use of asynchronous videos, extant literature has displayed its upsides [33]. Generally speaking, video tutorials tend to have a positive effect on students' performance. In this vein, videos improve student's motivation and autonomy [35] and allow them to regulate their workload by re-watching the classes whenever they want [36].

2.1. Impact of Video Tutorials on the Effectiveness

The effectiveness of videos depends on various factors [37]. In accordance with the cognitive theory of multimedia learning (CTML), which is how multimedia learning affects students' cognitive processes regarding the assimilation of knowledge [38], students need to be actively processing incoming information for learning to occur. In this sense, visual (such as an image on the screen) and verbal (the professor's voice) stimuli are addressed separately in two channels. It has been empirically proved that video and audio input simultaneously increase the student's engagement and therefore lead to better performance [39–41].

Furthermore, it is necessary to keep in mind that, given that the human capacity to attend to all the information presented is limited, too much information inhibits student's knowledge acquisition. Several authors [40,42] have highlighted that the instructor's presence does not necessarily bring about better engagement, satisfaction, or perceived learning in the students. Moreover, other authors [39,43] state that there is no optimal format for video tutorials.

In addition, the availability of video tutorials, where its visualization may be repeated (e.g., rewinding complex concepts), catching up on missed classes, is a positive advantage for students' learning process [44,45]. Furthermore, Koumi [46] say that "video by itself will not prove effective and will fail to achieve this potential if it is designed badly". Videos need to be pedagogically effective to achieve student learning. Thus, they have to be designed for cognitive engagement and constructive reflection [47]. However, as remarked in the empirical study done by Miner and Stefaniak [48], video should not replace direct teaching but should be a complement of it. In this line, Means, Toyama, Murphy, and Baki [49] demonstrated that a blended-learning method is more effective than entirely face-to-face classes. However, they found similar levels of effectiveness between purely online classes and face-to-face.

Based on this understanding, we propose the following hypotheses:

Hypothesis 1 (H1). *Format is positively related to Effectiveness.*

Hypothesis 2 (H2). *Usage is positively related to Effectiveness.*

The effectiveness of videos is based on deep knowledge [50,51]. The application of e-learning techniques in the mathematics field increases the memorization, comprehension, and internalization of contents [52]. In this way, it is important to highlight that "it is not enough for students to reproduce the knowledge faithfully; they must have a good command of the structure and existing interrelations and give coherence to the knowledge" [37].

In this sense, an in-depth understanding by using investigation methods and reflective and critical thinking is related to learning effectiveness [37]. According to Dede [53], an asynchronous method develops thoughtful and reflective thinking among students, improving the learning process [54]. Similarly, in order to a effectiveness in the learning process, it is essential to implement learning tasks of a challenging nature. Complex problems requiring divergent thinking can improve creativity and forge significant relationships and connections in learning [37].

Therefore, we propose the following hypotheses:

Hypothesis 3 (H3). *Depth is positively related to Effectiveness.*

Hypothesis 4 (H4). *Challenge is positively related to Effectiveness.*

2.2. Impact of Video Tutorials on Autonomy

Learner autonomy is the ability to control one's learning by defining one's goals and strategies for knowing and evaluating their development. In e-learning, where professors' intervention is not direct, motivation, environment, tasks, educators, and materials play an essential role in making students learn autonomously [55].

As we mentioned before, the use of the video tutorial format promotes students' autonomy [35]. Specifically, the video format gives students the possibility to choose when to participate, which is perceived as a tool that increases the students' degree of autonomy [56]. These findings are in line with [57], where it is indicated that videos are effective when allowing for autonomous learning, use vivid multimedia instruction, and enhance professor–learner communication [48].

Usually, e-learning methods have a positive impact because they favor self-regulation and allow students to develop their own learning processes [58]. As is said by Moreno-Guerrero et al. [52], "Autonomy is the dimension where there is a greater contrast when comparing the expository-traditional method with the e-learning method. This fact may be mainly due to the fact that the e-learning method favors self-regulation of learning". Learning autonomy allows each student to save study time by managing his/her rhythm and space to tailor to his/her educational needs. Thus, learning efficiency is related to autonomy [55].

Based on the aforementioned literature, we propose the following hypotheses:

Hypothesis 5 (H5). *Format is positively related to Autonomy.*

Hypothesis 6 (H6). *Effectiveness is positively related to Autonomy.*

Reflective and critical thinking leads to a deep formation. In this sense, students' inference abilities when using video tutorials, combined with self-assessment, where they may learn from their own errors, implies that depth learning is associated with improved motivation and autonomy [59].

Moreno-Guerrero et al. [52] also showed that these techniques favor students' inference, increasing their participation and interest in the contents while the professor becomes a guide. Moreover, they found that video tutorials directly influence autonomy and, in the same way, may lead to an acquisition of mathematical concepts and results. In this sense, autonomy permits students to coordinate the activities to be developed to achieve effective learning of the concepts, giving place to formative challenges.

In this way, the following hypotheses are proposed:

Hypothesis 7 (H7). *Challenge is positively related to Autonomy.*

Hypothesis 8 (H8). *Depth is positively related to Autonomy.*

2.3. Moderating Effects

Previous studies demonstrate that videos' effectiveness is influenced by a variety of characteristics: depth, challenge, format and usage. However, these studies do not capture the effectiveness's contribution to the relationship between previous constructs and students' autonomy. In this sense, we are referring to the moderating effect of effectiveness.

There is a large stream of literature demonstrating that variables such as gender [60,61], educational context [62], format [49], or study discipline [63] have a moderating effect on the effectiveness of online courses and blended learning.

In the same way, there is a large amount of meta-analysis research that included several possible moderators of videos' effectiveness. For example, Van Alten, Phielix, Janssen, and Kester [64] and Spanjers et al. [65] found that the inclusion of design characteristics such as quizzes, lecture activities, or small group assignments increases the effectiveness of sessions. Müller and Mildenberger [66] analyzed as moderating effects of the educational research, study condition (study, level), and methods, finding differences in the effectiveness between blended and conventional classes. Meanwhile, Means et al. [49] included several moderator variables explaining differences in outcomes between online learning and face-to-face environments. Müller and Mildenberger [66] also revealed similar moderating results. They did not find a significant effect of these variables. Vo et al. [63] tested the moderating effects of disciplines and the final evaluation method on students' outcomes. STEM disciplines had a more significant impact on students' outcomes.

Consequently, we hypothesized about the existence of moderating effects of effectiveness in the learning of Financial Mathematics.

Keeping in mind these studies, we postulate the following hypotheses:

Hypothesis 9 (H9). *Effectiveness positively moderates the link between Format and Autonomy.*

Hypothesis 10 (H10). *Effectiveness positively moderates the link between Depth and Autonomy.*

Hypothesis 11 (H11). *Effectiveness positively moderates the link between Challenge and Autonomy.*

3. Materials and Methods

The data used in the empirical part of this work are primary since they were obtained through a direct survey of students.

The authors are professors of the subject Mathematics of Financial Operations, corresponding to the second year of the Finance and Accounting degree at the University of Almeria, in Spain. The course is compulsory for students of this degree. In addition, it is offered as a third-year elective for students of the Mathematics degree in order to obtain a special recognition named "Mention in Mathematics and Finance". During the 2020–2021 academic year, the course was scheduled to be taught in blended learning, although it was actually taught entirely online due to the pandemic caused by COVID-19. However, following the initial spirit of blended learning, half of the online theoretical classes were taught in asynchronous mode, and the rest of the classes, including the practical classes, were in synchronous mode.

With the Camtasia 2020 software, the professors recorded video tutorials on the theoretical contents of the subject. At the end of the course, an anonymous questionnaire was sent to the students. Professors want to determine their opinion on the learning process with video tutorials. A total of six latent variables (constructs) were analyzed, all considered composites, and measured by between 2 and 5 manifest variables (indicators), which indicate different dimensions of the construct. Table 1 summarizes all the variables analyzed: constructs and corresponding indicators.

Table 1. Constructs and indicators.

Construct	Indicator	Description
Autonomy (Formative)	A1	Video tutorials favour my autonomous learning
	A2	Video tutorials help me to manage my study time better

A total of 21 indicators were considered, all measured on a Likert scale ranging from 1 to 7, as shown in Table 2. The Effectiveness construct was estimated in reflective mode, considering the covariation of its indicators. However, the rest of the constructs were estimated in a formative mode, considering a causal relationship of the indicators on the construct.

Table 2. Likert scale.

Value	Meaning
1	I fully disagree
2	I quite disagree
3	I disagree
4	Neither agree nor disagree (neutral)
5	I Agree
6	I quite agree
7	I fully agree

Three indicators were considered to measure the Autonomy achieved by the student with the use of video tutorials: strengthening of autonomous learning (A1), improvement in the management of study time (A2), and the possibility of resolving doubts without the need to attend face-to-face tutorials (A3). It should be noted that blended learning requires students to compensate for the professor's presence with their autonomous capacity in a more or less extensive part of the course.

The level of Effectiveness of the video tutorials translates into the degree of autonomy achieved by the student. If the video tutorials are useful for learning the subject (E1), if they help to understand the contents better (E2), and if they save time in the study (E3), they will be more effective.

In turn, the time the student has devoted to the video tutorials, both individually (U1) and in relation to peers (U2), determines their Usage, which will influence effectiveness.

If video tutorials are challenging for the learner, this should encourage students' autonomy. Video tutorials will be a Challenge when they help them to relate the subject matter meaningfully (C1), to coordinate the learning activities to be developed (C2), to seek alternative perspectives (C3), and to be creative in finding solutions (C4).

The Depth of the video tutorials will be conducive to learner autonomy. A video tutorial will be deep when it leads to questioning and the use of research methods (D1), helps to deepen the understanding of fundamental concepts and ideas (D2), relates fundamental concepts (D3), and develops critical and reflective thinking (D4).

Finally, the Format of the video tutorials, i.e., the way they are made, will also influence the student's autonomy. In this sense, the variety or not of the professors that appear in them (F1), the length of the subject matter covered in each video tutorial (F2), its duration (F3), its visual aspect (F4), and the clarity and precision of the explanations (F5) will be decisive in the student learning and, ultimately, in the autonomy achieved.

The constructs Depth and Challenge are based on the ATAE questionnaire for the analysis of assessment tasks [37]. The constructs Effectiveness, Format, and Autonomy were based on the survey conducted by Financial Mathematics professors in a previous course [67], while Use was newly created.

Figure 1 shows all the study variables developed and the relationships between them derived from the previously described hypotheses, namely the conceptual research model known as a nomogram. It should be noted that the model presented is absolutely original.

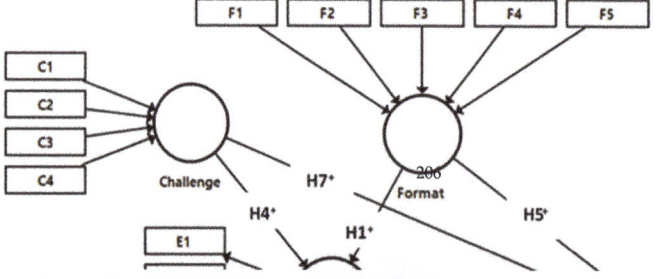

The relationships between each construct and its indicators constitute what is known as the *measurement model*. The relationships between the different constructs constitute the *structural model*. The assessment of both models was carried out using the statistical methodology of Partial Least Squares Structural Equation Modeling (PLS-SEM), implemented by the software *SmartPLS* (v.3.3.2.) [68].

PLS-SEM does not require the data to have a normal distribution, since it uses a non-parametric bootstrap technique to test the significance of the coefficients [69–71]. In this analysis, 10,000 samples with replacement were considered.

The PLS algorithm involves four stages. In the first stage, the structural model weights and latent variable scores are determined through a four-step iterative process [72–74]:

1. Preliminary outside estimation of the latent variables scores through the linear combination of their manifest variables:

$$\xi_{j,n} := \sum_{k_j} \tilde{w}_{k_j} x_{k_j,n},$$

where ξ_j is the latent variable, $j = 1, \cdots, J$; x_{k_i} is the manifest variable k of the latent variable i, $k = 1, \cdots, K$; $\tilde{w}_{k_j}$ is the estimated outer weight of the indicator x_{k_j}; n is the specific observation, $n = 1, \cdots, N$.

2. Inner weights estimation of the latent variable, by using the factor weighting scheme and according to the sign of the correlations between latent variables:

$$v_{j,i} = \begin{cases} \text{cov}(\xi_j, \xi_i), & \text{if } \xi_j \text{ and } \xi_i \text{ are adjacent} \\ 0, & \text{otherwise} \end{cases}$$

3. Internal estimation of latent variable scores by linear combination of their adjacent variables, by using the inner weights of the previous step:

$$\tilde{\xi}_j := \sum_i \beta_{j,i} \xi_i.$$

4. Outer weights estimation, which are calculated differently depending on whether the constructs are estimated in a formative or reflective mode:

$$\text{Formative (mode B) } \tilde{\xi}_{j,n} = \sum_{k_j} \tilde{w}_{k_j} x_{k_j,n} + \delta_{j,n}$$

$$\text{Reflective (mode A) } x_{k_j,n} = \tilde{w}_{k_j} \tilde{\xi}_{j,n} + e_{k_j,n}$$

where $\delta_{j,n}$ is the error term from a multiple regression and $e_{k_j,n}$ is the error term from a bivariate regression.

The algorithm is iterative and terminates when the weights in Step 4 experience a marginal variation between two consecutive iterations, or when the maximum number of iterations is reached.

Based on the latent variable scores obtained in Stage 1 and through a series of ordinary least squares regressions in Stages 2 and 3, the external weights and loadings of the manifest variables, the path coefficients, and a series of parameters, such as the direct and indirect effects, the coefficient of determination R^2 of the endogenous latent variables and different evaluation criteria of the model are determined.

Finally, in stage 4, the non-parametric bootstrapping procedure (sampling technique with replacement) is applied to test the significance of the model parameters.

The assessment of the measurement model requires different treatment of the reflective and formative constructs. To evaluate the reflective composite, it is necessary to contrast the reliability of each of its indicators as well as the reliability of the construct and its convergent and discriminant validity. Additionally, the evaluation of the formative composite involves studying the multicollinearity between the different indicators, as well as the relevance and significance of the weights of each indicator.

Composite reliability [72], which should range from 0.7 to 0.95, is the lower limit of internal consistency reliability of the reflective construct. This measure is determined by

$$\rho_c = \frac{\left(\sum_{k=1}^{K} l_k\right)^2}{\left(\sum_{k=1}^{K} l_k\right)^2 + \sum_{k=1}^{K} \text{var}(e_k)},$$

where l_k is the outer loading of the manifest variable k corresponding to a latent variable measured with K indicators; e_k is the measurement error of k; and $\text{var}(e_k)$ corresponds to the measurement error variance and it is calculated as $1 - l_k^2$.

Cronbach's alpha is the upper limit of internal consistency reliability

$$\text{Cronbach's } \alpha = \frac{K\bar{r}}{1 + (K-1)\bar{r}},$$

where $\bar{r}$ is the mean of the triangular correlation matrix.

The Dijkstra–Henseler's Rho usually stands between the two previous measures [75]:

$$\rho_A := (\hat{w}'\hat{w})^2 \frac{\hat{w}'(S - \text{diag}(S))\hat{w}}{\hat{w}'(\hat{w}\hat{w}' - \text{diag}(\hat{w}\hat{w}'))\hat{w}},$$

where $\hat{w}$ is the estimated weight vector of the construct and S is the empirical covariance matrix of the manifest variables.

The average variance extracted (AVE) is a measure of the convergent validity [76]:

$$\text{AVE} = \frac{\sum_{k=1}^{K} l_k^2}{K}.$$

It is considered acceptable when its value exceeds 0.5, which means that the construct explains more than 50% of its manifest variables' variance.

In the formative constructs, collinearity is analyzed through the variance inflation factor (VIF) of each indicator [77]:

$$\text{VIF}_k = \frac{1}{1 - R_k^2}.$$

If the VIF of an indicator is greater than 5, it indicates that there are problems of collinearity with other indicators of the construct, so it should be eliminated [78].

On the other hand, the assessment of the structural model involves determining the absence of collinearity problems between the constructs; evaluating the sign, magnitude, and statistical significance of the path coefficients; assessing the effect sizes; and, finally, determining the in-sample predictive power by means of the blindfolding procedure. Moreover, the out-of-sample predictive power will be analyzed by utilizing holdout samples [29].

The total number of students enrolled in the course was 162, of whom 23 students did not attend class and did not take the exams. Therefore, the actual sample was 139 students. Students were asked to answer the survey, completely voluntarily and anonymously, using the university's teaching platform through which the classes had been given. A total of 118 students responded, which represented an 85% response rate. Once the responses had been analyzed, 7 observations had to be eliminated, since inappropriate patterns of behavior were observed—for example, all responses had the same score. Consequently, the final sample was 111 observations, which represented 80% of the students who had taken the course, thus ensuring its representativeness.

The number of observations to reach acceptable levels of statistical power in PLS-SEM and a quality measurement model is 100 [78]. On the other hand, considering a mean effect size of 0.15, a significance level of 0.05, and a statistical power of 0.8, using the statistical software G*Power (v. 3.1.9.6., Kiel, Germany) [79] for the proposed model, the necessary sample size is 85. Therefore, since 111 responses were obtained in the study, the sample considered was adequate.

4. Results

Table 3 shows the main descriptive statistics of the sample. The students' opinion has been very mixed, as all but four of the indicators cover the full range between 1 and 7. The students' evaluation of the video tutorials is positive, since the average score exceeds 4. It is only below 4, with a value of 3.937, for their appreciation of the time of use (U2). The students consider that they have spent less time than their peers studying with video tutorials.

Table 3. Descriptive Statistics.

Construct	Indicator	Mean	Standard Deviation	Minimum	Maximum
Autonomy (Formative)	A1	4.982	1.395	1	7
	A2	4.766	1.571	1	7
	A3	4.216	1.365	1	7
Usage (Formative)	U1	4.568	1.313	1	7
	U2	3.937	1.247	1	7
Challenge (Formative)	C1	4.943	1.262	1	7
	C2	4.811	1.346	1	7
	C3	4.649	1.205	1	7
	C4	4.423	1.399	1	7
Effectiveness (Reflective)	E1	5.144	1.321	1	7
	E2	4.901	1.342	1	7
	E3	4.360	1.553	1	7
Depth (Formative)	D1	4.261	1.250	1	7
	D2	4.622	1.440	1	7
	D3	5.054	1.184	2	7
	D4	4.577	1.220	1	7
Format (Formative)	F1	4.342	1.679	1	7
	F2	5.036	1.287	2	7
	F3	4.937	1.232	1	7
	F4	5.622	1.163	3	7
	F5	5.054	1.199	2	7

Sample size: 111 interviews.

With an average of 5.622 and a minimum value of 3, the best-rated aspect was the visual aspect, considering that the video tutorials are attractive and clear. The highest-rated construct, overall, and with a smaller range of variation, was Format, indicating that the design of the video tutorials by the professors was appropriate.

Figure 2 depicts a global vision of the measurement and structural models.

4.1. Measurement Model

Table 4 shows the results corresponding to measurement model assessment.

First, Panel A displays the validation data for the reflective construct, Effectiveness. The construct indicators are valid when their loadings (λ) are ≥ 0.707 [80], which is verified in E1, E2, and E3 (Panel A1), and implies that the variance shared between the construct and its indicators is greater than the variance due to error. The commonality of the indicators (λ^2) indicates that their variation is explained between 66.59% and 75.69% by the construct.

On the other hand, construct reliability or internal consistency is guaranteed as long as Cronbach's alpha coefficient, composite reliability [81], and Dijkstra–Henseler's Rho (Panel A2) are ≥ 0.7 [75]. In this study, stricter reliability was found, with values >0.8, corresponding to more advanced research stages [82]. Furthermore, the 95% confidence intervals, obtained with a bootstrapping of 10,000 samples, verified that the lower limit is >0.7 and the upper limit is <0.95 [29].

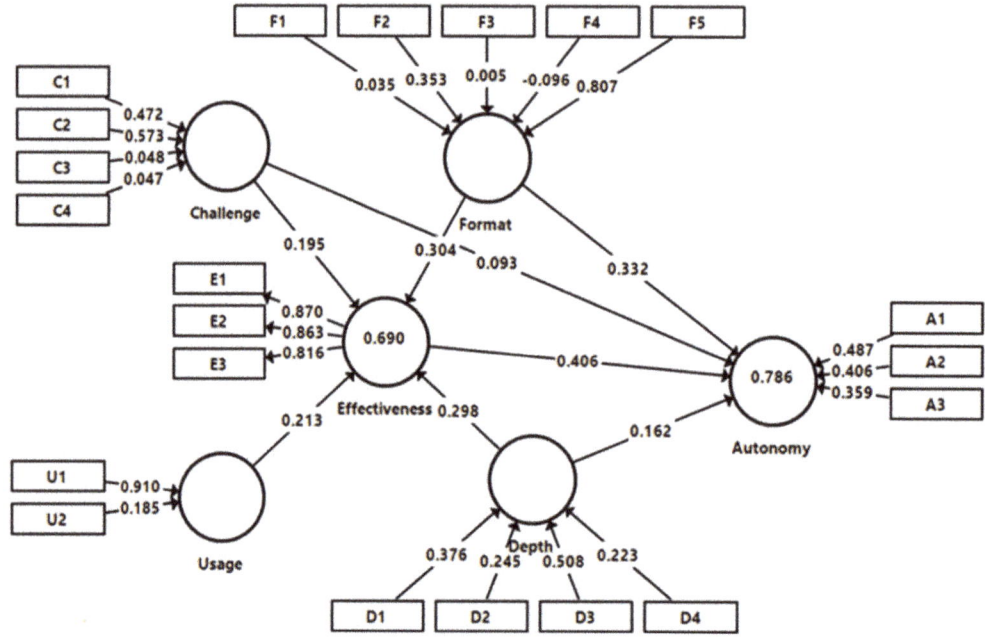

Figure 2. Model results.

In addition, convergent validity is determined by the average variance extracted (AVE), which must be ≥0.5 [83]. In this case, Effectiveness explains, on average, 72.2% of the indicators' variance.

Lastly, the discriminant validity of Effectiveness was tested by the Fornell–Larcker Criterion (Panel A3), according to which the square root of the AVE (0.850, highlighted in bold on the diagonal) must be greater than the construct correlation with the rest of the constructs (the values horizontally and vertically located on the same line) [23].

Second, Panel B reflects the evaluation results of the formative constructs. All variance inflation factors are <3, so there are no collinearity problems between the indicators [29,84]. It is important to note that the value of the weights indicates the indicator contribution to the corresponding construct. The six indicators with non-significant weights remained in the model, U2, C3, C4, and F3 had loadings >0.5; F1 and F4, whose loadings were <0.5, showed significant loadings [23]. The most highly rated indicators in each latent variable were A1, U1, C2, D3, and F5.

4.2. Structural Model

Table 5 presents the main parameters related to the assessment of the structural model. All the VIF values are ≤3, so the model did not present collinearity problems [29]. It should be noted that, as the path coefficients are shown as standardized values, their possible values range from +1 to −1. In this case, a one-tailed bootstrapping was considered for testing the hypotheses with their corresponding sign [85]. According to the latest requirements, the number of samples considered was 10,000 [86].

Table 4. Valuation of the Measurement Model.

Panel A. Reflective Construct (Effectiveness)				
Panel A1. Outer Loadings				
Indicator	Loading (λ)	CI 2.5%	CI 97.5%	p-Value
E1	0.870	0.808	0.913	0.000
E2	0.863	0.795	0.913	0.000
E3	0.816	0.735	0.871	0.000
Panel A2. Construct Reliability and Average Variance Extracted				
Criterion	Value	CI 2.5%	CI 97.5%	p-Value
Cronbach's Alpha	0.807	0.734	0.860	0.000
Dijkstra–Henseler's Rho	0.809	0.739	0.864	0.000
Composite Reliability	0.886	0.849	0.915	0.000
AVE	0.722	0.653	0.782	0.000

Panel A3. Discriminant Validity (Fornell-Larcker Criterion)						
Construct	Autonomy	Challenge	Depth	Effectiv.	Format	Usage
Challenge	0.739	n.a.				
Depth	0.751	0.751	n.a.			
Effectiveness	0.826	0.725	0.746	0.850		
Format	0.784	0.693	0.649	0.696	n.a.	
Usage	0.447	0.448	0.492	0.537	0.298	n.a.

Panel B. Formative Constructs							
Construct	Indicator	VIF	Weight	CI 2.5%	CI 97.5%	t-Stat.	Loading
Autonomy	A1	1.67	0.487 ***	0.316	0.644	5.746	0.878 **
	A2	1.36	0.406 ***	0.250	0.569	4.966	0.766 **
	A3	1.35	0.359 ***	0.200	0.504	4.652	0.730 **
Usage	U1	1.19	0.910 ***	0.647	1.059	8.386	0.986 **
	U2	1.19	0.185 ns	−0.189	0.546	0.983	0.556 **
Challenge	C1	1.77	0.472 ***	0.245	0.680	4.261	0.876 **
	C2	1.95	0.573 ***	0.354	0.763	5.540	0.919 **
	C3	1.89	0.048 ns	−0.203	0.292	0.377	0.608 **
	C4	2.04	0.047 ns	−0.202	0.298	0.367	0.646 **
Depth	D1	1.72	0.376 ***	0.165	0.605	3.335	0.687 **
	D2	1.73	0.245 **	0.007	0.445	2.201	0.727 **
	D3	1.69	0.508 ***	0.312	0.728	4.788	0.807 **
	D4	1.81	0.223 **	0.020	0.406	2.239	0.684 **
Format	F1	1.12	0.035 ns	−0.120	0.181	0.457	0.236 *
	F2	1.58	0.353 ***	0.084	0.593	2.686	0.744 **
	F3	1.60	0.005 ns	−0.213	0.238	0.043	0.534 **
	F4	1.35	−0.096 ns	−0.291	0.099	0.955	0.388 **
	F5	1.79	0.807 ***	0.566	0.982	7.585	0.945 **

n.a. denotes not applicable; ns denotes not significant; * p-value < 0.10; ** p-value < 0.05; *** p-value < 0.01.

The results showed that Format, Usage, and Depth positively influenced Effectiveness with a significance level of 99% (p = 0.001, 0.000, and 0.002, respectively), while Challenge did so at 95% (p = 0.024), which confirmed the hypotheses H1, H2, H3, and H4. Likewise, Format and Effectiveness positively influenced Autonomy, with a significance level of 99% (p = 0.000 in both cases), while Depth did so at 95% (p = 0.018) and Challenge did not present a significant influence (p = 0.149), so hypotheses H5, H6, and H8 were confirmed, but hypothesis H7 was not confirmed.

Concerning the mediation hypotheses, the indirect effects indicate that a partial mediation of Effectiveness between Depth and Autonomy was verified, with a significance level of 99% (p = 0.006), thus verifying hypothesis H10. Similarly, Effectiveness exerted a partial mediation, with a significance of 95% (p = 0.015), between Format and Autonomy, verifying hypothesis H9. In addition, with a significance of 90% (p = 0.061), Effectiveness fully mediated between Challenge and Autonomy, validating hypothesis H11.

Finally, Usage exerted a positive and significant influence on Autonomy through Effectiveness, with a significance of 99% (p = 0.003).

Table 5. Valuation of the Structural Model.

Panel A. Direct Effects

	Path	t	CI 5%	CI 95%	f^2	VIF
Challenge → Autonomy	0.093 ns	1.040	−0.053	0.240	0.014	2.930
Depth → Autonomy	0.162 **	2.103	0.042	0.295	0.043	2.894
Effectiveness → Autonomy	0.406 ***	3.522	0.228	0.609	0.266	2.909
Format → Autonomy	0.332 ***	3.430	0.161	0.479	0.224	2.300
Challenge → Effectiveness	0.195 **	1.971	0.040	0.366	0.043	2.831
Depth → Effectiveness	0.298 ***	3.238	0.137	0.439	0.106	2.680
Format → Effectiveness	0.304 ***	3.485	0.168	0.455	0.142	2.096
Usage → Effectiveness	0.213 ***	2.835	0.088	0.338	0.108	1.358

Panel B. Indirect Effects

	Effect	t	CI 5%	CI 95%
Challenge → Autonomy	0.079 *	1.546	0.013	0.180
Depth → Autonomy	0.121 ***	2.522	0.047	0.202
Format → Autonomy	0.124 **	2.182	0.050	0.234
Usage → Autonomy	0.087 ***	2.772	0.035	0.138

Panel C. Total Effects

	Effect	t	CI 5%	CI 95%
Challenge → Autonomy	0.172 **	1.956	0.033	0.323
Depth → Autonomy	0.283 ***	3.725	0.163	0.410
Effectiveness → Autonomy	0.406 ***	3.522	0.228	0.609
Format → Autonomy	0.455 ***	6.055	0.329	0.578
Usage → Autonomy	0.087 ***	2.772	0.035	0.138
Challenge → Effectiveness	0.195 **	1.971	0.040	0.366
Depth → Effectiveness	0.298 ***	3.238	0.137	0.439
Format → Effectiveness	0.304 ***	3.485	0.168	0.455
Usage → Effectiveness	0.213 ***	2.835	0.088	0.338

* p-value < 0.10; ** p-value < 0.05; *** p-value < 0.01; ns denotes not significant.

The coefficient of determination R^2 gives the explained variance of the latent dependent variables. Table 6 shows how much of that variance is explained by each preceding construct.

Table 6. Decomposition of the explained variance.

Dependent Variable	R^2	Antecedents Variables	Path Coefficients	Correlations	Explained Variance
Autonomy	0.786	Challenge	0.093	0.739	0.069
		Depth	0.162	0.751	0.122
		Effectiveness	0.406	0.826	0.336
		Format	0.332	0.784	0.260
Effectiveness	0.690	Challenge	0.195	0.725	0.142
		Depth	0.298	0.746	0.222
		Format	0.304	0.696	0.212
		Usage	0.213	0.537	0.114

Effectiveness was explained by 69%, and the constructs that contributed most were Depth (22.2%) and Format (21.2%), followed by Challenge (14.2%) and Usage (11.4%). Regarding the final dependent variable, Autonomy, the model explained 78.6% of its variance. In this case, the most influential construct was Effectiveness (33.6%), followed by Format (26%), Depth (12.2%), and Challenge (6.9%). Therefore, the explanatory power of the model is substantial [23,87].

The value of f^2 gives the effect size of the exogenous constructs on the endogenous constructs. By considering that $0.02 \leq f^2 < 0.15$ is a small effect, that $0.15 \leq f^2 < 0.35$ is a moderate effect, and that $f^2 \geq 0.35$ is a large effect [88], the latent variables Effectiveness and Format were found to exert a moderate effect on Autonomy (f^2 = 0.266 and 0.224, respectively), while the rest of the effects were found to be small.

Finally, in-sample predictive relevance of Effectiveness (reflective dependent construct) is given by the Stone–Geisser test [83], which yields a value of 0.475, indicating a prediction that could be considered high.

4.3. Out-of-Sample Prediction

The model analyzed also shows a high out-of-sample predictive power, as evidenced by the results in Table 7, which derive from a *holdout sample* procedure [89] implemented in SmartPLS. Therefore, the model is powerful enough to predict the results values of a new case, that is, cases not included in the sample analyzed. Indeed, this can be deduced from the fact that the PLS-LM difference is negative for all indicators.

Table 7. PLS predict valuation.

Panel A. Construct Prediction Summary								
	Q^2							
Autonomy	0.669							
Effectiveness	0.631							
Panel B. Indicator Prediction Summary								
	PLS			LM			PLS-LM		
	RMSE	MAE	Q^2	RMSE	MAE	Q^2	RMSE	MAE	Q^2
A1	0.991	0.766	0.502	1.095	0.854	0.393	−0.104	−0.088	0.109
A2	1.285	1.052	0.342	1.433	1.142	0.182	−0.148	−0.090	0.160
A3	1.054	0.866	0.419	1.250	0.994	0.182	−0.196	−0.128	0.237
E1	0.983	0.687	0.460	1.068	0.782	0.363	−0.085	−0.095	0.097
E2	0.990	0.772	0.467	1.073	0.855	0.373	−0.083	−0.083	0.094
E3	1.184	0.971	0.431	1.291	1.014	0.324	−0.107	−0.043	0.107

PLS: Partial Least Squares Path Model; LM: Linear Regression Model; RMSE: Root Mean Squared Error; MAE: Mean Absolute Error.

5. Discussion and Conclusions

As universities moved to teach in a blended or online environment, the use of video tutorials turned into a leading tool in higher education. The present research is a great step in the design and study of videos at the university level of Financial Mathematics. This research is motivated by the call for more extensive research into the effectiveness of video tutorials [33], [35], or [39]. To the best of our knowledge, previous studies that analyze the effectiveness of videos employing PLS-SEM methodology are scarce. Specifically, it is necessary to highlight the recent study of Ibarra-Sáiz and Rodríguez-Gómez [37], which makes an evaluation of the assessment system validated with PLS-SEM. Thus far, there is a gap in the empirical research because they have not used this methodology in the mathematics learning process during the COVID-19 era at the university level. Specifically, previous studies have not considered effectiveness as a moderating variable in the learning process. Following this argument, we highlight the articles [37,49,52] that deserve particular attention given that they are related to and support our paper's contribution.

As we mentioned before, to study the impact of video tutorials on learning efficiency, it is necessary to mention the study of Ibarra-Sáiz and Rodríguez-Gómez [37]. This paper is focused on analyzing the most effective way to learn. With respect to Format and Usage, in [37] it is claimed that it is not enough for students to reproduce the knowledge faithfully; they must have a good command of the structure and existing interrelations and give coherence to the knowledge. Regarding Depth, they claim that learning effectiveness is related to an in-depth understanding by using investigation methods and reflective and critical thinking. With respect to Challenge, complex problems requiring divergent thinking can improve creativity and forge significant relationships and connections in learning [37].

In this sense, our results are in line with their ones given that we confirm that in video tutorials, Format, Usage, Depth, and Challenge positively influence Effectiveness with a significance level higher than of 95% (p = 0.001, 0.000, 0.002, and 0.024, respectively). In particular, the descriptive statistics (see Table 3) reveal that the lowest score level given in the questionnaire corresponds to the U2 indicator (construct Usage), which leads us to infer that the students consider that they have spent less time than their peers with the video tutorials, recognizing that they do not dedicate enough time to study. However, in U1 (sufficient and suitable study time) they have an average of 4.568. Among all the constructs obtained, Format is the most prominent, which suggests that the students have satisfactorily evaluated the audio-visual presentation contained in the video tutorials, even though the mean of F1 (preference for recording by several professors rather than a single professor) is one of the lowest. This factor may result from students not being entirely satisfied with the video tutorials prepared by several professors, instead of a single one, or even by the same professor who previously taught the subject in the classroom mode. It might also be due to the existence of students who still miss the face-to-face mode or who have not been able to assimilate the transition from the traditional classroom to the online mode [19].

What is more, Format, Effectiveness, Depth and Autonomy, are also the constructs with the greatest degree of dispersion among the student responses. Such heterogeneity may be a direct consequence of the fact that the questionnaires were fulfilled entirely freely and anonymously, allowing the students to express themselves with complete freedom. Hence, the interactivity offered by the video tutorials has had particularly positive effects, one of the aspects to be encouraged in the post-COVID-19 teaching era as reported by Rey Lopez et al. [17] in pedagogical research in which the interviewees were the professors themselves. Moreover, Usage presents the lowest degree of acceptance among the students, which may be derived because the students prefer to interact with the professor through online streaming or simply because many prefer to write personally on the blackboard in the classroom [20,90].

On the other hand, to study the impact of video tutorials on autonomy, it is necessary to remark the research of Moreno-Guerrero et al. [52], where, as in our article, the effectiveness of a e-learning experience in the teaching of mathematics is studied (the difference resides in that this experience is in adult high school, while in our study it is at university). The present study's results for blended learning using video tutorials show that the Format, Effectiveness, and Depth positively influence Autonomy with a significance level of more than 95%. In this vein, our findings indicate that videos help to identify the key concepts of the subject and to coordinate the learning activities essential for its assimilation (see Panel B, Table 4). In line with our study, Moreno-Guerrero et al. [52] remark that the application of e-learning techniques in the mathematics field increases the memorization, comprehension, and internalization of contents. They focus on Autonomy, given that this dimension shows an enormous contrast when comparing the face-to-face method with the e-learning method, because the latter approach favors self-regulation of learning. Our results also show that students found the instructions to be clear and precise, considering that the time spent in their study was sufficient and adequate. Note that the considerations made by the students had already been foreseen by the professors beforehand, especially those that value that the explanations are clear and precise or do not give much importance to the length of each video. Both issues are justified by the fact that the professors designed each video in itself as a "knowledge pill" whose content summarized and condensed each of the subject's themes. In other words, the content of the videos was intended to be limited and condensed.

The model reveals the relationships between the different latent variables. For video tutorials to be effective in the teaching–learning process, it is essential that their content has depth and that the format (color, font, length, speed of explanation, etc.) in which they are made is appropriate. These results are in line with [47,57]. Although to a lesser extent, it is also important that the content is explained in a way that challenges the student. Finally, students rated the frequency of use as less critical. Therefore, we can conclude that students do not need to watch the video tutorials repeatedly for learning if they are well designed.

With respect to the desired objective of Autonomy in student learning, which was 78.6% explained, the most important variables are, in this order, Effectiveness and Format, as also indicated [35,57]. Depth also plays an essential role in achieving learner autonomy. However, the fact that the video tutorials challenge the students made a small contribution. In sum, we can state that the Format and Depth of the content have central importance in studying with video tutorials. Furthermore, in this particular case, the students positively valued the work done by the professors in the recorded video tutorials, considering that the format chosen was quite adequate.

Finally, our findings confirm the moderating effects of effectiveness in the learning of Financial Mathematics. The indirect effects of Effectiveness are new, since previous studies [60–66] just select gender, educational context, or study discipline as moderating variables. In this sense, we want to comment on Means et al.'s [49] contribution, where the effectiveness of online and blended learning through a meta-analysis of the empirical literature is analyzed. In this sense, the authors of [49] study several moderator variables explaining differences in outcomes between online learning and face-to-face environments, with the main results being in line with our paper outcome: a blended-learning method is more effective than entirely face-to-face classes. This is explained given that blending learning also involves additional learning time by using more instructional resources and course elements that encourage interactions among learners. This idea, initially proposed in [49], is developed and confirmed experimentally in our paper.

Another significant aspect to highlight is the predictive character of the model generated. In practice, the application of the PLS-SEM methodology only ensures that the adjusted model is exploratory or explanatory for the sample under consideration. Only in some cases is it guaranteed that the fitted model will have predictive relevance "out of the sample". However, given that values of Q2 are greater than 0 [23], and because the PLS-LM difference is negative for all indicators (Table 7), the model obtained can be considered entirely predictive. In line with the work of Karmila et al. [18], the results of this study demonstrate that the employment of video tutorials has served to improve three critical aspects of the teaching-learning process of the subject Mathematics of Financial Operations in the midst of the COVID-19 epoch: improving students' conceptual understanding skills and their learning outcomes while increasing their motivation and interest in the subject.

Similarly, several analogies with previous works can be observed. For example, the results of the four dimensions of the Adversity Quotient (AQ) developed by Anggraini and Mahmudi [91] among Indonesian students are analogous in terms of the degree of satisfaction manifested in the same way that its usability is also rated very positively. It also presents some parallelisms with Cassibba et al. [92], who applied a different methodological spectrum (Mathematics Teacher's Specialized Knowledge (MTSK)), implying that Sicilian teachers in charge of teaching mathematics did not experience a significant difference between online and traditional or classical teaching, a fact that is confirmed in our research from the students' perspective due to the high impact of the variables Effectiveness and Autonomy.

The relevance of this last variable is also noted by Fakhrunisa and Prabawanto [93] emphasizing the potential for independent and flexible learning, depending on the needs and circumstances of each student. Regarding the practical usefulness of this study, it should be taken into account that the quality and objectivity of the answers obtained from the learners, besides allowing us to obtain a PLS-SEM model of predictive nature, indirectly resulted in an analysis of the strength of the cognitive abilities of each student quite close to Hidayah et al. [94]. More specifically, it has mainly reinforced memory-related and visual-spatial skills (Effectiveness, Depth, and Format variables). It is essential to underline that, although students tend to prefer asynchronous online teaching [95], our methodological proposal in no way presupposes dispensing with it but rather complementing it, increasing the autonomous initiative of the e-learners in the field of mathematics.

Any innovative and groundbreaking methodological proposal, such as the one presented in this research work, can be implemented in other subjects, universities, and countries if the appropriate circumstances are fulfilled. Possible barriers include the lack of material or economic means, such as those detected by Mohammadi, Mohibbi, and Hedayati [96] in the establishment of HELMS (Higher Education Learning Management System) in Afghanistan, or the preparation, facilities, and infrastructure available for the teaching staff responsible for online teaching as reported by Kamsurya [97]. Another aspect to highlight for this approach to materialize in other classrooms is the need for coordination between schools or faculties, students, and professors [98], always bearing in mind the opinion and needs of the students [99].

Whatever the objections that may be raised, this approach to teaching develops the educational suggestions of international institutions such as UNESCO, since it implies the creation of methods, processes, and feedback analysis of each component involved in the educational process [100]. As Lockee [101] states, the new generations coming after the pandemic will be faced with an educational environment diametrically opposed to the classical one, in which online teaching will be predominant and will rely on methods and models that are more flexible with the reality of the times: COVID-19 has only accelerated this process that is yet to come.

The main limitation of this work lies in the sample used, since it corresponds to students of a given subject at a single university. In the future, it would be interesting to obtain data from students from other universities with the same video tutorials. It would also be desirable to extend the study to other subjects, in which case new video tutorials adapted to the subject would have to be developed. All this would make it possible to compare the results and, therefore, to give greater firmness to the conclusions reached.

Author Contributions: Conceptualization, methodology, and formal analysis, M.d.C.V.M.; data curation, M.d.C.V.M., P.A.M.-C., A.M.S.P. and M.d.C.M.V.; writing—original draft preparation, M.d.C.V.M., P.A.M.-C., A.M.S.P. and M.d.C.M.V.; writing—review and editing M.d.C.V.M., P.A.M.-C., A.M.S.P. and M.d.C.M.V.; project administration, M.d.C.V.M. All authors have read and agreed to the published version of the manuscript.

Funding: The APC was funded by VICERRECTORADO DE ORDENACIÓN ACADÉMICA DE LA UNIVERSIDAD DE ALMERÍA, Grupo Docente 20_21_4_05C (Docencia online de Matemáticas de las Operaciones Financieras).

Institutional Review Board Statement: Not applicable.

Informed Consent Statement: Not applicable.

Data Availability Statement: Not applicable.

Conflicts of Interest: The authors declare no conflict of interest.

References

1. Sasidharan, S.; Dhillon, H.S.; Singh, D.H.; Manalikuzhiyil, B. COVID-19: Pan(info)demic. *Turk. J. Anaesthesiol. Reanim.* **2020**, *48*, 438–442. [CrossRef]
2. Torjesen, I. Covid-19 will become endemic but with decreased potency over time, scientists believe. *BMJ* **2021**, *372*, n494. [CrossRef]
3. Shamekh, A.; Mahmoodpoor, A.; Sanaie, S. COVID-19: Is it the black death of the 21st century? *Health Promot. Perspect.* **2020**, *10*, 166–167. [CrossRef] [PubMed]
4. King, P.T.; Londrigan, S.L. The 1918 influenza and COVID-19 pandemics: The effect of age on outcomes. *Respirology* **2021**, *26*, 840–841. [CrossRef] [PubMed]
5. Cucinotta, D.; Vanelli, M. WHO Declares COVID-19 a Pandemic. *Acta Bio Med. Atenei Parm.* **2020**, *91*, 157–160.
6. Council of Ministers. Government Decrees State of Emergency to Stop Spread of Coronavirus COVID-19. 2020. Available online: https://www.lamoncloa.gob.es/lang/en/gobierno/councilministers/Paginas/2020/20200314council-extr.aspx (accessed on 5 August 2021).
7. Sun, K.L. Brief Report: The Role of Mathematics Teaching in Fostering Student Growth Mindset. *J. Res. Math. Educ.* **2018**, *49*, 330–335. [CrossRef]
8. Inglis, M.; Foster, C. Five Decades of Mathematics Education Research. *J. Res. Math. Educ.* **2018**, *49*, 462–500. [CrossRef]
9. Carr, M.E. Student and/or Teacher Valuing in Mathematics Classrooms: Where Are We Now, and Where Should We Go? In *Values and Valuing in Mathematics Education: Scanning and Scoping the Territory*; Clarkson, P., Seah, W.T., Pang, J., Eds.; Springer International Publishing: Cham, Switzerland, 2019.
10. Chirinda, B.; Ndlovu, M.; Spangenberg, E. Teaching Mathematics during the COVID-19 Lockdown in a Context of Historical Disadvantage. *Educ. Sci.* **2021**, *11*, 177. [CrossRef]

11. O'Sullivan, C.; an Bhaird, C.M.; Fitzmaurice, O.; Ní Fhloinn, E. *An Irish Mathematics Learning Support. Network (IMLSN) Report on Student Evaluation of Mathematics Learning Support: Insights from a Large Scale Multi-Institutional Survey*; Technical Report for National Centre for Excellence in Mathematics and Science Teaching and Learning (NCE-MSTL): Limerick, Ireland, September 2014.
12. Alabdulaziz, M.S. COVID-19 and the use of digital technology in mathematics education. *Educ. Inf. Technol.* **2021**, 1–25. [CrossRef]
13. Kalogeropoulos, P.; Roche, A.; Russo, J.; Vats, S.; Russo, T. Learning Mathematics From Home During COVID-19: Insights from Two Inquiry-Focussed Primary Schools. *Eurasia J. Math. Sci. Technol. Educ.* **2021**, *17*, em1957. [CrossRef]
14. Almarashdi, H.; Jarrah, A.M. Mathematics Distance Learning amid the COVID-19 Pandemic in the UAE: High School Students' Perspectives. *Int. J. Learn. Teach. Educ. Res.* **2021**, *20*, 292–307. [CrossRef]
15. Fitzmaurice, O.; Fhloinn, E.N. Alternative mathematics assessment during university closures due to COVID-19. *Ir. Educ. Stud.* **2021**, *40*, 187–195. [CrossRef]
16. Hodgen, J.; Taylor, B.; Jacques, L.; Tereshchenko, A.; Kwok, R.; Cockerill, M. *Remote Mathematics Teaching during COVID-19: Intentions, Practices and Equity*; UCL Institute of Education: London, UK, 2020.
17. Rey Lopez, S.; Bruun, G.R.; Mader, M.J.; Reardon, R.F. The Pandemic Pivot: The Impact of COVID-19 on Mathematics and Statistics Post-Secondary Educators. *Int. J. Cross-Discip. Subj. Educ.* **2021**, *12*, 4369–4378.
18. Karmila, D.; Putri, D.M.; Berlian, M.; Pratama, D.O.; Fatrima. The Role of Interactive Videos in Mathematics Learning Activities During the Covid-19 Pandemic. In *Proceedings of the International Conference on Educational Sciences and Teacher Profession (ICETeP 2020), Bengkulu, Indonesia, 7 November 2020*; Atlantis Press: Amsterdam, The Netherlands, 2021. [CrossRef]
19. Chisadza, C.; Clance, M.; Mthembu, T.; Nicholls, N.; Yitbarek, E. Online and face-to-face learning: Evidence from students' performance during the Covid-19 pandemic. *Afr. Dev. Rev.* **2021**, *33*, S114–S125. [CrossRef]
20. Pócsová, J.; Mojžišová, A.; Takáč, M.; Klein, D. The Impact of the COVID-19 Pandemic on Teaching Mathematics and Students' Knowledge, Skills, and Grades. *Educ. Sci.* **2021**, *11*, 225. [CrossRef]
21. Borba, M.C. The future of mathematics education since COVID-19: Humans-with-media or humans-with-non-living-things. *Educ. Stud. Math.* **2021**, 1–16. [CrossRef]
22. Glass, J.; Sue, V. Student preferences, satisfaction, and perceived learning in an online mathematics class. *MERLOT J. Online Learn. Teach.* **2008**, *4*, 325–338.
23. Hair, J.; Hult, G.; Ringle, C.; Sarstedt, M. *A Primer on Partial Least Squares Structural Equation Modeling (PLS-SEM)*, 2nd ed.; Sage Publications: Southend Oaks, CA, USA, 2017.
24. Hair, J.F.; Sarstedt, M.; Ringle, C.M.; Gudergan, S.P. *Advanced Issues in Partial Least Squares Structural Equation Modeling*; Sage Publications: Thousand Oaks, CA, USA, 2018.
25. Nitzl, C. The use of partial least squares structural equation modelling (PLS-SEM) in management accounting research: Directions for future theory development. *J. Account. Lit.* **2016**, *37*, 19–35. [CrossRef]
26. Wold, H. Estimation of principal components and related methods by iterative least squares. In *Multivariate Analysis*; Krishnaiah, P.R., Ed.; Academic Press: New York, NY, USA, 1966; pp. 391–420.
27. Wold, H. Nonlinear iterative partial least squares (NIPALS) modeling: Some current developments. In *Multivariate Analysis III*; Krishnaiah, P.R., Ed.; Academic Press: New York, NY, USA, 1973; pp. 383–407.
28. Hair, J.F.; Sarstedt, M.; Ringle, C.M.; Mena, J.A. An assessment of the use of partial least squares structural equation modeling in marketing research. *J. Acad. Mark. Sci.* **2012**, *40*, 414–433. [CrossRef]
29. Sarstedt, M.; Ringle, C.M.; Smith, D.; Reams, R.; Hair, J.F. Partial least squares structural equation modeling (PLS-SEM): A useful tool for family business researchers. *J. Fam. Bus. Strateg.* **2014**, *5*, 105–115. [CrossRef]
30. Avkiran, N.K. Rise of the Partial Least Squares Structural Equation Modeling: An Application in Banking. *Handb. Healthc. Logist.* **2018**, *267*, 1–29. [CrossRef]
31. Romo-González, J.R.; Tarango, J.; Machin-Mastromatteo, J.D. PLS SEM, a quantitative methodology to test theoretical models from library and information science. *Inf. Dev.* **2018**, *34*, 526–531. [CrossRef]
32. Hair, J.F.; Ringle, C.M.; Sarstedt, M. PLS-SEM: Indeed a Silver Bullet. *J. Mark. Theory Pract.* **2011**, *19*, 139–152. [CrossRef]
33. Noetel, M.; Griffith, S.; Delaney, O.; Sanders, T.; Parker, P.; Cruz, B.D.P.; Lonsdale, C. Video Improves Learning in Higher Education: A Systematic Review. *Rev. Educ. Res.* **2021**, *91*, 204–236. [CrossRef]
34. Commission of the European Communities. *The eLearning Action Plan. Designing Tomorrow's Education*; Commission of the European Communities: Brussels, Belgium, 2001.
35. Aelterman, N.; Vansteenkiste, M.; Haerens, L.; Soenens, B.; Fontaine, J.R.J.; Reeve, J. Toward an integrative and fine-grained insight in motivating and demotivating teaching styles: The merits of a circumplex approach. *J. Educ. Psychol.* **2019**, *111*, 497–521. [CrossRef]
36. McNulty, J.A.; Hoyt, A.; Chandrasekhar, A.J.; Gruener, G.; Price, R., Jr.; Naheedy, R. A Three-year Study of Lecture Multimedia Utilization in the Medical Curriculum: Associations with Performances in the Basic Sciences. *Med. Sci. Educ.* **2011**, *21*, 29–36. [CrossRef]
37. Ibarra-Sáiz, M.S.; Rodríguez-Gómez, G. Evaluating Assessment. Validation with PLS-SEM of ATAE Scale for the Analysis of Assessment Tasks. *Relieve Rev. ELectrón. Investig. EVal. Educ.* **2020**, *26*, 6.
38. Mayer, R.E. Cognitive Theory of Multimedia Learning. In *The Cambridge Handbook of Multimedia Learning*; Cambridge University Press: Cambridge, UK, 2014; pp. 31–48.

39. Lackmann, S.; Léger, P.-M.; Charland, P.; Aubé, C.; Talbot, J. The Influence of Video Format on Engagement and Performance in Online Learning. *Brain Sci.* **2021**, *11*, 128. [CrossRef]
40. Homer, B.D.; Plass, J.L.; Blake, L. The effects of video on cognitive load and social presence in multimedia-learning. *Comput. Hum. Behav.* **2008**, *24*, 786–797. [CrossRef]
41. Korving, H.; Hernández, M.; de Groot, E. Look at me and pay attention! A study on the relation between visibility and attention in weblectures. *Comput. Educ.* **2016**, *94*, 151–161. [CrossRef]
42. Wang, J.; Antonenko, P.D. Instructor presence in instructional video: Effects on visual attention, recall, and perceived learning. *Comput. Hum. Behav.* **2017**, *71*, 79–89. [CrossRef]
43. Ilioudi, C.; Giannakos, M.N.; Chorianopoulos, K. Investigating Differences among the Commonly Used Video Lecture Styles. *CEUR Workshop Proc.* **2013**, *983*, 21–26. [CrossRef]
44. Al-Samarrie, H. A Scoping Review of Videoconferencing Systems in Higher Education: Learning Paradigms, Opportunities, and Challenges. *Int. Rev. Res. Open Distrib. Learn.* **2019**, *20*, 121–140.
45. Moore, W.A.; Smith, A.R. Effects of video podcasting on psychomotor and cognitive performance, attitudes and study behaviour of student physical therapists. *Innov. Educ. Teach. Int.* **2012**, *49*, 401–414. [CrossRef]
46. Koumi, J. Potent Pedagogic Roles for Video. Available online: http://association.media-and-learning.eu/portal/resource/potent-pedagogic-roles-video (accessed on 5 August 2021).
47. Woolfitt, Z. The Effective Use of Videos in Medical Education. *Acad. Med.* **2015**, *1*, 45.
48. Miner, S.; Stefaniak, J.E. Learning via video in higher education: An exploration of instructor and student perceptions. *J. Univ. Teach. Learn. Pract.* **2018**, *15*, 2.
49. Means, B.; Toyama, Y.; Murphy, R.; Baki, M. The effectiveness of online and blended learning: A meta-analysis of the empirical literature. *Teach. Coll. Rec.* **2013**, *115*, 030303.
50. Entwistle, N.; McArthur, J. Perceptions of assessment and their influences on learning. In *Advances and Innovations in University Assessment and Feedback*; Edinburgh UP: Edinburgh, UK, 2014.
51. O'Donovan, B. How student beliefs about knowledge and knowing influence their satisfaction with assessment and feedback. *High. Educ.* **2016**, *74*, 617–633. [CrossRef]
52. Moreno-Guerrero, A.J.; Aznar-Díaz, I.; Cáceres-Reche, P.; Alonso-García, S. E-learning in the teaching of mathematics: An educational experience in adult high school. *Mathematics* **2020**, *8*, 840. [CrossRef]
53. Dede, C. Emerging influences of information technology on school curriculum. *J. Curric. Stud.* **2000**, *32*, 281–303. [CrossRef]
54. Veerman, A.; Veldhuis-Diermanse, E. Collaborative learning through computer-mediated communication in academic education. *Euro CSCL* **2001**, *2001*, 625–632.
55. Warni, S.; Aziz, T.A.; Febriawan, D. The use of technology in English as a foreign language learning outside the classroom: An insight into learner autonomy. *LLT J.* **2018**, *21*, 148–156.
56. Poot, R.; de Kleijn, R.A.M.; van Rijen, H.V.M.; van Tartwijk, J. Students generate items for an online formative assessment: Is it motivating? *Med. Teach.* **2017**, *39*, 315–320. [CrossRef]
57. Liaw, S.-S.; Huang, H.-M.; Chen, G.-D. Surveying instructor and learner attitudes toward e-learning. *Comput. Educ.* **2007**, *49*, 1066–1080. [CrossRef]
58. Akugizibwe, E.; Ahn, J.Y. Perspectives for effective integration of e-learning tools in university mathematics instruction for developing countries. *Educ. Inf. Technol.* **2019**, *25*, 889–903. [CrossRef]
59. García Pujals, A. The effect of formative assessment and instructional feedback on perception of learning, autonomy and motivation of German students of Spanish as a foreign language: A didactic proposal. Ph.D. Thesis, University of the Basque Country, Biscay, Spain, 14 June 2019. Available online: https://addi.ehu.es/handle/10810/35325 (accessed on 5 August 2021).
60. Zhang, Y.G.; Dang, M.Y. Understanding Essential Factors in Influencing Technology-Supported Learning: A Model toward Blended Learning Success. *J. Inf. Technol. Educ. Res.* **2020**, *19*, 489–510. [CrossRef]
61. Wongwatkit, C.; Panjaburee, P.; Srisawasdi, N.; Seprum, P. Moderating effects of gender differences on the relationships between perceived learning support, intention to use, and learning performance in a personalized e-learning. *J. Comput. Educ.* **2020**, *7*, 229–255. [CrossRef]
62. Strelan, P.; Osborn, A.; Palmer, E. The flipped classroom: A meta-analysis of effects on student performance across disciplines and education levels. *Educ. Res. Rev.* **2020**, *30*, 100314. [CrossRef]
63. Vo, H.M.; Zhu, C.; Diep, A.N. The effect of blended learning on student performance at course-level in higher education: A meta-analysis. *Stud. Educ. Eval.* **2017**, *53*, 17–28. [CrossRef]
64. Van Alten, D.C.D.; Phielix, C.; Janssen, J.; Kester, L. Effects of flipping the classroom on learning outcomes and satisfaction: A metaanalysis. *Educ. Res. Rev.* **2019**, *28*, 1–18. [CrossRef]
65. Spanjers, I.A.; Könings, K.; Leppink, J.; Verstegen, D.M.; de Jong, N.; Czabanowska, K.; van Merriënboer, J.J. The promised land of blended learning: Quizzes as a moderator. *Educ. Res. Rev.* **2015**, *15*, 59–74. [CrossRef]
66. Müller, C.; Mildenberger, T. Facilitating Flexible Learning by Replacing Classroom Time with an Online Learning Environment: A Systematic Review of Blended Learning in Higher Education. *Educ. Res. Rev.* **2021**, *34*, 100394. [CrossRef]
67. Valls Martínez, M.C.; Cruz Rambaud, S.; Muñoz Torrecillas, M.J.; Ramírez Orellana, A.; García Pérez, J. Presentaciones interactivas y videotutoriales en asignaturas de Finanzas y Contabilidad. In *VII Memoria Sobre Innovación Docente en la University of Almería (Curso Académico 2012–2013)*; Universidad de Almería, Servicio de Publicaciones: Almería, Spain, 2014.

68. Ringle, C.M.; Wende, S.; Becker, J.-M. *SmartPLS 3*; SmartPLS GmbH: Boenningstedt, Germany, 2015. Available online: http://www.smartpls.com (accessed on 5 August 2021).
69. Davison, A.C.; Hinkley, D.V. *Bootstrap Method and Their Application*; Cambridge University Press: Cambridge, UK, 1997.
70. Efron, B.; Tibshirani, R. Bootstrap Methods for Standard Errors, Confidence Intervals, and Other Measures of Statistical Accuracy. *Stat. Sci.* **1986**, *1*, 54–75. [CrossRef]
71. Debashis, K. Bootstrap Methods and Their Application. *Tecnnometrics* **2000**, *42*, 216–217.
72. Sarstedt, M.; Ringle, C.M.; Hair, J.F. Partial least squares structural equation modeling. In *Handbook of Market Research*; Homburg, C., Klarmann, M., Vomberg, A., Eds.; Springer: Cham, Switzerland, 2017.
73. Henseler, J.; Hubona, G.; Ray, P.A. Using PLS path modeling in new technology research: Updated guidelines. *Ind. Manag. Data Syst.* **2016**, *116*, 2–20. [CrossRef]
74. Henseler, J. On the convergence of the partial least squares path modeling algorithm. *Comput. Stat.* **2009**, *25*, 107–120. [CrossRef]
75. Dijkstra, T.K.; Henseler, J. Consistent Partial Least Squares Path Modeling. *MIS Q.* **2015**, *39*, 297–316. [CrossRef]
76. Henseler, J.; Ringle, C.M.; Sarstedt, M. A new criterion for assessing discriminant validity in variance-based structural equation modeling. *J. Acad. Mark. Sci.* **2015**, *43*, 115–135. [CrossRef]
77. Hair, J.F.; Risher, J.J.; Sarstedt, M.; Ringle, C.M. When to use and how to report the results of PLS-SEM. *Eur. Bus. Rev.* **2019**, *31*, 2–24. [CrossRef]
78. Reinartz, W.; Haenlein, M.; Henseler, J. An empirical comparison of the efficacy of covariance-based and variance-based SEM. *Int. J. Res. Mark.* **2009**, *26*, 332–344. [CrossRef]
79. Faul, F.; Erdfelder, E.; Buchner, A.; Lang, A.-G. Statistical power analyses using G*Power 3.1: Tests for correlation and regression analyses. *Behav. Res. Methods* **2009**, *41*, 1149–1160. [CrossRef]
80. Carmines, E.G.; Zeller, R.A. *Reliability and Validity Assessment*; Sage Publications: London, UK, 1979.
81. Werts, C.E.; Linn, R.L.; Jöreskog, K.G. Interclass Reliability Estimates: Testing Structural Assumptions. *Educ. Psychol. Meas.* **1974**, *34*, 25–33. [CrossRef]
82. Nunnally, J.; Bernstein, I.H. *Psychometric Theory*, 3rd ed.; McGraw-Hill: New York, NY, USA, 1994.
83. Fornell, C.; Larcker, D.F. Evaluating structural equation models with unobservable variables and measurement error. *J. Mark. Res.* **1981**, *18*, 39–50. [CrossRef]
84. Diamantopoulos, A.; Siguaw, J.A. Formative Versus Reflective Indicators in Organizational Measure Development: A Comparison and Empirical Illustration. *Br. J. Manag.* **2006**, *17*, 263–282. [CrossRef]
85. Kock, N. One-Tailed or Two-Tailed P Values in PLS-SEM? *Int. J. e-Collab.* **2015**, *11*, 1–7. [CrossRef]
86. Streukens, S.; Leroi-Werelds, S. Bootstrapping and PLS-SEM: A step-by-step guide to get more out of your bootstrap results. *Eur. Manag. J.* **2016**, *34*, 618–632. [CrossRef]
87. Chin, X.W. The partial least squares approach to structural equation modeling. In *Modern Methods for Business Research*; Marcoulides, G., Ed.; Lawrence Erlbaum Associates: London, UK, 1998.
88. Cohen, J. *Statistical Power Analysis for the Behavioral Sciences*, 2nd ed.; Routledge Academic: New York, NY, USA, 1988.
89. Shmueli, G.; Ray, S.; Estrada, J.M.V.; Chatla, S.B. The elephant in the room: Predictive performance of PLS models. *J. Bus. Res.* **2016**, *69*, 4552–4564. [CrossRef]
90. Busto, S.; Dumbser, M.; Gaburro, E. A Simple but Efficient Concept of Blended Teaching of Mathematics for Engineering Students during the COVID-19 Pandemic. *Educ. Sci.* **2021**, *11*, 56. [CrossRef]
91. Anggraini, T.W.; Mahmudi, A. Exploring the students' adversity quotient in online mathematics learning during the Covid-19 pandemic. *J. Res. Adv. Math. Educ.* **2021**, *6*, 221–238. [CrossRef]
92. Cassibba, R.; Ferrarello, D.; Mammana, M.F.; Musso, P.; Pennisi, M.; Taranto, E. Teaching Mathematics at Distance: A Challenge for Universities. *Educ. Sci.* **2020**, *11*, 1. [CrossRef]
93. Fakhrunisa, F.; Prabawanto, S. Online Learning in COVID-19 Pandemic: An Investigation of Mathematics Teachers' Perception. In Proceedings of the 2020 The 4th International Conference on Education and E-Learning, Yamanashi, Japan, 6–8 November 2020; pp. 207–213.
94. Hidayah, I.N.; Sa'Dijah, C.; Subanji, S. The students' cognitive engagement in online mathematics learning in the pandemic Covid-19 era. In Proceedings of the 4th International Conference on Mathematics and Science Education (ICoMSE) 2020: Innovative Research in Science and Mathematics Education in The Disruptive Era, Malang, Indonesia, 25–26 August 2021.
95. Libasin, Z.; Azudin, A.R.; Idris, N.A.; Rahman, M.S.A.; Umar, N. Comparison of Students' Academic Performance in Mathematics Course with Synchronous and Asynchronous Online Learning Environments during COVID-19 Crisis. *Int. J. Acad. Res. Prog. Educ. Dev.* **2021**, *10*, 492–501.
96. Mohammadi, M.K.; Mohibbi, A.A.; Hedayati, M.H. Investigating the challenges and factors influencing the use of the learning management system during the Covid-19 pandemic in Afghanistan. *Educ. Inf. Technol.* **2021**, *26*, 5165–5198. [CrossRef] [PubMed]
97. Kamsurya, R. Learning Evaluation of Mathematics during the Pandemic Period COVID-19 in Jakarta. *Int. J. Pedagog. Dev. Lifelong Learn.* **2020**, *1*, ep2008. [CrossRef]
98. Yohannes, Y.; Juandi, D.; Diana, N.; Sukma, Y. Mathematics Teachers' Difficulties in Implementing Online Learning during the COVID-19 Pandemic. *J. Hunan Univ. Nat. Sci.* **2021**, *48*, 1–12.
99. Wardani, E.R.; Mardiyana; Saputro, D.R.S. Online Mathematics Learning during the Covid-19 Pandemic. *J. Phys. Conf. Ser.* **2021**, *1808*, 012044. [CrossRef]

100. ECLAC-UNESCO. Education in the Time of COVID-19. Available online: https://repositorio.cepal.org/bitstream/handle/11362/45905/1/S2000509_en.pdf (accessed on 5 August 2021).
101. Lockee, B.B. Online education in the post-COVID era. *Nat. Electron.* **2021**, *4*, 5–6. [CrossRef]

Article

Spatio-Temporal Patterns of CO_2 Emissions and Influencing Factors in China Using ESDA and PLS-SEM

Bin Wang [1,*], Qiuxia Zheng [2], Ao Sun [3], Jie Bao [4] and Dianting Wu [1]

1. Faculty of Geographical Science, Beijing Normal University, Beijing 100875, China; wudianting@bnu.edu.cn
2. Urban and Regional Development Research Department, Chongqing Academy of Economics Research, Chongqing 401147, China; ylsfzheng@163.com
3. School of Geography, South China Normal University, Guangzhou 510631, China; ray_suen@m.scnu.edu.cn
4. School of Business, Anhui University, Hefei 230601, China; 15015@ahu.edu.cn
* Correspondence: wangbin2017@mail.bnu.edu.cn

Abstract: Controlling carbon dioxide (CO_2) emissions is the foundation of China's goals to reach its carbon peak by 2030 and carbon neutrality by 2060. This study aimed to explore the spatial and temporal patterns and driving factors of CO_2 emissions in China. First, we constructed a conceptual model of the factors influencing CO_2 emissions, including economic growth, industrial structure, energy consumption, urban development, foreign trade, and government management. Second, we selected 30 provinces in China from 2006 to 2019 as research objects and adopted exploratory spatial data analysis (ESDA) methods to analyse the spatio-temporal patterns and agglomeration characteristics of CO_2 emissions. Third, on the basis of 420 data samples from China, we used partial least squares structural equation modelling (PLS-SEM) to verify the validity of the conceptual model, analyse the reliability and validity of the measurement model, calculate the path coefficient, test the hypothesis, and estimate the predictive power of the structural model. Fourth, multigroup analysis (MGA) was used to compare differences in the influencing factors for CO_2 emissions during different periods and in various regions of China. The results and conclusions are as follows: (1) CO_2 emissions in China increased year by year from 2006 to 2019 but gradually decreased in the eastern, central, and western regions. The eastern coastal provinces show spatial agglomeration and CO_2 emission hotspots. (2) Confirmatory analysis showed that the measurement model had high reliability and validity; four latent variables (industrial structure, energy consumption, economic growth, and government management) passed the hypothesis test in the structural model and are the determinants of CO_2 emissions in China. Meanwhile, economic growth is a mediating variable of industrial structure, energy consumption, foreign trade, and government administration on CO_2 emissions. (3) The calculated results of the R^2 and Q^2 values were 76.3% and 75.4%, respectively, indicating that the structural equation model had substantial explanatory and high predictive power. (4) Taking two development stages and three main regions as control groups, we found significant differences between the paths affecting CO_2 emissions, which is consistent with China's actual development and regional economic pattern. This study provides policy suggestions for CO_2 emission reduction and sustainable development in China.

Keywords: CO_2 emissions; ESDA; PLS-SEM; China

Citation: Wang, B.; Zheng, Q.; Sun, A.; Bao, J.; Wu, D. Spatio-Temporal Patterns of CO_2 Emissions and Influencing Factors in China Using ESDA and PLS-SEM. *Mathematics* 2021, 9, 2711. https://doi.org/10.3390/math9212711

Academic Editor: Bahram Adrangi

Received: 25 September 2021
Accepted: 21 October 2021
Published: 25 October 2021

Publisher's Note: MDPI stays neutral with regard to jurisdictional claims in published maps and institutional affiliations.

Copyright: © 2021 by the authors. Licensee MDPI, Basel, Switzerland. This article is an open access article distributed under the terms and conditions of the Creative Commons Attribution (CC BY) license (https://creativecommons.org/licenses/by/4.0/).

1. Introduction

Climate change is a major global challenge faced by humanity today and has attracted extensive attention from the international community [1]. Countries have committed to reducing greenhouse gases to address climate change caused by greenhouse gases such as carbon dioxide (CO_2) [2,3]. In 2016, 178 parties signed the Paris Agreement, which became the third landmark international legal text to tackle climate change, following the UN Framework Convention on Climate Change in 1992 and the Kyoto Protocol in 1997, shaping global climate governance patterns. The Paris Agreement obliges all parties to commit to a long-term greenhouse gas emission reduction strategy for the mid-21st century until 2020 to promote early and significant reductions in global emissions. At the 75th UN General Assembly in September 2020, China proposed reaching its carbon peak by 2030 and striving to achieve carbon neutrality by 2060. A carbon peak is reached when carbon emissions (mainly CO_2) no longer increase. Carbon neutrality refers to a state of net-zero emissions after efforts to reduce or offset CO_2 emissions. China's carbon peak and carbon neutrality targets are consistent with the Paris Agreement and key to achieving the 1.5 °C global temperature decrease target.

China is at a critical stage of industrialisation and urbanisation. CO_2 emissions are continuing to increase, and many energy conservation and emission-reduction activities could be implemented. China has made continuous attempts to construct low-carbon cities, establish a circular economy, conserve resources, encourage an environment-friendly society, and pursue other paths to green development, with some positive results [4,5]. During the 14th Five-Year Plan (2021–2025), China will enter a critical period in which the country will focus on CO_2 reduction, promote the synergy and efficiency of pollution and CO_2 reduction, facilitate a comprehensive green transformation in economic and social development, realise a qualitative change in the improvement of ecological and environmental quality, and promote major changes in the way society functions. China will pursue its long-term goals of peak carbon and carbon neutrality; promote economic and social transformation; and accelerate the building of a green, low-carbon economic system. It will also continue to play a positive and constructive leading role in the international community, promote the establishment of a new cooperation mechanism for global climate governance, and strive to build a community with a common future for humanity [6–8].

In 2020, China's CO_2 emissions totalled 9894 billion tons, the highest in the world, and thus China came under great pressure to reduce emissions. China has a vast land area, and there are great differences between regions in the natural environment, development stage, population distribution, industrial structure, urbanisation, and other characteristics. The main factors that affect CO_2 emissions and their degree of influence are also different [9,10]. Therefore, it is of great theoretical and practical significance to explore the spatio-temporal patterns and driving factors of CO_2 emissions in different regions within China to formulate energy conservation and emission reduction policies to achieve the goals of peak carbon and carbon neutrality.

The determinants of CO_2 emissions are diverse, and the relationships between these driving forces are complex. Therefore, it is of great significance for China's ability to further control CO_2 emissions in the future to identify the influencing factors of CO_2 emissions and further clarify the causal relationships between these driving forces. First, we identified the driving factors affecting CO_2 emissions and constructed a conceptual model. We then analysed the spatio-temporal evolution of CO_2 emissions in China by using exploratory spatial data analysis (ESDA). We analysed the effect paths of the driving factors affecting CO_2 emissions using partial least squares structural equation modelling (PLS-SEM) methods. Finally, we propose policy suggestions for China's sustainable development.

2. Literature Review and Hypotheses

Increased CO_2 emissions result from economic growth, industrial structure, energy consumption, and other factors. Clarifying the key factors that affect CO_2 emission for research is of critical importance. However, most studies have focused on the relationship between specific factors and CO_2 emissions, while few studies have systematically explored the factors and their degree of influence on CO_2 emissions. Therefore, this study attempts to construct a theoretical framework concerning CO_2 emissions and explores the direction and intensity of each driving factor. We discuss the factors influencing CO_2 emissions in six dimensions: economic growth, industrial structure, energy consumption, urban development, foreign trade, and government management. We then hypothesise that these factors influence CO_2 emissions and construct a conceptual model (Figure 1).

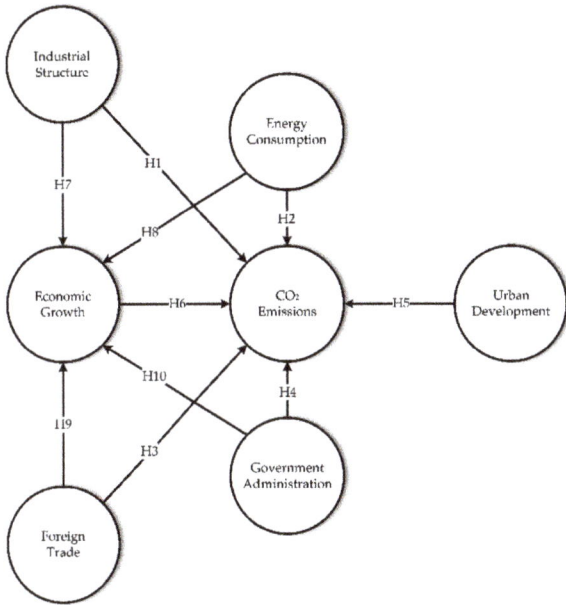

Figure 1. Proposed conceptual model.

2.1. Industrial Structure

With adjustments and developments of industrial structures, the CO_2 emissions of an economic system also change. The theory of industrial structural evolution holds that, during the process of economic development, the labour force shifts from primary industries dominated by agriculture to secondary industries dominated by factory production of goods, which is the main industry for CO_2 emissions. The higher the proportional contribution of the secondary industry to the total output value, the greater the threat to the environment and the higher CO_2 emissions. In particular, the higher the proportion of energy-intensive industries, the greater the CO_2 emissions. As the economy continues to expand, the labour force moves to tertiary industries dominated by services. In this process, the total CO_2 emissions will continue to increase, but the growth rate will decelerate. Additionally, different industrial structures have different impacts on CO_2 emissions. Some scholars have studied the influencing factors on CO_2 emissions in different industrial sectors, such as transportation, agriculture, industry, and tourism, and results vary from study to study [11,12]. Zheng et al. explored interregional differences in industry, construction, and transportation; warehousing industries had a significantly positive impact on interregional differences in CO_2 emissions from 2007 to 2016 [13]. Li et al. suggested that governments should increase the proportion of high-tech industries through technological progress, vigorously develop resource-saving and environment-friendly tertiary industries, and develop a low-carbon economy by promoting clean production technology [14]. Many scholars have found that industrial structure upgrading has a significant positive impact on economic growth, thereby promoting CO_2 emission reduction [15,16]. On the basis of the aforementioned literature, we propose the following hypothesis:

Hypothesis 1 (H1): *Industrial structure contributes to more CO_2 emissions.*

2.2. Energy Consumption

Economic growth heavily depends on energy consumption, which is a direct source of CO_2 emissions. Energy consumption quantity, intensity, and structure have some impact on CO_2 emissions. Economic development is inseparable from the use of all kinds of energy. The larger the economy grows, the more energy is consumed, and the more CO_2 emissions rise. Wang et al. concluded that reducing energy intensity could, to a large extent, reduce CO_2 emission intensity. By optimising industrial structure, CO_2 emission intensity could also be inhibited [10]. Energy intensity is the energy consumed per unit of output and is used to measure energy efficiency. The higher the level of technological progress, the less energy is wasted [17]. The structure of energy consumption has an important impact on CO_2 emissions. The carbon content of energy used in economic construction determines CO_2 emissions; the higher the carbon content, the greater the CO_2 emissions, and vice versa [18]. Xu et al. analysed the dynamic relationships between CO_2 emissions, economic growth, and the consumption of various fossil fuels in China and found that increasing coal and petroleum consumption significantly promoted CO_2 emissions. However, natural gas offers a cleaner substitute for other fossil fuels [19]. Thus, the impact of energy consumption on CO_2 emissions is clear. China is in a period of highly industrialised development, and coal-fired power generation is still the main form of power generation. Coal-fired power plants burn much fossil fuel and emit a large amount of CO_2, although the Chinese government has regulated the power industry as a key sector [20]. However, compared with European and American countries, China still has a long way to go in terms of energy conservation and emission reduction. Therefore, the sustainable development of electric energy is the main path of a low-carbon economy. In this way, the following hypothesis is proposed:

Hypothesis 2 (H2): *Energy consumption contributes to more CO_2 emissions.*

2.3. Foreign Trade

Engaging in international trade may be a factor in changing CO_2 emissions. Foreign trade is one of the principal ways to transfer CO_2 emissions through an international division of labour [21]. The expansion of foreign trade, as well as the structure and volume of foreign trade, may be factors in a country's CO_2 emissions. It is undeniable that China's foreign trade products are characterised by high energy consumption and low added value. The effect of import and export on CO_2 emissions is intuitively positive. However, relevant studies have found that the impact of foreign trade on CO_2 emissions may be positive [22], negative [23], or irrelevant [24], depending on both the foreign trade structure and trade volume [25] and also to the research period and region. Nevertheless, industrial transfer is one way of transferring CO_2 emissions. According to the foreign direct investment (FDI) theory, to maximise profits and reduce environmental costs, investors implement cross-border pollution transfer to transfer high-emission industries to lower-income regions, thus realising CO_2 emissions transfer. China is in a period of rapid economic development and must introduce a large amount of foreign capital for economic construction, which has a significant impact on CO_2 emissions. As the energy price is low, and the regulation of polluting industries is lower than that of higher-income countries, many export industries with high energy consumption and carbon emissions rapidly develop and become production sites for capital investment countries to transfer pollution [26]. Additionally, FDI has a technological spillover effect on CO_2 emissions that can improve the technical level of the region and reduce CO_2 emissions. Therefore, FDI affects the CO_2 emissions of the host country. On the basis of the aforementioned literature, we propose the following hypothesis:

Hypothesis 3 (H3): *Foreign trade contributes to more CO_2 emissions.*

2.4. Government Administration

Government intervention may be a key factor influencing CO_2 emissions. Related studies have found that governments affect CO_2 emissions in many ways. First, governments formulate corresponding energy transition and environmental policies to constrain the CO_2 emissions of enterprises and society [17,27]. Second, governments can increase financial expenditure on environmental governance [28] and scientific and technological investment, as well as collect pollution control fees and taxes from enterprises [29] to reduce levels of environmental pollution. Third, local governments may take the initiative to relax environmental regulations to achieve regional economic growth and thereby attract investment, which results in the deterioration of regional environmental quality and an increase in CO_2 emissions. Local government decision-making competition has three effects on CO_2 emissions: market distortion, investment bias, and a race to the bottom of environmental policies.

Hypothesis 4 (H4): *Government administration contributes to reduced CO_2 emissions.*

2.5. Urban Development

Urban development is an important factor affecting CO_2 emissions. First, a country's level of urbanisation is closely related to its economic development. The larger the scale of urban development, the more CO_2 emissions increase. However, improvements in urbanisation levels can decrease CO_2 emissions and thus realise low-carbon development [30]. In the early stages of urban development, Sun found that the expansion of the scope of urbanisation promotes improvement in CO_2 emission efficiency. After the urbanisation level reaches a critical point, the economic growth rate falls behind the growth rate in CO_2 emissions [31]. Second, a large number of energy resources is necessary for infrastructure construction during the process of urbanisation, which leads to an increase in CO_2 emissions. Zhou et al. found that spatial urbanisation was positively associated with CO_2 emissions due to the new infrastructure construction and conversion of existing land [32]. Third, during urbanisation, many people migrate from rural to urban areas, and this substantial increase in the urban population also increases CO_2 emissions [33]. Improvements in economic development and urbanisation can help achieve low-carbon development in an urban agglomeration [34].

Hypothesis 5 (H5): *Urban development contributes to more CO_2 emissions.*

2.6. Economic Growth

Economic growth is an important index for measuring the economic development of a country or region. According to the environmental Kuznets curve (EKC), when a country has a low level of economic development and uses few energy resources, its CO_2 emissions are low. However, with the acceleration of industrialisation, more fossil energy is needed, the degree of environmental deterioration worsens, and CO_2 emissions rise. When economic development reaches a certain level, the degree of environmental pollution and CO_2 emissions gradually decrease. Dinda et al. analysed the EKC hypothesis and postulated an inverted-U-shaped relationship between different pollutants and per capita income; that is, environmental pressure increases up to a certain level as income increases and decreases thereafter [35]. Fang et al. verified the panel EKC between economic growth and CO_2 emissions in China from 1995 to 2016 and found that, as the economy developed and GDP per capita increased, more energy was consumed, and the amount of environmental pollution increased [36]. Mardani et al. further found that CO_2 emissions were stimulated at higher or lower levels as economic growth increased or decreased. Conversely, a potential reduction in emissions harms economic growth [37].

China faces the dual pressures of economic growth and environmental protection [36]. On the one hand, China needs to reduce CO_2 emissions to jointly cope with the global climate crisis with the international community. On the other hand, China should also pay attention to domestic economic growth and social progress to enhance its national strength and improve its resilience to external environmental changes. Economic growth is a task and goal that cannot be ignored. Therefore, economic growth plays a connecting role between industrial structure, energy consumption, trade growth, government management, and carbon emissions. In recent years, China has witnessed rapid economic growth, increasing energy consumption, and a rapid increase in CO_2 emissions. In the future, China will need enough space for CO_2 emissions to reach the level of higher-income countries. Thus, as the economy continues to grow, CO_2 emissions will continue to accumulate.

Therefore, we propose the following hypotheses:

Hypothesis 6 (H6): *Economic growth contributes to more CO_2 emissions.*

Hypothesis 7 (H7): *Economic growth mediates the relationship between industrial structure and CO_2 emissions.*

Hypothesis 8 (H8): *Economic growth mediates the relationship between energy consumption and CO_2 emissions.*

Hypothesis 9 (H9): *Economic growth mediates the relationship between foreign trade and CO_2 emissions.*

Hypothesis 10 (H10): *Economic growth mediates the relationship between government administration and CO_2 emissions.*

3. Materials and Methods

3.1. Study Area

This study selected 30 provinces, autonomous regions, and municipalities as the research areas to acquire relevant statistical data, excluding Hong Kong, Macao, Taiwan, and Tibet. These areas can be divided into three main regions: eastern, central, and western.

- Eastern region: Beijing, Tianjin, Hebei, Liaoning, Shanghai, Jiangsu, Zhejiang, Fujian, Shandong, Guangdong, Guangxi, and Hainan.
- Central region: Shanxi, Henan, Anhui, Jilin, Heilongjiang, Jiangxi, Hubei, Hunan, and Inner Mongolia.
- Western region: Chongqing, Sichuan, Guizhou, Yunnan, Shaanxi, Gansu, Ningxia, Qinghai, and Xinjiang.

The CO_2 emissions of 30 provinces in China showed an overall upward trend from 2006 to 2019, increasing from 8.32 billion in 2006 to 13.92 billion in 2019, with an annual growth rate of 4.03%.

3.2. Data Collection

This study selected 420 data observations from 30 provinces, autonomous regions, and municipalities from 2006 to 2019 as research samples. The main data in this paper include CO_2 emissions and influencing factors. The former were calculated from fossil energy consumption data published in the China Statistical Yearbook of Energy (2006–2019), while the latter were mainly derived from the China Statistical Yearbook (2006–2019). Missing values were supplemented using linear interpolation with IBM SPSS Statistics 25.0 software (IBM Inc., Armonk, NY, USA).

(1) Data on CO_2 emissions. The estimates of CO_2 emissions of provinces in China are taken from the IPCC Guidelines for the National Greenhouse Gas Emission Inventory. According to China Energy Statistical Yearbook, primary energy consumption can be divided into eight categories: coal, coke, crude oil, fuel oil, gasoline, kerosene, diesel, and natural gas. As there are so many energy sources, the total CO_2 emissions from combustion should be the sum of the CO_2 emissions from each energy source.

$$CE_{it} = \sum_{i=1}^{n} E_{ijt} \times \omega_j \qquad (1)$$

where CE_{it} represents the CO_2 emissions generated by energy combustion in province i in year t, E_{ijt} represents the burned amount of energy j in province i, ω_j represents the CO_2 emissions coefficient of energy j combustion, and n represents the eight kinds of energy.

(2) Data for influencing factors. Following the principles of systematicity, representativeness, and availability of index selection, this study used 16 specific indicators from the six dimensions of economic development level, industrial structure, energy consumption, urban development, foreign trade, and government management, using the following specific descriptions:

(i) Economic growth. Gross domestic product (GDP) and per capita GDP represent the level of economic growth, which can measure the impact of the economy on CO_2 emissions. Household consumption level is selected to reflect the economic conditions of people's lives, which is an important reflection of national economic growth.

(ii) Industrial structure. The proportion of the value added by the secondary industry to GDP can represent the industrial structure of the main source of CO_2. The proportion of industrial added value in the GDP reflects that China is still in the development stage of industrialisation at present; its main industries have high energy consumption, and that consumption will not be significantly reduced in a short time.

(iii) Energy consumption. With China's economic development and social progress, the production and consumption of electric energy are increasing. The power industry has become a major contributor to fossil fuel consumption and CO_2 emissions. Annual carbon emissions from electricity generation are close to 50% of the country's total energy CO_2 emissions. Therefore, it is reasonable to take electric energy production and consumption as two indicators to measure CO_2 emissions.

(iv) Urban development. The urban employed population reflects the agglomeration of the urban population. Urban fixed-asset investment reflects the input of urban factors. Traffic is an important source of urban CO_2 emissions, and the number of civilian vehicles can be used as an indicator to measure the impact of traffic on CO_2 emissions.

(v) Foreign trade. The total FDI and number of FDI enterprises can reflect the investment in foreign capital; the total export–import volume is used to reflect foreign trade. As FDI and total export–import volume are denominated in U.S. dollars, the U.S. GDP deflator is used to calculate the real value for the year 2019.

(vi) Government management. The impact of the Chinese government's intervention on CO_2 emissions can be considered from two perspectives: financial spending and administrative management. Scientific and technological input is selected as the financial index, reflecting the government's ability to curb CO_2 emissions by improving scientific and technological levels. Administrative management indicators include the accepted number of domestic patent applications and the authorised number of domestic patent applications, which are used to measure the impact of government and enterprises on CO_2 emissions from the perspective of science and technology.

3.3. Methods

Quantitative studies of CO_2 emissions have been a hot topic in recent years. Previous studies widely used include the Kaya identity [38–40]; the IPAT model [41–43]; the arithmetic mean divisia index (AMDI); the logarithmic mean divisia index (LMDI) [44–47]; and the stochastic impacts by regression on population, affluence, and technology (STIRPAT) model [48–51]. This study integrated ESDA and PLS-SEM to study the spatio-temporal differences and influencing factors of CO_2 emissions in China. The spatial correlation and hotspot analysis of the ESDA method intuitively reflected the spatial distribution and spatial agglomeration of CO_2 emissions. The PLS-SEM method was used to analyse the influence paths, intensity, and significance of the factors influencing CO_2 emissions.

3.3.1. ESDA

ESDA is an analytical method for exploring the spatial relevance of geographical phenomena from the perspective of spatial analysis; it is also suitable for studying the spatial agglomeration of CO_2 emissions [52,53].

Spatial Autocorrelation

Global and local autocorrelation can reveal the spatial connections and differences between research units, which are often expressed by Moran's I. Global autocorrelation can describe the spatial correlation pattern of a whole research area; the local autocorrelation can reflect the spatial agglomeration characteristics of each unit within the region and identify high-value agglomeration and low-value agglomeration at different spatial locations. This method is suitable for representing the spatial agglomeration characteristics of CO_2 emissions in the research area [54,55]. The calculation formula is

$$I = \frac{n}{\sum_{i=1}^{n}(x_i - \bar{x})^2} \cdot \frac{\sum_{i=1}^{n}\sum_{j=1}^{n} w_{ij}(x_i - \bar{x})(x_j - \bar{x})}{\sum_{i=1}^{n}\sum_{j=1}^{n} w_{ij}} \quad (2)$$

$$I_L = \frac{n(x_i - \bar{x})\sum_{j=1}^{n} w_{ij}(x_j - \bar{x})}{\sum_{i=1}^{n}(x_i - \bar{x})^2} \quad (3)$$

where I represents the global Moran index, and I_L represents the local Moran index. n is the number of provinces, and x_i and x_j are the CO_2 emission values of province i and province j, respectively. w_{ij} is the adjacency matrix between province i and province j. At the significance level, the value of the global Moran I index ranges from -1 to 1. If the value of the index is greater than 0, there is a positive spatial correlation between the two provinces; if the value is less than 0, there is a negative spatial correlation; if it is equal to 0, there is no spatial correlation. The local spatial association mode can be divided into high–high (H-H), high–low (H-L), low–high (L-H), and low–low (L-L). The global Moran index statistics can only test the global spatial correlation but cannot determine the specific spatial agglomeration region, whereas the local Moran index and the Getis–Ord G_i^* local statistics can solve this problem.

Hotspots Analysis

On the basis of the Getis–Ord G_i^* value, ARCGIS software can be used to automatically draw a spatial clustering graph of high and low values with statistical significance, which reveals the hotspots and coldspots of regional CO_2 emissions [56]. The formulas are as follows:

$$G_i^* = \frac{\sum_{i=1}^{n} w_{i,j}x_j - \bar{X}\sum_{j=1}^{n} w_{i,j}}{S\sqrt{\frac{n\sum_{j=1}^{n} w_{i,j}^2 - \left(\sum_{j=1}^{n} w_{i,j}\right)^2}{n-1}}} \quad (4)$$

$$S = \sqrt{\frac{\sum_{j=1}^{n} x_j^2}{n} - \bar{X}^2} \quad (5)$$

$$\bar{X} = \frac{\sum_{j=1}^{n} x_j}{n} \quad (6)$$

where x_j is the attribute value of province j, ω_{ij} is the spatial weight of province i and province j, and n is the total number of provinces. G_i^* is the score of the z value, which reflects the spatio-temporal agglomeration characteristics of high and low CO_2 emissions.

3.3.2. PLS-SEM

PLS-SEM is used to estimate a complex causal relationship model with latent variables; it is widely used in economics, sociology, and other fields [57]. Therefore, this study is a new attempt to explore the influence of various driving factors on CO_2 emissions using PLS-SEM. The PLS-SEM model has the following three advantages: first, multiple latent variables are introduced to better reflect the path, orientation, and intensity between latent variables and CO_2 emissions; second, the complex relationships between latent variables and their causal effects can be further clarified; third, comparative analyses between control groups better reflect the temporal difference and spatial heterogeneity. PLS-SEM must generally follow four steps: model construction, hypothesis formulation, confirmatory factor analysis, and structural model analysis [58–60]. We also adopted multigroup analysis to understand the differences between factors affecting CO_2 emissions in different periods and regions. As both the total sample and the control sample size exceeded 100, the PLS-SEM technology of SmartPLS 3.3.3 software was suitable for testing the causal relationships proposed in the conceptual model.

Multigroup analysis (MGA) allows for testing whether pre-defined data groups have significant differences in their group-specific parameter estimates (e.g., outer weights, outer loadings, and path coefficients) [61]. In recent years, the MGA method has also been widely applied in tourism [61,62], marketing [63,64], and other fields. This study adopted MGA to analyse the difference in the impact paths on CO_2 emissions in different periods and regions. SmartPLS 3.3.3 provides outcomes of four different approaches that are based on bootstrapping results from every group.

4. Results

4.1. Spatio-Temporal Evolution Patterns and Agglomeration Characteristics of CO_2 Emissions

4.1.1. Spatio-Temporal Patterns at Different Scales

At the national scale, CO_2 emissions have been rising steadily in China, from 277.43 million tons in 2006 to 463.97 million tons in 2019, with an annual growth rate of 4.03%. From the perspective of the development stage, CO_2 emissions grew rapidly from 2006 to 2011 but slowed from 2012 to 2019, with annual growth rates of 7.63% and 1.71%, respectively (Figure 2).

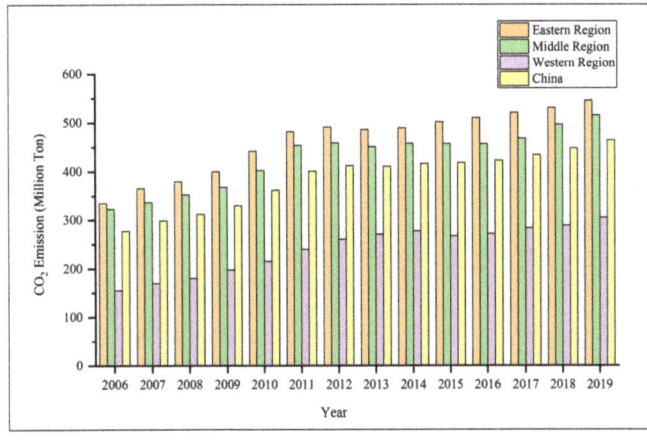

Figure 2. Averages of CO_2 emissions in three main regions in China from 2006 to 2019.

At the regional scale, the average CO_2 emissions of the eastern, central, and western regions were 334.95 million tons, 322.68 million tons, and 155.49 million tons, respectively, in 2006. The average CO_2 emissions of the eastern, middle, and western regions increased to 535.34 million tons, 515.03 million tons, and 304.43 million tons in 2019, with annual growth rates of 3.82, 3.66, and 5.30%, respectively. The order of mean CO_2 emission is eastern region > central region > western region (Figure 2).

At the provincial scale, CO_2 emissions in nine provinces were higher than the national average of CO_2 emissions (277.43 million tons) in 2006, and CO_2 emissions in 11 provinces were more than the national average of CO_2 emissions (463.97 million tons) in 2019. Shandong province has always ranked first in CO_2 emissions in China, whereas Qinghai province always ranked last from 2006 to 2019. The growth rate of CO_2 emission in Beijing decreased year by year, showing a negative growth rate. The growth rate of CO_2 emissions in Shanghai and Henan remained at relatively low levels, 0.69 and 1.31%, respectively, while the growth rates for Ningxia and Xinjiang, although they were located in the western region, were 10.82 and 10.05%, respectively (see Figure 3).

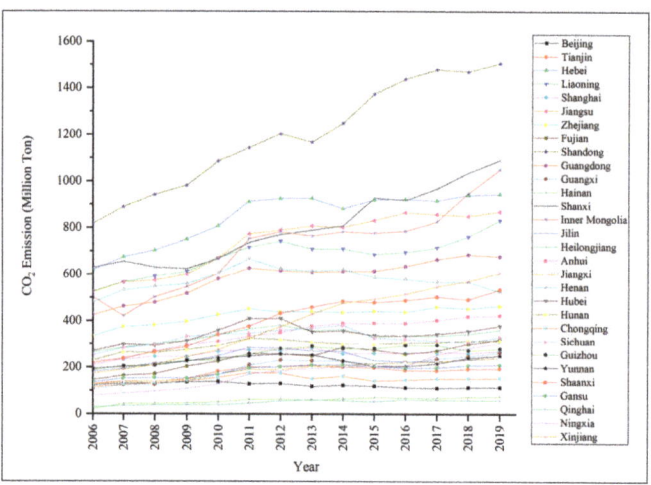

Figure 3. CO_2 emissions for 30 provinces in China from 2006 to 2019.

4.1.2. Spatial Agglomeration Characteristics

To display the spatial difference characteristics of the CO_2 emissions, we calculated the spatial correlation and hotspot analysis of CO_2 emissions for 30 provinces in China from 2006 to 2019, and the z values were divided into five levels using the natural breakpoint method.

The spatial correlation of CO_2 emissions in China was calculated with Geoda 1.18 software using Formulas (2) and (3) (see Figure 4).

The results show that the global Moran's I was 0.087, and the statistical significance test at 5% indicated that the spatial distribution of CO_2 emissions in each province was significantly positive; that is, the regions with high CO_2 emissions were relatively concentrated in terms of their locations. Shandong, Hebei, Henan, Shanxi, Liaoning, Jiangsu, Zhejiang, and Inner Mongolia were H-H agglomeration areas, indicating that local CO_2 emissions and those of neighbouring provinces were high. Beijing, Shanghai, Tianjin, and Anhui were L-H agglomeration areas, indicating that local CO_2 emissions were low and those of the surrounding provinces were high. Xinjiang was an H-L agglomeration area, indicating that local CO_2 emissions were high and those of the surrounding provinces were low.

Figure 4. Local autocorrelation of CO_2 emissions in China from 2006 to 2019.

The hotspots and coldspots of regional CO_2 emissions were plotted with ArcGIS 10.6 software using Formulas (4)–(6) (see Figure 5).

Figure 5. Hotspots of CO_2 emissions in China from 2006 to 2019.

The results show that most of the hotspots were distributed in the eastern coastal areas. Liaoning and Inner Mongolia ranked highest, followed by the Beijing–Tianjin–Hebei and the Yangtze River Delta regions (Shanghai–Jiangsu–Zhejiang). Shandong, Shanxi, and Henan ranked in the central. Hainan, as an isolated province, was a coldspot for CO_2 emissions.

4.2. Measurement Model

4.2.1. Reliability Test

Reliability and construct validity are the keys to establishing the measurement model. Cronbach's alpha, composite reliability (CR), and average variance extracted (AVE) were obtained using PLS algorithms [65]. Table 1 shows all factor loadings of Cronbach's alpha, CR, and AVE.

Table 1. Factor loadings, Cronbach's alpha, composite reliability (CR), and average variance extracted (AVE).

Latent Variables	Indicators	Factor Loading	*t*-Value	Cronbach's Alpha	CR	AVE
	Economic growth					
EG1	GDP	0.864	81.788	0.847	0.899	0.747
EG2	Household consumption level	0.850	43.414			
EG3	Per capita GDP	0.879	54.032			
	Industrial structure					
IS1	Industrial added value as a percentage of GDP	0.986	375.863	0.968	0.984	0.969
IS2	Secondary industry as a percentage of GDP	0.983	297.067			
	Energy consumption					
EC1	Electric energy production	0.975	345.139	0.953	0.977	0.955
EC2	Consumption of electric power	0.979	500.524			
	Urban development					
UD1	Total fixed-asset investment	0.961	188.443	0.948	0.967	0.906
UD2	Urban employment	0.925	96.994			
UD3	Number of civilian motor vehicles	0.968	235.794			
	Foreign trade					
FT1	Number of foreign-invested enterprises	0.983	227.454	0.965	0.977	0.935
FT2	Total volume of foreign trade	0.970	285.679			
FT3	Foreign direct investment	0.948	123.828			
	Government administration					
GA1	Accepted number of domestic patent applications	0.987	413.46	0.975	0.984	0.953
GA2	Authorised number of domestic patent applications	0.986	273.979			
GA3	Expenditures in science and technology of local government	0.956	105.985			

Note: CR = composite reliability; AVE = average variance extracted.

(1) Cronbach's alpha was calculated to ensure composite reliability. By convention, Cronbach's alpha should be greater or equal to 0.800 for a good scale and 0.700 for an acceptable scale. Table 1 shows that all Cronbach's alpha values ranged from 0.899 to 0.984, which indicates that the measurement indexes had high reliability. (2) The CR was used to examine internal consistency. The higher the CR value, the higher the internal consistency of the plane; 0.700 is an acceptable threshold. We found the model to be sufficiently reliable and internally consistent, as shown in Table 1. (3) The AVE reflected the average commonality of each latent factor and was used to establish convergent validity. The AVE should be above 0.500 for all latent variables, whereby at least 50% of measurement variance is explained. Table 1 shows that all AVE values ranged from 0.747 to 0.969. In conclusion, the model had high reliability and construct validity, as evidenced by the verification of Cronbach's alpha, CR, and AVE.

4.2.2. Validity Test

Convergent and discriminant validities were utilised to examine the model's construct validity [65]. (1) Factor loading measures convergent validity and must be greater than 0.500. The factor loadings were greater than 0.850 (see Table 1), indicating that the measured variables had high convergence validity. (2) The square root of each construct's AVE was greater than the bivariate correlation with the other constructs, which shows that the model had high convergent validity. Table 2 shows the square roots of the AVEs and all the correlations. (3) The HTMT ratio is an important indicator used to evaluate discriminant validity by applying a PLS algorithm [66]. The HTMT ratio for the two latent variables was below 0.900. The HTMT values are shown in brackets in Table 3. In conclusion, the model has high convergent and discriminant validity, found through the verification of Cronbach's alpha, CR, and AVE.

Table 2. Results of the hypothesis model.

Hypothesis	Path	Path Coefficient	t-Value	p-Value	95% BCa Confidence Intervals	f^2	Support
H1	IS→CO_2	0.122 ***	4.661	0.000	[0.080, 0.165]	0.048	Yes
H2	EC→CO_2	0.995 ***	21.654	0.000	[0.916, 1.066]	0.858	Yes
H3	FT→CO_2	−0.012	0.205	0.419	[−0.116, 0.107]	0.000	No
H4	GA→CO_2	−0.528 ***	6.372	0.000	[−0.639, −0.380]	0.130	Yes
H5	UD→CO_2	0.001	0.011	0.496	[−0.093, 0.102]	0.000	No
H6	EG→CO_2	0.223 ***	4.013	0.000	[0.133, 0.307]	0.058	Yes
H7	IS→EG	−0.157 ***	4.651	0.000	[−0.229, −0.111]	0.068	Yes
H8	EC→EG	0.279 ***	4.311	0.000	[0.181, 0.399]	0.113	Yes
H9	FT→EG	0.302 ***	4.660	0.000	[0.200, 0.414]	0.068	Yes
H10	GA→EG	0.334 ***	3.270	0.001	[0.144, 0.472]	0.061	Yes

Note: BCa: bias corrected and accelerated bootstrap. Significance level: *** $p < 0.001$. EG = economic growth; IS = industrial structure; EC = energy consumption; UD = urban development; FT = foreign trade; GA = government administration.

Table 3. Discriminant validity.

	EG	IS	EC	UD	FT	GA
EG	**0.865**					
IS	−0.083 (0.300)	**0.984**				
EC	0.641 (0.599)	0.294 (0.305)	**0.977**			
UD	0.695 (0.662)	0.177 (0.188)	0.864 (0.900)	**0.952**		
FT	0.756 (0.796)	−0.012 (0.071)	0.586 (0.606)	0.566 (0.595)	**0.967**	
GA	0.788 (0.809)	−0.007 (0.070)	0.689 (0.710)	0.771 (0.803)	0.873 (0.900)	**0.976**

Note: The square roots of AVEs are shown on the diagonal in bold. The HTMT ratios are in brackets. EG = economic growth; IS = industrial structure; EC = energy consumption; UD = urban development; FT = foreign trade; GA = government administration.

4.3. Structural Model

4.3.1. Path Coefficients and Significance

The bootstrapping method is a non-parametric statistical procedure that detects the statistical significance of various PLS-SEM results, including path coefficients, Cronbach's alpha, HTMT, R^2 values, and the effect size f^2. Here, the number of sub-samples was set to 5000 and the t-test results were guaranteed to be significant at the level of 0.05 [65]. Table 2 shows the standardised path coefficients, t-values, and results. The results show that the path coefficients of the four latent variables (economic growth, industrial structure, energy consumption, and government administration) passed the significance test, and the four hypotheses were acceptable. This reveals that four factors had an impact on China's total CO_2 emissions from 2006 to 2019. However, urban development and foreign trade did not pass the hypothesis test in this model, which does not mean that these two latent variables did not contribute to carbon emissions. On the contrary, these two variables had an impact on CO_2 emissions through the latent variable of economic growth. An f^2 above 0.02, 0.15, or 0.35 is considered a small, medium, or large effect, respectively. Additionally, the direct and indirect effects were analysed to verify the multiple mediation model (see Figure 6).

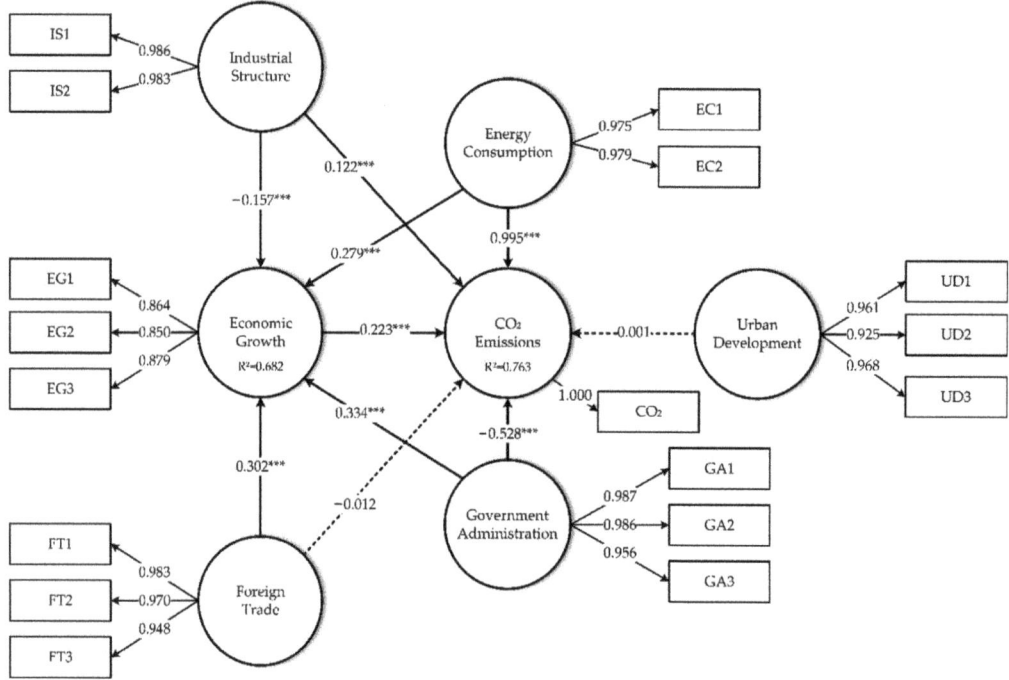

Figure 6. Results of PLS-SEM. Note: The dotted line indicates that the path fails the hypothesis test, and the solid line indicates that the path passes the hypothesis test. *** $p < 0.001$.

4.3.2. Direct, Indirect, Total, and Mediation Effects

Table 4 shows the direct, indirect, and total effects among latent variables and CO_2 emissions. Moreover, the positive effect of industrial structure on CO_2 emissions is more direct (0.122) than indirect (−0.035) through economic growth. The effect of energy consumption on CO_2 emissions is more direct (0.955) than indirect (0.060) through economic growth. The effect of foreign trade on CO_2 emissions is entirely indirect (0.067); the direct effect is not statistically significant. The effect of government administration on CO_2 emissions is also more direct (−0.528) than indirect (0.077).

Table 4. Results of direct, indirect, and mediating effects of economic growth.

Path	Direct Effect	Indirect Effect	t-Value	p-Value	95% Bca Confidence Intervals	Mediation Effect
IS→CO_2	0.122 ***					Complementary
IS→EG→CO_2		−0.035 ***	3.313	0.000	[−0.054, −0.019]	(partial mediation)
EC→CO_2	0.995 ***					Complementary
EC→EG→CO_2		0.060 ***	4.305	0.000	[0.400, 0.088]	(partial mediation)
FT→CO_2	−0.012					Indirect-only
FT→EG→CO_2		0.067 ***	3.114	0.001	[0.035, 0.103]	(full mediation)
GA→CO_2	−0.528 ***					Complementary
GA→EG→CO_2		0.077 *	1.955	0.025	[0.025, 0.138]	(partial mediation)

Note: Bca = bias corrected and accelerated bootstrap. Significance level: *** $p < 0.001$, * $p < 0.05$. IS = industrial structure; EC = energy consumption; FT = foreign trade; GA = government administration; UD = urban development; EG = economic growth; CO_2 = CO_2 emissions.

Economic growth is a mediating variable of the effect of industrial structure, energy consumption, foreign trade, and government administration on CO_2 emissions. Therefore, H7, H8, H9, and H10 are supported. Table 4 shows the mediation effects of economic growth in the model.

4.3.3. Predictive Power

R^2 and Q^2 are important metrics for evaluating the predictability of a model. The R^2 value was obtained using a PLS calculation; Q^2 was calculated using blindfolding [65]. (1) The determination coefficient R^2 value was used as an indicator of the overall predictive strength of the model. Falk and Miller consider that R^2 value should be greater than 0.100 [67]; Chin [68] recommends 0.670, 0.330, and 0.100 (substantial, moderate, and weak, respectively); while Hair et al. [65] consider 0.750, 0.500, and 0.250 (substantial, moderate, and weak, respectively). The R^2 value ($R^2 = 0.763 > 0.750$) indicated that 76.3% of CO_2 emissions could be explained by the causality of six latent variables, showing that the model had substantial explanatory power. In addition, the R^2 value of economic growth was 0.682, indicating 68.2% of economic growth can be explained by industrial structure, energy consumption, foreign trade, and government administration. (2) The Stone–Geisser Q^2 value is a criterion for evaluating construct cross-validated redundancy. The larger the Q^2 value, the stronger the prediction correlation. The Q^2 value of CO_2 emissions was 0.754, indicating that the model had high predictive relevance.

4.4. Multigroup Analysis

The PLS-MGA p-values revealed significant group differences in the MGA. Table 5 and Figure 7 show the MGA for different regions and periods in China.

On the basis of the analysis of the temporal and spatial evolution characteristics of CO_2 emissions, we divided the study period into two control groups: 2006–2011 was the first growth stage, and 2012–2019 was the second growth stage. For the growth stage groups, H3, H4, and H6 differed significantly, showing that the impact of foreign trade, government administration, and economic growth on CO_2 emission for the first stage compared to the second stage group. Meanwhile, H7, H9, and H10 differed significantly, indicating that the impact of industrial structure, foreign trade, and government administration on economic growth in the first stage is stronger than in the second stage.

The pairwise comparison of the three regions in China revealed differences in the influencing factors of CO_2 emissions between the regions.

(1) For the eastern and central groups, H3, H4, and H6 differed significantly, indicating that the impact of foreign trade, government administration, and economic growth on CO_2 emissions in the eastern region is stronger than in the central region.
(2) For the eastern and western groups, H1, H2, H3, and H5 differed significantly, indicating that the impact of industrial structure, energy consumption, foreign trade, and urban development on CO_2 emissions in the eastern region is stronger than in the western region. Meanwhile, H8 differed significantly.
(3) For the central and western groups, H1, H2, and H5 differed significantly, indicating that industrial structure, energy consumption, and urban development have stronger impacts on CO_2 emissions for the central than for the western region. Meanwhile, H7, H8, and H10 differed significantly. Likewise, we analysed the confidence intervals that allow us to verify if a path coefficient is significantly different from 0 as another way to assess the significance.

Table 5. Results of multigroup analysis on different periods and regions in China.

Hypothesis	Diff.	p-Value	Part (A): First Stage (n = 180) vs. Second Stage (n = 240).							
			Stage 1 (2006–2011)				Stage 2 (2012–2019)			
			PC Mean	t-Value	p-Value	95% BCa CI	PC Mean	t-Value	p-Value	95% BCa CI
H1	−0.002	0.478	0.077 ***	2.475	0.007	[−0.028, 0.129]	0.080 **	2.421	0.008	[0.026, 0.132]
H2	−0.022	0.411	0.956 ***	9.454	0.000	[−0.797, 1.13]	0.978 ***	18.685	0.000	[0.888, 1.057]
H3	−0.449	**0.001**	−0.307 ***	3.216	0.001	[−0.499, −0.174]	0.134	1.172	0.121	[−0.029, 0.333]
H4	0.299	**0.036**	−0.446 ***	3.651	0.000	[−0.626, −0.252]	−0.741 ***	6.097	0.000	[−0.93, −0.545]
H5	−0.016	0.453	0.083	0.706	0.240	[−0.113, 0.237]	0.095	1.391	0.082	[−0.013, 0.2]
H6	0.201	**0.025**	0.420 ***	5.481	0.000	[−0.292, 0.54]	0.227 ***	3.432	0.000	[0.121, 0.327]
H7	0.134	**0.029**	0.052	0.923	0.178	[−0.045, 0.126]	−0.082 *	1.776	0.038	[−0.179, −0.016]
H8	−0.143	0.109	0.031	0.499	0.309	[−0.088, 0.172]	0.171 *	2.166	0.015	[0.055, 0.327]
H9	−0.314	**0.011**	0.207 *	2.139	0.016	[−0.049, 0.38]	0.541 ***	5.442	0.000	[0.381, 0.696]
H10	0.480	**0.001**	0.667 ***	6.046	0.000	[−0.466, 0.819]	0.172	1.325	0.093	[−0.065, 0.359]

Hypothesis	Diff.	p-Value	Part (B): Eastern Region (n = 168) vs. Central Region (n = 140).							
			Eastern Region				Central Region			
			PC Mean	t-Value	p-Value	95% BCa CI	PC Mean	t-Value	p-Value	95% BCa CI
H1	−0.010	0.433	0.034	1.050	0.147	[−0.018, 0.102]	0.052	0.988	0.162	[−0.038, 0.126]
H2	0.016	0.469	0.929 ***	6.125	0.000	[0.656, 1.125]	0.882 ***	11.954	0.000	[0.765, 1.010]
H3	0.400	**0.000**	−0.099	1.026	0.152	[−0.225, 0.070]	−0.500 ***	6.346	0.000	[−0.613, −0.359]
H4	−0.424	**0.001**	−0.638 ***	5.434	0.000	[−0.812, −0.442]	−0.200 ***	3.639	0.000	[−0.287, −0.108]
H5	−0.110	0.309	0.215	1.542	0.062	[−0.004, 0.497]	0.360 **	2.348	0.009	[0.134, 0.610]
H6	0.278	**0.009**	0.226 *	2.144	0.016	[0.062, 0.365]	−0.071	0.949	0.171	[−0.196, 0.046]
H7	0.105	0.093	−0.148 **	2.839	0.002	[−0.273, −0.084]	−0.263 ***	4.874	0.000	[−0.353, −0.173]
H8	0.043	0.422	0.512 **	2.604	0.005	[0.227, 0.891]	0.483 ***	7.164	0.000	[0.368, 0.586]
H9	−0.037	0.378	0.190 **	2.460	0.007	[0.080, 0.339]	0.232 **	2.222	0.013	[0.045, 0.386]
H10	0.141	0.239	0.298	1.619	0.053	[−0.020, 0.567]	0.150 **	1.683	0.046	[0.015, 0.300]

Hypothesis	Diff.	p-Value	Part (C): Eastern Region (n = 168) vs. Western Region (n = 112).							
			Eastern Region				Western Region			
			PC Mean	t-Value	p-Value	95% BCa CI	PC Mean	t-Value	p-Value	95% BCa CI
H1	0.161	**0.01**	0.034	2.148	0.147	[−0.016, 0.103]	−0.129 *	1.050	0.016	[−0.207, −0.020]
H2	−0.400	**0.026**	0.928 ***	8.889	0.000	[0.648, 1.132]	1.362 ***	6.033	0.000	[1.087, 1.541]
H3	−0.557	**0.000**	−0.098	2.651	0.152	[−0.224, 0.074]	0.544 **	1.027	0.004	[0.205, 0.719]
H4	−0.465	0.076	−0.637 ***	0.503	0.000	[−0.811, −0.440]	−0.103	5.459	0.307	[−0.655, 0.386]
H5	1.137	**0.003**	0.217	2.233	0.061	[0.004, 0.499]	−1.025 *	1.550	0.013	[−1.519, −0.183]
H6	0.164	0.163	0.225 *	0.329	0.013	[0.070, 0.360]	−0.007	2.222	0.371	[−0.198, 0.216]
H7	−0.119	0.071	−0.148 **	0.730	0.003	[−0.279, −0.086]	−0.036	2.803	0.233	[−0.152, 0.041]
H8	0.388	**0.040**	0.515 **	1.346	0.005	[0.218, 0.887]	0.152	2.583	0.089	[−0.021, 0.302]
H9	0.125	0.273	0.189 **	0.373	0.006	[0.080, 0.337]	0.090	2.488	0.355	[−0.281, 0.336]
H10	−0.367	0.068	0.296	3.831	0.051	[−0.021, 0.569]	0.632 ***	1.636	0.000	[0.396, 0.947]

Hypothesis	Diff.	p-Value	Part (D): Central Region (n = 140) vs. Western Region (n = 112).							
			Central Region				Western Region			
			PC Mean	t-Value	p-Value	95% BCa CI	PC Mean	t-Value	p-Value	95% BCa CI
H1	0.171	**0.013**	0.053	2.126	0.166	[−0.042, 0.127]	−0.129 *	0.969	0.017	[−0.211, −0.025]
H2	−0.416	**0.002**	0.882 ***	9.022	0.000	[0.765, 1.007]	1.357 ***	11.907	0.000	[1.091, 1.543]
H3	−0.957	N.A.	−0.501 ***	2.680	0.000	[−0.604, −0.354]	0.542 **	6.403	0.004	[0.214, 0.719]
H4	−0.041	0.473	−0.200 ***	0.499	0.000	[−0.288, −0.106]	−0.118	3.614	0.309	[−0.655, 0.405]
H5	1.246	**0.001**	0.361 **	2.235	0.009	[0.121, 0.600]	−1.011 *	2.348	0.013	[−1.561, −0.245]
H6	−0.114	0.238	−0.070	0.332	0.171	[−0.199, 0.045]	−0.003	0.951	0.370	[−0.204, 0.211]
H7	−0.223	**0.003**	−0.262 ***	0.727	0.000	[−0.355, −0.176]	−0.035	4.848	0.234	[−0.147, 0.042]
H8	0.345	**0.006**	0.482 ***	1.333	0.000	[0.369, 0.592]	0.153	7.060	0.091	[−0.026, 0.301]
H9	0.162	0.223	0.230 *	0.372	0.013	[0.050, 0.390]	0.085	2.241	0.355	[−0.281, 0.328]
H10	−0.508	**0.005**	0.152 *	3.776	0.044	[0.009, 0.292]	0.634 ***	1.702	0.000	[0.388, 0.958]

Note: BCa CI = bias corrected and accelerated bootstrap confidence intervals. Significance level: * $p < 0.05$, ** $p < 0.01$, *** $p < 0.001$. Bold font: PLS-MGA p-value below 5% and above 95% indicate significant values. Diff. = path coefficient differences; PC = path coefficient; N.A. = not available. IS = industrial structure; EC = energy consumption; FT = foreign trade; GA = government administration; UD = urban development; EG = economic growth; CO_2 = CO_2 emissions.

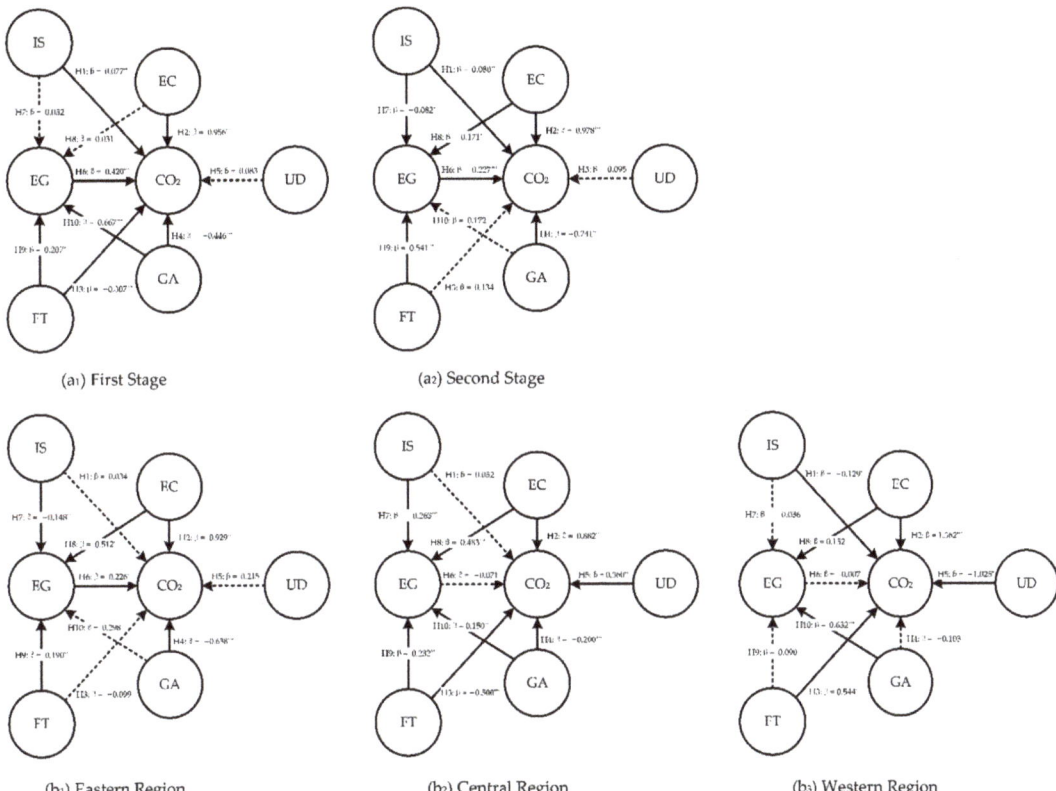

Figure 7. MGA of different periods and regions. Note: The dotted line indicates that the path fails the hypothesis test, and the solid line indicates that the path passes the hypothesis test. * < 0.05; p ** $p < 0.01$; *** $p < 0.001$. IS = industrial structure; EC = energy consumption; FT = foreign trade; GA = government administration; UD = urban development; EG = economic growth; CO_2 = CO_2 emissions.

5. Discussion

5.1. Theoretical Implications

First, Figures 2–4 show that spatio-temporal differences in China's CO_2 emissions have a spatial scale effect. On the national scale, China's total CO_2 emissions show a trend of gradual and steady growth. This is because China is in a period of rapid economic growth, and its CO_2 emissions are thus increasing rapidly. At the regional scale, the CO_2 emissions of the three main regions show decreasing distribution in the east, central, and west regions, which is consistent with the actual situation of China's regional economic development [10]. At the provincial scale, provinces with high CO_2 emissions were mainly distributed in the eastern coastal region, indicating that provinces with high CO_2 emissions tend to be clustered together. Therefore, when formulating policies and development plans, the government should consider the spatial spillover effects of CO_2 emissions and the relevant influencing factors between neighbouring provinces.

Second, through confirmatory and structural model analyses, we found that the measurement model of the influencing factors for CO_2 emissions in China had high reliability and validity. Economic growth, industrial structure, energy consumption, and government management had a significant impact on carbon emissions. However, urbanisation development and foreign trade fail to pass the hypothesis test. These two latent variables had an impact on CO_2 emissions through the mediating effects of economic growth. The R^2 was

76.3%, and the Q^2 value was 75.4%, indicating that the structural equation model had a substantial explanatory and high predictive power.

Third, the path coefficients and significance of the six latent variables on CO_2 emissions were analysed. The results are as follows:

(1) H1: Industrial structure contributes to more CO_2 emissions, which is consistent with the research results of Long et al. [69]. China is at the central stage of an industrialised economy; the development of secondary industries requires energy and resources. The proportion of secondary industries in China is larger than that of higher-income countries. Secondary industries have a high energy demand, and therefore their CO_2 emissions are also high.

(2) H2: Energy consumption contributes to more CO_2 emissions, which is consistent with the results of Meng et al. [20]. With China's economic development, the production and consumption of electric energy are growing, and the electric power industry has become the main sector of fossil fuel consumption and CO_2 emission. China's energy consumption has always been based on coal; thermal power generation has long been the main component of power energy products, which will inevitably lead to the increase of CO_2 emissions.

(3) H3: Foreign trade fails to pass the hypothesis test. With the acceleration of China's integration into economic globalisation, China's foreign trade volume increases, and a large number of foreign-funded enterprises enter into China, resulting in the rapid development of the local economy and thus contributing more CO_2 dioxide. This result supports hypothesis H9.

(4) H4: Government administration contributes to reduced CO_2 emissions, which is consistent with the research results of Zheng et al. [70]. Government intervention reduces CO_2 emissions. The Chinese government can effectively reduce CO_2 emissions by implementing a variety of measures that are consistent with the national goal of energy conservation and emission reduction. Meanwhile, the government encourages enterprises to innovate independently and improve energy efficiency to reduce their CO_2 emissions. Therefore, the Chinese government has a restraining role in CO_2 emissions.

(5) H5: Urban development fails to pass the hypothesis test. China is in a period of rapid urbanisation development, and the contribution of increased urbanisation to CO_2 emissions exists. As a matter of fact, with the influx of migrants into cities, the construction of infrastructure, such as buildings and transportation, is accelerating, which will bring about the rapid development of the urban economy and inevitably increase CO_2 emissions.

(6) H6: Economic growth has a positive impact on CO_2 emissions, which is consistent with the findings of Cui et al. [16] and Li et al. [71]. China is currently at the central stage of economic development and undergoing large-scale economic growth and rapid development. Therefore, energy consumption and CO_2 emissions are high. This is consistent with the spatio-temporal evolution pattern of China's CO_2 emissions.

Economic growth is an important intervening variable in the model. The mediation of economic growth is complementary (partial mediation) between industrial structure and CO_2 emissions; the mediation of economic growth is complementary (partial mediation) between energy consumption and CO_2 emissions; the mediation of economic growth is indirect-only (full mediation) between foreign trade and CO_2 emissions; the mediation of economic growth is complementary (partial mediation) between government administration and CO_2 emissions. These results are consistent with China's multiple goals of pursuing both environmental protection and economic and social progress.

Fourth, through the MGA of different stages and regions, we found differences in the effect intensity of different influencing factors on CO_2 emissions at different development stages and different regions. Compared with the second stage, the impact of foreign trade, government administration, and economic growth on CO_2 emissions in the first stage was significantly different. This is because of China's accession to the World Trade

Organization in the early 21st century, which significantly accelerated China's participation in economic globalisation and opening-up and increased its import and export trade. During the 11th Five-Year Plan period (2006–2010), the central government promulgated the implementation of energy-saving target responsibility and evaluation, which has achieved some results. Progress has also been made toward the overall goal of building a resource-conserving and environment-friendly society, as proposed in the 12th Five-Year Plan (2011–2015) and 13th Five-Year Plan (2016–2020).

Fifth, through an MGA of different regions, we found differences in the effect intensity of various influencing factors on CO_2 emissions in different regions; this result is consistent with the current state of China's regional economic development [46].

(1) The impact of foreign trade (H3), government administration (H4), and economic growth (H6) on CO_2 emissions in the eastern region was stronger than that in the central region. The eastern coastal region took the lead in implementing the reform and opening-up strategy, and its economy started early and developed rapidly. In addition, the eastern region is a densely populated urban area of China; as such, it is also the core region for CO_2 emissions. In recent years, the eastern region has taken the lead in implementing the energy conservation and emission reduction strategy. For example, the CO_2 emissions for Beijing showed negative growth, while those for Shanghai and Tianjin showed low growth.

(2) The impact of industrial structure (H1), energy consumption (H2), foreign trade (H3), and urban development (H5) on CO_2 emissions in the eastern region was stronger than in the western region. Compared with the eastern region, there is a huge gap in economic development, industrial structure, and urban development in the western region, which is also consistent with the actual situation of China's regional development. The western region is a relatively backward region regarding economic development and has lower CO_2 emissions than the eastern and central regions. However, with the implementation of development and the opening-up strategy in recent years, emissions in the western region have rapidly increased, especially in Ningxia and Xinjiang, where the growth rates of CO_2 emissions are among the highest in the entire country. The current development in the western region still follows the model for inefficient, blind, and energy-intensive development; thus, the growth rate of CO_2 emissions is accelerating. Additionally, the western region is a key and difficult area in China's ecological construction due to its vast territory and fragile environment. Overall, CO_2 emissions in the west are problematic.

(3) The impacts of industrial structure (H1), energy consumption (H2), and urban development (H5) on CO_2 emissions in the central region were stronger than that in the western region. The central region is the main energy- and resource-producing area in China, and its industrial structure is dominated by secondary industries. Faced with prosperity in the east and development in the west, the central region has experienced economic collapse. In recent years, the central region has taken over industries with excess capacity from the east, and the rise of urban clusters has led to continuous CO_2 emissions.

5.2. Practical Implications

Given the above analysis, China's energy conservation and emission reduction measures can be proposed in six dimensions.

- China should improve the quality of economic growth through industrial upgrading, energy restructuring, and technological progress to change the economic growth model; it should also realise an extensive development model to intensify development model change.
- Per the law of industrial structure development at the present stage, China should strive to develop tertiary industries, promote the upgrading of secondary industries, advocate a circular economy, and encourage the development of green and environmental protection industries.

- China should continue to increase the proportion of clean energy in energy consumption; develop clean energy, such as wind power, solar energy, hydropower, and nuclear energy; and develop clean energy technologies and improve energy efficiency.
- China should remain on the path of green urbanisation and raise people's awareness about green consumption.
- The Chinese government should formulate a strict environmental access system, improve environment-related rules and regulations, and control high pollution and high energy consumption projects.
- The government has increased investment in science and technology and environmental protection and has adopted preferential policies to encourage local enterprises to engage in independent innovation.
- Energy conservation and emission reductions are gradual processes that should be based on actual economic growth stages and regional patterns to promote the transformation and sustainable development of different regions.

6. Conclusions

This study analysed the spatial and temporal pattern of CO_2 emissions in China and clarified the determinants of CO_2 emissions and their internal relationships. First, six driving factors that affect CO_2 emissions were clarified, including economic growth, industrial structure, energy consumption, urban development, foreign trade, and government management, and a conceptual model was constructed. Second, mathematical statistics, spatial autocorrelation, and hotspot analysis were used to analyse the spatial and temporal patterns of CO_2 emissions in China. Third, the reliability, validity, path coefficient, and prediction ability of the structural equation model for CO_2 emissions were tested using 420 samples from 30 provinces in China during 14 years from 2006 to 2019. Fourth, multi-group analysis was used to analyse the differences between the impact paths of different regions and stages. Finally, we propose targeted policy suggestions for China's future CO_2 emission reduction to provide a reference for achieving the goals of peak carbon and carbon neutrality. The main conclusions are as follows:

First, China's CO_2 emissions have different spatial and temporal patterns. On the national scale, China's total CO_2 emissions show a trend of gradual and steady growth. At the regional scale, the CO_2 emissions of the three main regions show a decreasing distribution pattern of east, central, and west. At the provincial scale, the eastern coastal area is a hotspot for high CO_2 emissions. These results are consistent with the actual status of China's regional economic development.

Second, this study constructed a structural equation model on drivers of CO_2 emissions, including economic growth, industrial structure, energy consumption, urban development, foreign trade, and government management. The conceptual model was tested with 420 samples from 30 provinces in China from 2006 to 2019. We found that the R^2 value was 76.3% and the Q^2 value was 75.4%, indicating that the proposed model had a substantial explanatory power and a high predictive power.

Third, through path coefficient analysis and hypothesis testing, we found that latent variables had different degrees of influence on CO_2 emissions. As China's secondary industries still occupy an important position in GDP and consume a large amount of energy, the industrial structure contributes to more CO_2 emissions. Because coal has always been the main energy consumption in China, electricity consumption and production burn a large amount of fossil fuel and emit a large amount of CO_2. Therefore, energy consumption contributes to higher CO_2 emissions. Government administration can effectively control CO_2 emissions through positive and effective measures. Foreign trade does not pass the hypothesis test but has an impact on CO_2 emissions through the mediating role of economic growth. Urban development does not pass the hypothesis test, but China's urbanisation is an important factor that cannot be ignored. As China's current economic development is still on the rise, the impact of economic growth on CO_2 emissions is significantly positive.

Meanwhile, economic growth is a mediating variable of the impact of industrial structure, energy consumption, foreign trade, and government administration on CO_2 emissions.

Fourth, through the calculation of the MGA methods, we found that latent variables in different development stages and regions have different effects on CO_2 emissions. By comparing the two development stages, we found significant differences in the impact of foreign trade, government administration, and economic growth on CO_2 emissions. This is consistent with the extent of China's reform and opening-up and the government's concept and practice of green development. Through a comparison of the three main regions, we found that the impact of foreign trade, government administration, and economic growth on CO_2 emissions in the eastern region was stronger than that in the central region; the impact of industrial structure, energy consumption, foreign trade, and urban development on CO_2 emissions in the eastern region was stronger than that in the western region; and the impact of industrial structure, energy consumption, and urban development on CO_2 emissions in the central region was stronger than that in the western region. These calculation results are consistent with the actual situation of China's regional economic development.

There are some limitations in this paper, and further research suggestions are proposed. First, the factors that affect CO_2 emissions are diverse, and more latent variables should be considered, such as renewable energy production, macro-control policies, technological progress, and residents' consumption awareness. Second, the selection of indicators has an impact on the intensity of CO_2 emissions, and therefore other indicators should be considered in the future. Third, more causal relationships between latent variables and their mediating effects should be explored in depth.

Author Contributions: Conceptualisation, B.W., A.S. and J.B.; methodology, B.W., A.S. and Q.Z.; software, A.S. and B.W.; validation, B.W. and Q.Z.; formal analysis, B.W. and A.S.; data curation, D.W.; writing—original draft preparation, B.W.; writing—review and editing, B.W. and Q.Z.; visualisation, A.S. and B.W.; supervision, J.B. and D.W.; project administration, J.B. and D.W.; funding acquisition, J.B. and D.W. All authors have read and agreed to the published version of the manuscript.

Funding: This work was supported by the National Natural Science Foundation of China (grant no. 41801142, 41771128), the Science and Technology Basic Resources Investigation Program of China's 'Multidisciplinary Joint Expedition for China-Mongolia-Russia Economic Corridor' (grant no. 2017FY101300, 2017FY101302), and the Collaborative Innovation Centre Project of Geopolitical Environment and Frontier Development in Southwest China (grant no. KJHX2018494).

Institutional Review Board Statement: Not applicable.

Informed Consent Statement: Not applicable.

Data Availability Statement: Data are made available upon request.

Conflicts of Interest: The authors declare no conflict of interest.

References

1. Sachs, J.D.; Schmidt-Traub, G.; Mazzucato, M.; Messner, D.; Nakicenovic, N.; Rockström, J. Six transformations to achieve the sustainable development goals. *Nat. Sustain.* **2019**, *2*, 805–814. [CrossRef]
2. Baruch-Mordo, S.; Kiesecker, J.M.; Kennedy, C.M.; Oakleaf, J.R.; Opperman, J.J. From Paris to practice: Sustainable implementation of renewable energy goals. *Environ. Res. Lett.* **2018**, *14*, 024013. [CrossRef]
3. Kokotovic, F.; Kurecic, P.; Mjeda, T. Accomplishing the sustainable development goal 13—Climate action and the role of the European Union. *Interdiscip. Descr. Complex Syst.* **2019**, *17*, 132–145. [CrossRef]
4. Yang, L.; Li, Y. Low-carbon city in China. *Sustain. Cities Soc.* **2013**, *9*, 62–66. [CrossRef]
5. He, J. Global low-carbon transition and China's response strategies. *Adv. Clim. Chang. Res.* **2016**, *7*, 204–212. [CrossRef]
6. He, J.; Yu, Z.; Zhang, D. China's strategy for energy development and climate change mitigation. *Energy Policy* **2012**, *51*, 7–13.
7. He, J. Situation and measures of China's CO_2 emission mitigation after the Paris agreement. *Front. Energy* **2018**, *12*, 353–361. [CrossRef]
8. Kuhn, B.M. China's commitment to the sustainable development goals: An analysis of push and pull factors and implementation challenges. *Chin. Political Sci. Rev.* **2018**, *3*, 359–388. [CrossRef]
9. Yang, S.; Zhao, D.; Wu, Y.; Fan, J. Regional variation in carbon emissions and its driving forces in China: An index decomposition analysis. *Energy Environ.* **2013**, *24*, 1249–1270. [CrossRef]

10. Wang, Y.; Zheng, Y. Spatial effects of carbon emission intensity and regional development in China. *Environ. Sci. Pollut. Res.* **2021**, *28*, 14131–14143. [CrossRef]
11. Sun, L.; Wu, L.; Qi, P. Global characteristics and trends of research on industrial structure and carbon emissions: A bibliometric analysis. *Environ. Sci. Pollut. Res.* **2020**, *27*, 44892–44905. [CrossRef]
12. Guo, F.; Meng, S.; Sun, R. The evolution characteristics and influence factors of carbon productivity in China's industrial sector: From the perspective of embodied carbon emissions. *Environ. Sci. Pollut. Res.* **2021**, *28*, 50611–50622. [CrossRef]
13. Zheng, H.; Gao, X.; Sun, Q.; Han, X.; Wang, Z. The impact of regional industrial structure differences on carbon emission differences in China: An evolutionary perspective. *J. Clean. Prod.* **2020**, *257*, 120506. [CrossRef]
14. Li, L.; Hong, X.; Peng, K. A spatial panel analysis of carbon emissions, economic growth and high-technology industry in China. *Struct. Chang. Econ. Dyn.* **2019**, *49*, 83–92. [CrossRef]
15. Zhou, D.; Zhang, X.; Wang, X. Research on coupling degree and coupling path between China's carbon emission efficiency and industrial structure upgrading. *Environ. Sci. Pollut. Res.* **2020**, *27*, 25149–25162. [CrossRef] [PubMed]
16. Cui, E.; Ren, L.; Sun, H. Analysis on the regional difference and impact factors of CO_2 emissions in China. *Environ. Prog. Sustain. Energy* **2017**, *36*, 1282–1289. [CrossRef]
17. Wang, D.; Liu, X.; Yang, X.; Zhang, Z.; Wen, X.; Zhao, Y. China's energy transition policy expectation and its CO_2 emission reduction effect assessment. *Front. Energy Res.* **2021**, *8*, 8. [CrossRef]
18. Wang, S.; Wang, J.; Li, S.; Fang, C.; Feng, K. Socioeconomic driving forces and scenario simulation of CO_2 emissions for a fast-developing region in China. *J. Clean. Prod.* **2019**, *216*, 217–229. [CrossRef]
19. Xu, B.; Zhong, R.; Hochman, G.; Dong, K. The environmental consequences of fossil fuels in China: National and regional perspectives. *Sustain. Dev.* **2019**, *27*, 826–837. [CrossRef]
20. Meng, M.; Jing, K.; Mander, S. Scenario analysis of CO_2 emissions from China's electric power industry. *J. Clean. Prod.* **2017**, *142*, 3101–3108. [CrossRef]
21. Wang, S.; Wang, X.; Tang, Y. Drivers of carbon emission transfer in China—An analysis of international trade from 2004 to 2011. *Sci. Total Environ.* **2020**, *709*, 135924. [CrossRef]
22. Zhang, L.; Xiong, L.; Cheng, B.; Yu, C. How does foreign trade influence China's carbon productivity? Based on panel spatial lag model analysis. *Struct. Chang. Econ. Dyn.* **2018**, *47*, 171–179. [CrossRef]
23. Chen, Y.; Wang, Z.; Zhong, Z. CO_2 emissions, economic growth, renewable and non-renewable energy production and foreign trade in China. *Renew. Energy* **2019**, *131*, 208–216. [CrossRef]
24. Tan, S.; Zhang, M.; Wang, A.; Zhang, X.; Chen, T. How do varying socio-economic driving forces affect China's carbon emissions? New evidence from a multiscale geographically weighted regression model. *Environ. Sci. Pollut. Res.* **2021**, *28*, 41242–41254. [CrossRef]
25. Jiang, M.; An, H.; Gao, X.; Liu, S.; Xi, X. Factors driving global carbon emissions: A complex network perspective. *Resour. Conserv. Recycl.* **2019**, *146*, 431–440. [CrossRef]
26. Cao, Z.; Wei, J. Industrial distribution and LMDI decomposition of trade-embodied CO_2 in China. *Dev. Econ.* **2019**, *57*, 211–232. [CrossRef]
27. Pan, X.; Pan, X.; Li, C.; Song, J.; Zhang, J. Effects of China's environmental policy on carbon emission efficiency. *Int. J. Clim. Chang. Strat. Manag.* **2019**, *11*, 326–340. [CrossRef]
28. He, L.; Yin, F.; Zhong, Z.; Ding, Z. The impact of local government investment on the carbon emissions reduction effect: An empirical analysis of panel data from 30 provinces and municipalities in China. *PLoS ONE* **2017**, *12*, e0180946. [CrossRef] [PubMed]
29. Lin, B.; Jia, Z. The energy, environmental and economic impacts of carbon tax rate and taxation industry: A CGE based study in China. *Energy* **2018**, *159*, 558–568. [CrossRef]
30. Zhang, F.; Jin, G.; Li, J.; Wang, C.; Xu, N. Study on dynamic total factor carbon emission efficiency in China's urban agglomerations. *Sustainability* **2020**, *12*, 2675. [CrossRef]
31. Sun, W.; Huang, C. How does urbanization affect carbon emission efficiency? Evidence from China. *J. Clean. Prod.* **2020**, *272*, 122828. [CrossRef]
32. Zhou, C.; Wang, S.; Wang, J. Examining the influences of urbanization on carbon dioxide emissions in the Yangtze River Delta, China: Kuznets curve relationship. *Sci. Total Environ.* **2019**, *675*, 472–482. [CrossRef]
33. Han, X.; Cao, T.; Sun, T. Analysis on the variation rule and influencing factors of energy consumption carbon emission intensity in China's urbanization construction. *J. Clean. Prod.* **2019**, *238*, 117958. [CrossRef]
34. Xu, L.; Du, H.; Zhang, X. Driving forces of carbon dioxide emissions in China's cities: An empirical analysis based on the geodetector method. *J. Clean. Prod.* **2021**, *287*, 125169. [CrossRef]
35. Dinda, S. Environmental kuznets curve hypothesis: A survey. *Ecol. Econ.* **2004**, *49*, 431–455. [CrossRef]
36. Fang, D.; Hao, P.; Wang, Z.; Hao, J. Analysis of the influence mechanism of CO_2 emissions and verification of the environmental kuznets curve in China. *Int. J. Environ. Res. Public Health* **2019**, *16*, 944. [CrossRef]
37. Mardani, A.; Streimikiene, D.; Cavallaro, F.; Loganathan, N.; Khoshnoudi, M. Carbon dioxide (CO_2) emissions and economic growth: A systematic review of two decades of research from 1995 to 2017. *Sci. Total Environ.* **2019**, *649*, 31–49. [CrossRef] [PubMed]

38. Duro, J.A.; Padilla, E. International inequalities in per capita CO_2 emissions: A decomposition methodology by Kaya factors. *Energy Econ.* **2006**, *28*, 170–187. [CrossRef]
39. Yuan, J.; Xu, Y.; Hu, Z.; Zhao, C.; Xiong, M.; Guo, J. Peak energy consumption and CO_2 emissions in China. *Energy Policy* **2014**, *68*, 508–523. [CrossRef]
40. Green, F.; Stern, N. China's changing economy: Implications for its carbon dioxide emissions. *Clim. Policy* **2017**, *17*, 423–442. [CrossRef]
41. Dietz, T.; Rosa, E.A. Effects of population and affluence on CO_2 emissions. *Proc. Natl. Acad. Sci. USA* **1997**, *94*, 175–179. [CrossRef] [PubMed]
42. Feng, K.; Hubacek, K.; Guan, D. Lifestyles, technology and CO_2 emissions in China: A regional comparative analysis. *Ecol. Econ.* **2009**, *69*, 145–154. [CrossRef]
43. Wang, C.; Wang, F.; Zhang, X.; Yang, Y.; Su, Y.; Ye, Y.; Zhang, H. Examining the driving factors of energy related carbon emissions using the extended STIRPAT model based on IPAT identity in Xinjiang. *Renew. Sustain. Energy Rev.* **2017**, *67*, 51–61. [CrossRef]
44. Wang, C.; Chen, J.; Zou, J. Decomposition of energy-related CO_2 emission in China: 1957–2000. *Energy* **2005**, *30*, 73–83. [CrossRef]
45. Hatzigeorgiou, E.; Polatidis, H.; Haralambopoulos, D. CO_2 emissions in Greece for 1990–2002: A decomposition analysis and comparison of results using the arithmetic mean divisia index and logarithmic mean divisia index techniques. *Energy* **2008**, *33*, 492–499. [CrossRef]
46. Xu, S.-C.; He, Z.-X.; Long, R.-Y. Factors that influence carbon emissions due to energy consumption in China: Decomposition analysis using LMDI. *Appl. Energy* **2014**, *127*, 182–193. [CrossRef]
47. Shao, S.; Yang, L.; Gan, C.; Cao, J.; Geng, Y.; Guan, D. Using an extended LMDI model to explore techno-economic drivers of energy-related industrial CO_2 emission changes: A case study for Shanghai (China). *Renew. Sustain. Energy Rev.* **2016**, *55*, 516–536. [CrossRef]
48. Zheng, H.; Hu, J.; Guan, R.; Wang, S. Examining determinants of CO_2 emissions in 73 cities in China. *Sustainability* **2016**, *8*, 1296. [CrossRef]
49. Li, W.; Wang, W.; Wang, Y.; Qin, Y. Industrial structure, technological progress and CO_2 emissions in China: Analysis based on the STIRPAT framework. *Nat. Hazards* **2017**, *88*, 1545–1564. [CrossRef]
50. Zheng, D.C.; Liu, W.X.; Li, X.X.; Lin, Z.Y.; Jiang, H. Research on carbon emission diversity from the perspective of urbanization. *Appl. Ecol. Environ. Res.* **2018**, *16*, 6643–6654. [CrossRef]
51. Chen, J.; Lian, X.; Su, H.; Zhang, Z.; Ma, X.; Chang, B. Analysis of China's carbon emission driving factors based on the perspective of eight major economic regions. *Environ. Sci. Pollut. Res.* **2021**, *28*, 8181–8204. [CrossRef]
52. Li, J.; Cheng, J.; Diao, B.; Wu, Y.; Hu, P.; Jiang, S. Social and economic factors of industrial carbon dioxide in China: From the perspective of spatiotemporal transition. *Sustainability* **2021**, *13*, 4268. [CrossRef]
53. Sha, W.; Chen, Y.; Wu, J.; Wang, Z. Will polycentric cities cause more CO_2 emissions? A case study of 232 Chinese cities. *J. Environ. Sci.* **2020**, *96*, 33–43. [CrossRef]
54. Chen, J.; Wang, L.; Li, Y. Research on the impact of multi-dimensional urbanization on China's carbon emissions under the background of COP21. *J. Environ. Manag.* **2020**, *273*, 111123. [CrossRef] [PubMed]
55. Sun, Z.-Q.; Sun, T. The impact of multi-dimensional urbanization on China's carbon emissions based on the spatial spillover effect. *Pol. J. Environ. Stud.* **2020**, *29*, 3317–3327. [CrossRef]
56. Zhang, X.; Zhao, Y. Identification of the driving factors' influences on regional energy-related carbon emissions in China based on geographical detector method. *Environ. Sci. Pollut. Res.* **2018**, *25*, 9626–9635. [CrossRef] [PubMed]
57. Sarstedt, M.; Cheah, J.-H. Partial least squares structural equation modeling using SmartPLS: A software review. *J. Mark. Anal.* **2019**, *7*, 196–202. [CrossRef]
58. Ben Jabeur, S.; Sghaier, A. The relationship between energy, pollution, economic growth and corruption: A partial least squares structural equation modeling (PLS-SEM) Approach. *Econ. Bull.* **2018**, *38*, 1927–1946.
59. Wei, Y.; Zhu, X.; Li, Y.; Yao, T.; Tao, Y. Influential factors of national and regional CO_2 emission in China based on combined model of DPSIR and PLS-SEM. *J. Clean. Prod.* **2019**, *212*, 698–712. [CrossRef]
60. Soltani, M.; Rahmani, O.; Ghasimi, D.S.; Ghaderpour, Y.; Pour, A.B.; Misnan, S.H.; Ngah, I. Impact of household demographic characteristics on energy conservation and carbon dioxide emission: Case from Mahabad city, Iran. *Energy* **2020**, *194*, 116916. [CrossRef]
61. Li, W.; Zhao, S.; Ma, J.; Qin, W. Investigating regional and generational heterogeneity in low-carbon travel behavior intention based on a PLS-SEM approach. *Sustainability* **2021**, *13*, 3492. [CrossRef]
62. Wang, B.; Li, J.; Sun, A.; Wang, Y.; Wu, D. Residents' green purchasing intentions in a developing-country context: Integrating PLS-SEM and MGA methods. *Sustainability* **2019**, *12*, 30. [CrossRef]
63. Luo, L.; Qian, T.Y.; Rich, G.; Zhang, J.J. Impact of market demand on recurring hallmark sporting event spectators: An empirical study of the Shanghai Masters. *Int. J. Sports Mark. Spons.* **2021**. [CrossRef]
64. Schirmer, N.; Ringle, C.M.; Gudergan, S.P.; Feistel, M.S.G. The link between customer satisfaction and loyalty: The moderating role of customer characteristics. *J. Strateg. Mark.* **2018**, *26*, 298–317. [CrossRef]
65. Hair, J.F.; Hult, G.T.M.; Ringle, C.M.; Sarstedt, M. *A Primer on Partial Least Squares Structural Equation Modeling (PLS-SEM)*, 3rd ed.; Sage Publications: Thousand Oaks, CA, USA, 2021.

66. Henseler, J.; Ringle, C.M.; Sarstedt, M. A new criterion for assessing discriminant validity in variance-based structural equation modeling. *J. Acad. Mark. Sci.* **2015**, *43*, 115–135. [CrossRef]
67. Falk, R.F.; Miller, N.B. *A Primer for Soft Modeling*; University of Akron Press: Akron, OH, USA, 1992.
68. Chin, W.W. The partial least squares approach to structural equation modeling. *Mod. Methods Bus. Res.* **1998**, *295*, 295–336.
69. Long, R.; Shao, T.; Chen, H. Spatial econometric analysis of China's province-level industrial carbon productivity and its influencing factors. *Appl. Energy* **2016**, *166*, 210–219. [CrossRef]
70. Zeng, L.; Lu, H.; Liu, Y.; Zhou, Y.; Hu, H. Analysis of regional differences and influencing factors on China's carbon emission efficiency in 2005–2015. *Energies* **2019**, *12*, 3081. [CrossRef]
71. Li, R.; Wang, Q.; Liu, Y.; Jiang, R. Per-capita carbon emissions in 147 countries: The effect of economic, energy, social, and trade structural changes. *Sustain. Prod. Consum.* **2021**, *27*, 1149–1164. [CrossRef]

Article

Using PLS-SEM to Analyze the Effect of CSR on Corporate Performance: The Mediating Role of Human Resources Management and Customer Satisfaction. An Empirical Study in the Spanish Food and Beverage Manufacturing Sector

Fernando Gimeno-Arias [1], José Manuel Santos-Jaén [2,*], Mercedes Palacios-Manzano [2] and Héctor Horacio Garza-Sánchez [3]

1. Department of Management and Finance, University of Murcia, 30100 Murcia, Spain; fernando.gimeno@um.es
2. Department of Accounting and Finance, University of Murcia, 30100 Murcia, Spain; palacios@um.es
3. Department of Accounting and Finance, University of Nuevo Leon, Nuevo León 66455, Mexico; hector.garzasc@uanl.edu.mx
* Correspondence: jmsj1@um.es; Tel.: +34-86-888-7922

Abstract: Although in recent decades corporate social responsibility (CSR) has been subjected to numerous studies in management and marketing literature about its impact on business results, the mechanism by which it affects performance has not been established. There is a lack of consensus when it comes to explaining how CSR actions are related to firm performance. Our research helps to understand this relationship through mediating effects such as CSR-oriented human resource management and customer satisfaction because employees and customers are critical stakeholders of companies and contribute directly to the determination of the corporate results. Through a study on a sample of small and medium-sized Spanish food and beverage manufacturing companies, and by using partial least squares structural equation modelling (PLS-SEM), we found that CSR does indeed impact business performance when CSR actions are mainly oriented towards more efficient management of human resources and customer satisfaction. In this way, the results lead us to conclude that depending on the stakeholder to which these actions are oriented, a specific orientation of the company's CSR policy can be more efficient in corporate performance.

Keywords: corporate social responsibility; corporate performance; human resources management; customer satisfaction; partial least squares structural equation modelling (PLS-SEM)

1. Introduction

Corporate social responsibility (CSR) is considered crucial for business success [1] and a strategic business necessity in order to achieve competitive advantage [2,3]. A large amount of research has shown that CSR orientation is the key to stimulating long-term stability, growth, and sustainable performance in a dynamic and changing environment [4].

CSR seems to have rather an unclear impact on corporate performance as no true causality has yet been proven. Despite the large number of studies undertaken to investigate this issue, the results are not conclusive [5,6]. There is a strong consensus that companies that incorporate CSR into their strategy actions increase their value creation [7–11]. However, the effects of CSR on corporate performance are not sufficiently specified, as there is a diversity of conclusions [12–16].

The lack of consensus might reflect model specification problems, such as omissions of intangible resources [2,5]. In this sense, [17] has stressed the importance of developmental models incorporating omitted variables and test mediating mechanisms to establish causal links between CSR practices and business performance. Although the most recent literature in this field showcases models that incorporate mediation mechanisms with variables, such

as firm reputation [6], individual-level organizational identification, employees' innovative job performance [18], or customer loyalty [19], there remains a wide field of research due to the diversity of variables that are part of the strategic management of the firms. The progressive incorporation of variables with mediating effects will help to understand and reach a consensus on how CSR actions and performance are related within the company.

The academic community has highlighted the lack of studies promoting CSR practices and have demanded more research in this area, especially from SMEs (small and medium-sized enterprises). With this purpose in mind, we have considered it necessary to examine whether there is a relationship between CSR activities and business performance and to study if this relationship could be mediated by other factors, such as good human resource management [20,21] or customer satisfaction [22,23], through the incorporation of CSR into business strategy. In this sense, the presence of research that has explored mediating effects [19,23–32] leads to work on these issues.

Given the state of knowledge, we consider it essential to deepen research to understand how certain aspects of management omitted in previous studies mediate the impact of CSR on the organization's performance. With this purpose, this paper proposes and tests a model of the relationship between CSR activities, firm performance, human resource management and customer satisfaction in a sample of Spanish SMEs. According to the International Council for Small Business (ICSB) SMEs represent 90% of the business activity and therefore, from the analysis of this type of company, conclusions can be drawn that can be generalized to all companies. Thus, our research objective is to investigate whether the mediating effects serve as connectors between CSR and business performance. This article proposes to answer the following questions: Does CSR influence the performance of SMEs in the food and beverage manufacturer sector? Is this influence mediated by human resource management or customer satisfaction? In order to draw our conclusions, we obtained data from a sample of Spanish companies by means of a questionnaire. We investigated the relationships present in the companies between CSR, human resource management, customer satisfaction, and firm performance.

To process the data and study the adequacy of the proposed behavioral model, we apply structural equation modeling (SEM) by using the method partial least squares (PLS) because "using PLS-SEM to estimate common factor models is much less of an issue than (incorrectly) using factor-based SEM to estimate composite models" [33]. Moreover, not only does PLS-SEM work well with small samples [34], it has also proven to be the most appropriate technique to apply in order to estimate multiple relationships between latent constructs, above all, if they involve mediation [33]. The empirical results show that the relationship between CSR and performance does not occur directly but is mediated by human resource management and customer satisfaction.

We contribute to the field of study of mathematics by illustrating its multiple applications as a transversal science. The use of existing knowledge in statistics-mathematics in various areas allows us to conduct research with greater scientific rigor. Specifically, in the social sciences, the use of complex statistical techniques, such as PLS-SEM, has revolutionized research methodology in recent decades. It has provided the field with better-founded conclusions, better interconnections with reality and greater universality of their conclusions. On the other hand, this research provides a contribution to management literature by considering the CSR as a key business performance tool in SMESs in the food and beverage manufacturer sector, through the analysis of the relationship between CSR practices and performance, and by incorporating the mediating effects of human resource management and customer satisfaction. The inclusion of these two mediating effects seeks to draw the inconclusive results concerning the effect of CSR on firm performance in the research [35]. These results align with previous studies [19,21,22,26], although our research considers them as a whole and not in isolation.

The paper is structured as follows. After this introduction, Section 2 contains the literature review and proposes the hypotheses development. Section 3 describes the research method, the variables and model to test the hypotheses. In Section 4, the results are provided. Finally, Section 5 shows the main results and consequences and presents the conclusions.

2. Literature Review and Hypothesis

In recent decades, stakeholder theory [36] has become one of the dominant paradigms in business management research, through strategic interpretation of firms, which positions stakeholder management as a means of achieving the objectives of shareholders and corporate managers [37]. Stakeholder theory reconsidered the traditional view focused on agents directly related to the firm's output-shareholders, customers, employees, and suppliers-introducing a new perspective on management understanding through the extended web of the stakeholder [9]. Thus, the firms' stakeholders are defined as "any group or individual that can affect or is affected by an organization" [7], and these include shareholders, creditors, employees, customers, suppliers, public interest groups, and governmental bodies [38]. This type of management makes it possible to exploit the existing relationship between stakeholders and business objectives, such as profitability, stability, and growth [11].

One of the most interesting consequences of the stakeholder theory is the incorporation of CSR into business strategy, as a means of simultaneously achieving corporate and stakeholder objectives from a social perspective [8,9,11,38] namely "the conscience of the corporation is a logical and moral extension of the consciences of its principals" [37]. In this way, corporate management is guided by incentives to maximize shareholder value, and the individual or collective interests of stakeholders such as employees, customers, suppliers, and local communities within which corporations operate [11]. So, CSR includes firm actions aimed at efficient management to ensure the sustainability of the economy, the environment, and society in general [26].

From this perspective, scholars ask themselves about the benefits of incorporating CSR into the firm strategy and its impact on corporate performance. Does CSR have a direct impact on performance [12–16,39], and which company elements can model the impact of CSR on performance? The literature points CSR as a VRIO element (valuable, rare, difficult to imitate, and non-substitutable in the corporate organization) and therefore a source of a sustained competitive advantage [19,40,41], as it is mainly applied in human resources [42]. Thus, another conceptual framework of the influence of CSR on performance can be found in the Resource-Based View (RBV) Theory.

Therefore, we find that Corporative Social Responsibility (CSR) is a set of firms actions and policies (economic, social, and environmental) directed to interact with their stakeholders [26], i.e., "philanthropic in nature and not necessarily related directly to the operational business of a firm" [10]. We can find in the literature an extended definition published by the European Commission in 2001 stating that "firms voluntarily interact with their stakeholders and integrate social and environmental concerns to corporate management" [43]. The CSR idea is that for-profit organizations have a responsibility towards society [26]. These definitions regarding CSR lead us to think that disinterested actions in firms' decisions may even incur a source of cost [6]. Nevertheless, CSR actions can contribute to the competitive advantage and superior performance of the corporations [10].

In this sense, CSR is based on the win–win concept through the evidence that social values are linked with corporate performance because prosperous societies are a business-friendly environment [43].

CSR is a complex concept because it "is an umbrella term overlapping with some yet synonymous with other concepts of business–society relations" [44]. CSR requires investments in resources, and the Return on Investments (ROI) is only noticeable in the long term and not necessarily in financial terms [10]. However, performance is a concept generally based on financial aspects as Return On Assets (ROA) and Tobin's Q [43]. These dimensional differences explain the difficulty of empirically demonstrating a, clear direct causal effect of CSR practices on performance, particularly in situations where practices are not necessarily related directly to a firm's operational business [10].

On the other hand, the economic objectives of CSR practices are not exclusively performance based. They could be varied and complex as a consequence of multiple categories of CSR related to the interests of the stakeholders. According to European Commission, "CSR that supports economic growth and prosperity is a means by which firms can pre-empt economic responsibility issues that might arise in their interactions with customers, suppliers and shareholders in the Marketplace". Under this standpoint, CSR practices could contribute to the economic development of companies through indirect effects related to stakeholders [43].

Finally, the firm's performance is an indicator of its capacity to achieve its goals and performance includes both financial and non-financial measures [45]. However, CSR is a strategic action of the firm that supports the economic aspects of the company and social cohesion, equity, integrity, and environmental responsibility [10]. Some authors consider CSR actions as "marketing techniques that enable companies to simultaneously pursue economic and social goals" [43]. The complexity of the CSR construct has very different objectives because it aims to satisfy other stakeholders and not all of them have the same interest in the firm's performance [18]. From a strategic point of view, CSR contributes to increasing a firm's value through superior competitive advantage [40], which may be related to human, organizational, or relational elements [5]. With this in mind, the relationship between CSR and corporate performance is mediated by the type of CSR action and stakeholder it impacts, and these arguments lead us to not expect a direct relationship between CSR actions and firm performance, and to formulate the following hypothesis:

Hypothesis 1 (H1). *CSR practices do not directly affect corporate performance.*

Although we do not find a direct cause–effect relationship between CSR practices and firm performance, the literature suggests that the adoption of CSR actions is related to a source of labor productivity and sales growth [22], elements necessary to achieve corporate performance. Labor productivity and sales growth are types of performance directly related to two concrete stakeholders: employees and customers. In this sense, specific works suggest a direct link between CSR-employee behavior [20] and CSR-customer satisfaction [22]. When the firms adopt CSR practices, it contributes to building closer relationships with their stakeholders [43], and particularly in human resources management, customer satisfaction or firm reputation [18].

In this way, through CSR "firms voluntarily interact with their stakeholders and integrate social and environmental concerns to corporate management" [43]. The interaction with one of their stakeholders, such as employees, could be a strategy of human resources that generate "structural cohesion, and employee-generated synergy that propels a company forward" [25]. The European Commission suggests in its 2003 report that proactive CSR motivates the workforce by offering training and development opportunities. CSR practices increase employees' pride of belonging; they consider being a member of the firm is valuable and identify themselves with a socially responsible organization [19,26,29]. These situations show that companies can influence their employees' attitudes. The behavior for their job contribute to organization's purpose and better performance [46].

According to the relationship described, CSR could be a powerful tool for manage human resources through promoting employees' positive attitudes, trust, and commitment to the firm, corporate identity, or job satisfaction [21]. CSR actions oriented to a human resource could be integrated into human resources management (HRM) because they "influence and shape the skills, attitudes, and behaviors of individuals, so that they can do their job better and achieve the objectives of the organization" [46]. The literature in human resources has demonstrated that adequate management strategy oriented to workers is a practical action to attain organizational objectives [27] because it favors the corporate identification of employees who tend to make extra efforts to defend corporative interests [28] and influences employees' behavior, thus promoting superior worker performance and, as a consequence, a greater level of firm performance [25,46].

Human resources are key competitive factors in the firm because they are rare, valuable, inimitable, and non-substitutable [40], and are "organizational systems designed to achieve a competitive advantage through people" [42]. This competitive advantage is built by adequately managing human resources and linking this effort with the strategic goals of the firm; therefore, HRM are a source of a competitive advantage [25]. Researchers in the field of human resources report explanations about the effects of HRM on firm performance, finding a positive influence [25,45]. In this sense, the relationship between CSR and firm performance is mediated through HRM, which lead us to make the following hypothesis:

Hypothesis 2 (H2). *The effect of CSR on firm performance is mediated by human resource management (HRM).*

This H2 Hypothesis is subdivided into the following three:

Hypothesis 2$_a$ (H2$_a$). *CSR practices have a positive effect on HRM.*

Hypothesis 2$_b$ (H2$_b$). *HRM has a positive effect on corporate performance.*

Hypothesis 2$_c$ (H2$_c$). *CSR practices indirectly affect corporate performance through HRM.*

Numerous studies have established empirical evidence of the relationship between CSR and customer satisfaction [23]. The literature shows CSR actions as "a competitive strategy for corporations to increase profits, customer satisfaction, customer loyalty, corporate reputation, and positive attitudes towards the company's brands" [19]. Although it is difficult to establish a direct relationship between CSR and performance, there is a broad consensus in the literature in positioning customer satisfaction as a precursor to improved performance [19,29,31,41,47].

Satisfaction in business relationships is a fundamental aspect of the marketing literature [48] because relationships that contribute to business success are characterized by high levels of satisfaction among the parties [49–51]. In addition, the presence of satisfaction in the commercial relationship fosters stronger bonds [52], generates loyalty [48], provokes a desire for continuity [53], and encourages participation in collaborative activities [52]. These characteristics define satisfaction as "an affective state developed based on the evaluation of the relationship with a particular exchange partner or the degree of fulfillment of the expectations of each partner in an exchange relationship" [48]. The positive affective state arises as a consequence of the assets of the relationship [54], through a cognitive process that compares the expectations of the parties concerned with regards to their performance, both in their tangible aspects (profitability, growth, income) and in the intangible aspects reflected in their emotional dimension [52]. When assessing satisfaction, this double dimension (tangible and intangible) implies its division into economic and non-economic satisfaction [50,52,53]. Economic satisfaction is derived from the achievement of performance objectives [50] and is defined as "evaluation of the economic outcomes that flow from the relationship with its partner such as sales volume, margins, and discounts" [55].

On the other hand, non-economic satisfaction or "social satisfaction" is defined as "evaluation of the psychosocial aspects of its relationship, in that the interactions with the exchange partner are fulfilling, gratifying, and facile" [55]. This meaning of satisfaction focuses on a positive affective response to the psychological aspects of the customer relationship [50], the contacts are appreciated on a personal level and the client considers the firm concerned, respectful, and open to the exchange of ideas [53]. There is a broad consensus in the literature that economic satisfaction is the antecedent of social satisfaction [52,53,55] because high levels of economic satisfaction will have an emotional impact, provoking in the agents a collaborative and constructive response to any contingency, which will positively affect the increase in social satisfaction [52], although most of the studies focus on the social aspects [50].

Theoretical and empirical evidence show CSR actions of the firms are perceived by their customers positively and lead to higher satisfaction and loyalty [23]. Hence, CSR is a driver of client satisfaction [30]. In this case, social satisfaction is built on the customers' psychosocial perceptions [50]. Marketing literature points to customer satisfaction as a key driver of a firm's long-term profitability and higher market value [23] because "customers are among the most important stakeholders" [19]. Supported by the literature, we can establish a clear positive relationship between customer satisfaction and firm performance because many companies use customer satisfaction as an indicator of performance [31], the situation described in the following hypothesis:

Hypothesis 3 (H3). *The effect of CSR practices on firm performance is mediated by Customer Satisfaction (CS).*

This H3 Hypothesis is subdivided into the following three:

Hypothesis 3$_a$ (H3$_a$). *CSR practices have a positive effect on CS.*

Hypothesis 3$_b$ (H3$_b$). *CS has a positive effect on corporate performance.*

Hypothesis 3$_c$ (H3$_c$). *CSR practices indirectly affect corporate performance through CS.*

The literature shows as CSR activities in the firm oriented to human resources management and the perception that employees have about CSR is considered a positive influence on their attitudes and behaviors, that it leads to worker satisfaction [27–29]. The strategies oriented to human resources add value to the firm because the workers obtain different skills, abilities, and capabilities [45]. These higher abilities of employees are associated with job satisfaction, higher productivity, and better decision-making which contribute to customer satisfaction and as a result, the likelihood of better organizational performance [24].

In this way, employees who are more engaged with the organization improve their customers' experiences, especially those in direct contact with them. This better customer service directly relates to improving organizational performance in terms of sales, market share, and profitability [56]. So, it is expected that the increased employee motivation resulting from the company's CSR actions will lead to increased customer satisfaction [32,57]. Following these considerations, we can formulate a hypothesis:

Hypothesis 4 (H4). *The effect of CSR on customer satisfaction (CS) is mediated by human resources management (HRM).*

This H4 Hypothesis is subdivided into the following two:

Hypothesis 4$_a$ (H4$_a$). *HRM practices have a positive effect on CS.*

Hypothesis 4$_b$ (H4$_b$). *CSR practices indirectly affect CS through HRM.*

As we have discussed, human resources management originated by the company's CSR policies has an impact on customer satisfaction, which fosters higher corporate performance. In this sense, based on Barney [40], "from the stakeholder's point of view, it is always the focus on the interest of the individual stakeholder that influences the organization's performance" [19]. The individual stakeholder with a more direct impact on firm performance is the customer, and customer satisfaction is one of the most widely used performance indicators by business analysts [19,30,58].

Based on the above reasoning, more satisfied customers promote higher company reputation, more sales growth, greater competitive advantage, and higher levels of firm performance [31]. For those employees who have direct contact with the customer, the result of HRM involves "helping achieve business objectives, adapt to change, meet customers' needs, and increase financial performance through the deliverable of effective strategy execution" [59] So, we can establish the following hypothesis:

Hypothesis 5 (H5). *Human resources management (HRM) has indirect effects on corporate performance through customer satisfaction (CS).*

The integration of the partial relationships described above leads to the establishment of a general model. In this way, we have described the influence that CSR has on employee motivation as part of HRM [19,25,26,29,60], as well as the influence CSR has on CS [19,21–23,31,41,47].

On the other hand, we have found a positive influence of HRM on firm performance [25,27,28,45,46] and CS on firm performance [19,23,49–51]. We have also found a positive relationship between HRM and CS [31,59].

Using this syllogism, we can establish a model of relationships using the following hypothesis:

Hypothesis 6 (H6). *CSR practices have indirect effects on corporate performance through human resources management (HRM) and customer satisfaction (CS).*

Based on the above, Figure 1 represents the model for this research.

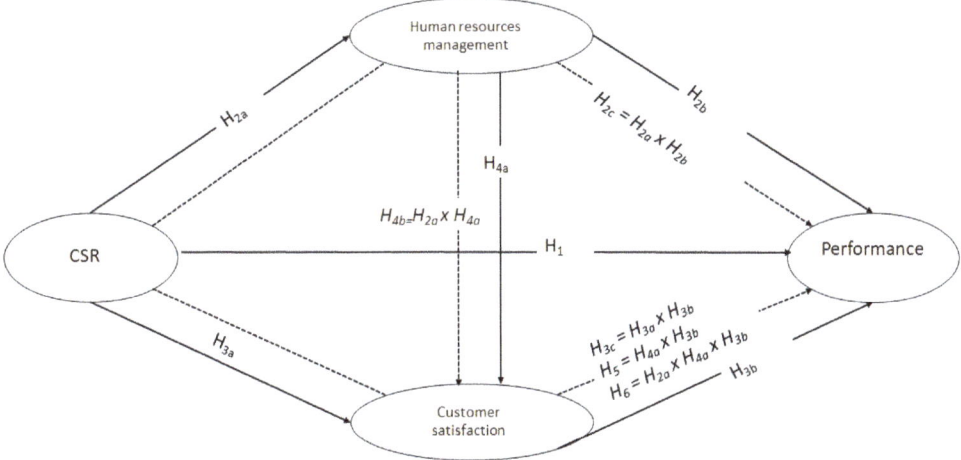

Figure 1. Research model. Note: The path lines in $H2_c$, $H3_c$, $H4_b$, $H5$ and $H6$ represent mediation relationships. Source: Authors.

3. Methodology
3.1. Research Design and Data Collection

The study was carried out using a sample of 166 SMEs in the food and beverage manufacturer Spanish sector. The food and beverage manufacturing sector in Spain is comprised of 30,573 companies, which implies a sampling error of 7.59 (with a confidence level of 95%). In Spain, the food and beverage manufacturing sector accounts for 27% of the total number of manufacturing companies, according to data obtained from the National Statistics Institute (INE), the governmental institution that provides statistical data. Therefore, the conclusions obtained in the sector study can be extrapolated to other manufacturing companies. These companies were randomly selected using the SABI database, the most detailed database of company information in Spain with data on more than 2.6 million companies (financial, directors and contacts, corporate structures, audit reports, and a great deal of other relevant information). For this purpose, a telephone questionnaire was administered to managers of these companies during the spring of 2018. The telephone interview allows (1) selecting the key respondent for each company, and (2) resolving any doubts that the questionnaire may raise guaranteeing its face validity. The distribution of the companies according to their size is shown in Table 1.

Table 1. Sample distribution.

Sector Industry	Total of Companies		Micro Companies		Small Companies		Medium Companies	
	Number	Percent of Total	Number	Percent of Total	Number	Percent of Total	Number	Percent of Total
Food and beverage manufacturing	166	100%	60	36.1%	72	43.4%	34	20.5%

Source: Authors.

In order to reduce the social acceptance bias, the anonymity of the responses was ensured at all times [61]. In addition, to ensure that the questionnaire was easy to understand, a pre-test was carried out with 5 companies. Based on the sample size obtained and a confidence level of 95%, the sampling error obtained is 7.8%. Moreover, with the power of 0.95, an effect size of 0.15, and 3 predictors, the calculated minimum sample size for this study was 119. Therefore, our sample of 166 companies is more than sufficient to validate all the effects found in the research [62].

Despite the fact that all data have been obtained from the same source, the results (see Section 4) show that there are no multicollinearity problems, ruling out a problem related to common method bias [63]. In the same way, a possible problem related to non-response bias has also been ruled out. For this purpose, the sample was divided into two groups, one with 75% of the first responses and the other with the rest. The results of the ANOVA test show that there were no significant differences between the two groups.

3.2. Variables

The four constructs have been measured with a Likert-type scale with five levels of answers, from 1 = "Strongly disagree" to 5 = "Strongly agree" for CSR, customer satisfaction and human resources management, and 1 = "Unimportant" to 5 = "Very important" for performance. The questions used in the survey can be found in Table A1 (Appendix A).

3.2.1. CSR

According to Galbreath and Shum [64], there are several good ways to measure CSR. Our research used five indicators adapted from the literature [65–70]. These indicators have been previously validated in other studies in the field [71]. A variable has been obtained that incorporates various CSR practices such as: (1) helping the community, (2) developing transparent management, and (3) protecting the environment.

3.2.2. Human Resources Management

Based on the previous literature and keeping in mind that human resources management is affected by external and internal factors [72], we have used seven indicators adapted from the literature [73–77]. As a result of this, in our questionnaire, we asked: (1) the equity of salary, (2) professional growth, (3) employees' opportunities to be involved in decision making, (4) hiring criteria, (5) investment money and time in training, and (6) the existence of permanent training.

3.2.3. Customer Satisfaction

In order to understand how customer satisfaction has evolved, a variable composed of three indicators adapted from previous literature has been created [19,23,31,32,47]. For this purpose, companies were asked about the evolution of their corporate image and reputation, the quality of their products and the satisfaction of their customers.

3.2.4. Performance

Performance was evaluated with a scale of three items based on previous research [4,78,79]. The scale created has been used to measure the financial and non-financial performance of companies, in aspects such as profitability, sales growth, and customer satisfaction, in comparison with the performance of their competitors. The use of this type of measurement, rather than relying on accounting data, allows us to measure a company's success more efficiently [78].

3.3. Data Analysis

Due to the high correlation between the indicators that make up the same construct or latent variable, the model comprises four composite variables performed in mode A [80]. According to Hair et al. [33], to estimate composite models, PLS-SEM is better than SEM. Therefore, we considered PLS-SEM, a variance-based structural equation modeling technique [81], the most appropriate for analyzing the model [82–84]. Other reasons that led us to choose this technique were that PLS-SEM is the most appropriate to apply to estimate multiple relationships between latent constructs above all if they involve mediation [33] and also due to this technique working very well when the sample is not very large [34]. Therefore, the model proposed in the research has been analyzed using Smart-PLS software 3.3 [85]. Following Henseler's indications, a bootstrapping procedure with 10,000 subsamples was performed to test the hypotheses [86].

4. Results

4.1. Analysis of the Measurement Model

In this model, it has been assumed that all latent variables are antecedents of their indicators. That is, they are reflective variables. For this reason, in order to evaluate the measurement model, indicators and construct reliability, convergence validity and discriminant validity, as well as analysis were carried out. The results are presented in Table 2.

Table 2. Outer model results for 18 indicators corresponding to four constructs.

	Mean	SD	Loading	t-Student *	Q_B^2	α	ρA	ρC	AVE
CSR						0.837	0.859	0.884	0.604
CSR.1	3.663	1.009	0.731	12.890					
CSR.2	3.705	0.894	0.800	20.291					
CSR.3	3.753	0.984	0.814	22.210					
CSR.4	3.849	1.022	0.718	12.000					
CSR.5	3.934	0.851	0.818	29.526					
Human Resources Management					0.146	0.873	0.889	0.903	0.574
HRM.1	3.211	1.140	0.597	8.035	0.079				
HRM.2	3.657	1.010	0.636	9.345	0.090				
HRM.3	3.482	1.004	0.778	12.329	0.107				
HRM.4	3.434	1.078	0.770	14.063	0.154				
HRM.5	3.783	0.906	0.831	30.353	0.242				
HRM.6	3.861	0.835	0.831	22.965	0.197				
HRM.7	3.825	0.931	0.821	22.534	0.155				
Customer Satisfaction					0.259	0.731	0.735	0.847	0.649
CUS.1	3.958	0.853	0.790	24.419	0.487				
CUS.2	4.187	0.773	0.801	19.607	0.130				
CUS.3	4.120	0.710	0.825	20.238	0.161				
Performance					0.442	0.878	0.878	0.925	0.805
PERF.1	3.982	0.832	0.870	40.647	0.467				
PERF.2	3.886	0.888	0.933	69.331	0.445				
PERF.3	3.789	0.842	0.886	40.144	0.414				

Significance and standard deviations (SD) performed by 10,000 repetitions bootstrapping procedure. Q_B^2: cross-validated redundancies index performed by a 9-step distance-blindfolding procedure. α: Chronbach's alpha; ρ$_A$: Dijkstra–Henseler's composite reliability; ρ$_C$: Jöreskog's composite reliability; AVE: average variance extracted; *: All loadings are significant at the 0.001 level. Source: Authors.

The reliability of the constructs was analyzed through factor loadings, Cronbach's Alpha, composite reliability, the Dijkstra–Henseler rho ratio, and the average variance extracted (AVE) [87].

The reliability of the indicators was examined through their loadings. All but two of the factor loadings exceed the established minimum value of 0.7 [88]. Furthermore, these two loads have values close to this minimum level, so they can also be accepted [89]. Therefore, the reliability of the indicators has been demonstrated.

The values for Cronbach's alpha, composite reliability, and the Dijkstra–Henseler rho ratio range from 0.731 to 0.925, demonstrating the reliability of the constructs [88]. Convergent validity was measured using average variance extracted (AVE). All values are greater than 0.5, confirming the internal consistency of the reflective scales [90].

The Fornell–Larcker criterion [91] was used to check the discriminant validity. For this purpose, the correlations between each pair of constructs were checked to ensure they did not exceed the square root of the AVE of each of the constructs, as can be seen in the results shown in Table 3 below. In addition, the heterotrait–monotrait (HTMT) ratio of correlations [92] was estimated. All values are below the conservative threshold of 0.850, confirming the adequate discriminant validity for all latent variables.

Table 3. Discriminant validity.

		I	II	III	IV
I	CSR	**0.777**	*0.567*	*0.587*	*0.576*
II	Human resources management	0.518	**0.757**	*0.606*	*0.534*
III	Customer satisfaction	0.637	0.497	**0.805**	*0.849*
IV	Performance	0.508	0.457	0.472	**0.897**

HTMT ratio over the diagonal (italics). Fornell–Lacker criterion: square root of AVE in diagonal (bold) and construct correlations below the diagonal. Source: Authors.

For further analysis, this research evaluated the quality by finding that the normed fit index (NFI) and the standardized root mean square residual (SRMR) and (NFI) did not exceed the value of 0.09 and 0.08 [86,93]. These results clearly show an adequate fit of the model.

Finally, through the Q_B^2 statistical test (a cross-validated redundancy index), carried out by the blindfolding method [94], the predictive relevance of the independent latent variables has been evaluated. The findings in Table 2 reveal that all Q_B^2 are positive, confirming the satisfactory explanatory qualities of the model [95].

4.2. Path Analysis

The structural model analysis begins by checking for the possible existence of a multicollinearity problem by analyzing the variance inflation factor (VIF). The results in Table 4 show how the values fluctuate from 1 to 1.85. Therefore, there is no multicollinearity problem in the structural model [96].

A further bootstrapping (10,000 resamples) procedure was then carried out with the aim to calculate *t*-values and percentile confidence intervals [97]. The coefficient of determination (R^2) of the dependent variables, the algebraic sign and magnitude, as well as the effect size (f^2) of the standardized regression coefficients were measured (Hair et al., 2017). The results are presented in Table 4 and Figure 2.

Table 4. Structural model and hypotheses testing.

	Path	SD	t-Value	f^2	95CI	VIF	H	Supported
Direct effects								
CSR → Performance	0.133	0.084	0.130	0.000	[−0.132;0.142]	1.85	H1	Yes
CSR → Human resources management	0.518	0.066	7.795 ***	0.367	[0.412;0.630]	1.00	$H2_a$	Yes
Human resources management → Performance	0.133	0.066	2.024 *	0.028	[0.030;0.247]	1.46	$H2_b$	Yes
CSR → Customer satisfaction	0.519	0.063	8.234 ***	0.355	[0.415;0.621]	1.37	$H3_a$	Yes
Customer satisfaction → Performance	0.671	0.070	9.551 ***	0.579	[0.563;0.793]	1.80	$H3_b$	Yes
Human resources management → Customer satisfaction	0.228	0.069	3.297 ***	0.068	[0.112;0.337]	1.37	$H4_a$	Yes
Indirect effects						VAF		
Individual indirect effects								
CSR → Human resources management → Performance	0.069	0.036	1.926 *		[0.016–0.133]	13.58%	$H2_c$	Yes
CSR → Customer satisfaction → Performance	0.349	0.060	5.518 ***		[0.255–0.463]	68.70%	$H3_c$	Yes
CSR → Human resources management → Customer satisfaction	0.118	0.041	2.904 **		[0.056–0.189]	18.52%	$H4_b$	Yes
Human resources management → Customer satisfaction → Performance	0.153	0.048	3.120 **		[0.037–0.129]	53.50%	H5	Yes
CSR → Human resources management → Customer satisfaction → Performance	0.079	0.028	2.806 **		[0.001–0.016]	15.55%	H6	Yes
Global indirect effects								
CSR → Customer satisfaction	0.118	0.049	2.904 **		[0.056–0.189]	18.52%		
CSR → Performance	0.497	0.059	8.486 ***		[0.414–0.605]	97.83%		
Human resources management → Performance	0.015	0.049	3.12 **		[0.074–0.233]	5.35%		
Total effect								
CSR → Performance	0.508	0.057	8.851 ***		[0.414–0.603]			
Human resources management → Performance	0.286	0.079	3.641 ***		[0.158–0.416]			
CSR → Customer satisfaction	0.637	0.049	13.062 ***		[0.563–0.793]			

R^2 adjusted [99% CI in brackets]: Customer satisfaction: 0.437 [0.340; 0.549]; Human resources management: 0.264 [0.164; 0.393]; Performance: 0.560 [0. 496; 0.648]. Blindfolding Q^2 index as shown in Table 3; Standardized path values reported. SD: Standard deviation; f^2: size effect index; 95CI: 95% Bias Corrected confidence interval; VIF: Inner model variance inflation factors; VAF: Variance Accounted Formula × 100 represents the proportion mediated. Significance, standard deviations, 95% bias-corrected CIs were performed by 10,000 repetitions bootstrapping procedure; *: $p < 0.05$; **: $p < 0.01$; ***: $p < 0.001$. Only total effects that differ from direct effects are shown. Source: Authors.

The findings reveal that CSR does not directly affect performance, as the effect found, and although positive, it is not significant ($\beta = 0.133$), rejecting H_1. However, a positive and direct influence of CSR on human resources management and customer satisfaction has been found ($\beta = 0.518$ *** and $\beta = 0.519$ *** respectively), thus verifying $H2_a$ and $H3_a$. Likewise, a positive and significant influence of human resources management on performance and customer satisfaction has also been found ($\beta = 0.133$ * and $\beta = 0.228$ *** respectively), thus verifying $H2_b$ and $H4_a$. Finally, the results also show a positive and significant influence of customer satisfaction on performance ($\beta = 0.671$ ***), verifying $H3_b$.

R^2 is used as a measure to analyze the model's predictive power as it shows how the variance of a variable can be explained by those variables that predict it in the model. The higher the R^2 value, the greater the predictive power of the model. As can be seen in the results, the variance explained is 43.7% for customer satisfaction, 26.4% for Human resources management and 56.0% for performance, exceeding the minimum value of 10% established by Falk & Miller [98].

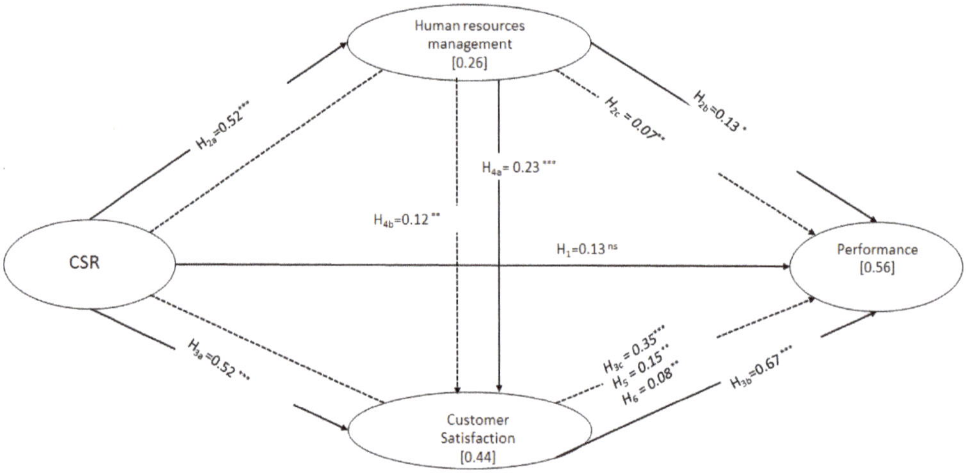

Figure 2. Results. *: $p < 0.05$; **: $p < 0.01$; ***: $p < 0.001$; ns: not significant. Source: Authors.

According to Cohen [62], the contribution of each independent variable to R^2 values of a dependent variable is to measure through f^2. Levels of f^2 of 0.02, 0.15, and 0.35 indicate a small, medium, or large effect, respectively. In addition, Chin [99] established a minimum value for f^2 of 0.02. As can be seen from the results, this minimum value is exceeded in all cases, except for the relationship between CSR and performance. Based on the above, the results show that customer satisfaction has a substantial effect on performance. There is also a significant effect of CSR on human resource management and customer satisfaction.

4.3. Mediating Effects

The results in Table 4 also show the indirect effects and the variance accounted for (VAF) [100]. VAF shows the size of the indirect effect relative to the total effect. As can be seen, the findings reveal how the indirect effects of CSR on performance through human resources management and customer satisfaction are both positive and significant ($\beta = 0.069$ * and $\beta = 0.349$ *** respectively, plus a sequential indirect effect $\beta = 0.079$ **). Regarding the VAF, the indirect effect of CSR on performance is about 97.83% of the total effect, with 13.58% through human resources management, 68.70% through customer satisfaction, and an additional 15.55% sequentially. Since the direct effect is not significant and indirect effects are significant, and the proportions mediated are prominent, a full mediation is suggested, supporting H2$_c$, H3$_c$ and H6. Furthermore, the results show that the indirect effect of CSR on customer satisfaction is positive and significant ($\beta = 0.118$ **). The proportion mediated by human resources management is 18.52% (VAF) of the total effect of CSR on customer satisfaction, supporting H4$_b$. Finally, customer satisfaction partially mediates between human resources management and performance ($\beta = 0.153$ **), with a 53.50% (VAF) of the total effect of human resources management on performance, supporting H5.

In sum, the findings demonstrate that the relationship between CSR and performance is fully mediated by human resources management and customer satisfaction. For CSR practices to have a positive influence on company performance, they must be partly aimed at improving employee and customer satisfaction.

4.4. Evaluation of the Predictive Performance

According to Shmueli [101], the predictive performance of a model is its ability to generate new predictions. Therefore, predictive validity (out-of-sample prediction) shows how a given outcome variable can be predicted from a given set of measures of a variable [102].

The model's predictive capacity has been evaluated through a cross-validation with holdout samples [95] by running the PLS predict algorithm with SmartPLS [103].

As can be observed in Table 5, the model shows predictive power for all the constructs, since their Q^2 value is above 0. A similar conclusion is achieved when comparing the RMSE or MAE results of the PLS-SEM with those of a linear regression model (LM) model. In almost all results obtained, PLS-SEM produces lower errors and higher Q2, demonstrating the model's predictive performance [89].

Table 5. PLS predict assessment.

	Construct Prediction Summary							
	Q^2							
HUMAN RESOURCES MANAGEMENT	0.250							
CUSTOMER SATISFACTION	0.391							
PERFORMANCE	0.243							
	INDICATOR PREDICTION SUMMARY							
	PLS			LM			PLS-LM		
Indicator	RMSE	MAE	Q^2	RMSE	MAE	Q^2	RMSE	MAE	Q^2
HRM.1	1.103	0.895	0.071	1.124	0.901	0.036	−0.021	−0.006	0.035
HRM.2	0.970	0.743	0.091	0.994	0.762	0.047	−0.024	−0.019	0.044
HRM.3	0.961	0.752	0.096	0.979	0.772	0.062	−0.018	−0.020	0.034
HRM.4	1.000	0.805	0.147	1.006	0.807	0.138	−0.006	−0.002	0.009
HRM.5	0.797	0.608	0.236	0.813	0.622	0.206	−0.016	−0.014	0.030
HRM.6	0.755	0.584	0.191	0.767	0.602	0.165	−0.012	−0.018	0.026
HRM.7	0.861	0.649	0.155	0.864	0.655	0.149	−0.003	−0.006	0.006
CUS.1	0.612	0.497	0.492	0.496	0.330	0.667	0.116	0.167	−0.175
CUS.2	0.743	0.586	0.089	0.734	0.588	0.111	0.009	−0.002	−0.022
CUS.3	0.675	0.531	0.108	0.662	0.531	0.143	0.013	0.000	−0.035
PERF.1	0.733	0.581	0.235	0.744	0.596	0.211	−0.011	−0.015	0.024
PERF.2	0.802	0.640	0.195	0.815	0.661	0.168	−0.013	−0.021	0.027
PERF.3	0.782	0.629	0.151	0.764	0.619	0.189	0.018	0.010	−0.038

PLS: Partial least squares path model; LM: Linear regression model; RMSE: Root mean squared error; MAE: Mean absolute error. Q^2: PLS-predict index performed with 10 k-fold and 10 repetitions. Source: Authors.

4.5. Robustness Checks

4.5.1. Endogeneity

In order to rule out the existence of endogeneity problems, we applied the Gaussian copula approach developed by Park and Gupta [104], which is suitable to identify endogeneity issues [105]. In order to do this, we previously checked that the variables that could generate an endogeneity problem are nonnormally distributed. We carried out, using Stata v.16, the skewness/kurtosis test for normality, the Shapiro–Wilk W test for normal data and the Shapiro–Francia W' test for normal data on the independent variable scores of CSR, human resources management and customer satisfaction. The findings show that none of the variables has a normally distributed score. These results allow us to continue with the Gaussian copula approach [106].

The results in Table 6 show that the effect of the Gaussian copula is not a significant equation. Therefore, endogeneity is not an issue in this model.

Table 6. Gaussian Copula Approach.

		Path		Copula	
Relationship	Model	β	t	β	t
Perf ← CSR	1	0.679	9.69 ***		
Perf ← CSR	2	0.666	9.42 ***	0.091	0.34
Cus ← CSR	1	0.227	3.33 ***		
Cus ← CSR	2	0.244	3.62 ***	0.718	2.46
HRM ← CSR	1	0.517	7.76 ***		
HRM ← CSR	2	0.818	3.37 ***	−0.432	−1.29

***: $p < 0.001$.

We checked for endogeneity arising from the existence of omitted variables when trying to explain the dependent variable [107]. For this purpose and according to Antonakis et al. [108], we introduced as control variables size, age and the percentage of the company's capital owned by the family ownership. Once the control variables have been introduced, we run the PLS algorithm again and we can see that the results obtained are identical to those obtained without the control variables. Thus, we can establish that the omitted variables are controlled in this model.

4.5.2. Heterogeneity

To identify unobserved heterogeneity in PLS path models, we carried out the FIMIX-PLS procedure [109]. We established a maximum number of iterations of 5000 and 10 repetitions. Taking into account the size of the sample, an effect size of 0.15 and a power level of 80%, we established two segments. Subsequently, FIMIX-PLS was run for 1 to 2 segments.

Table 7 shows the results obtained, which are ambiguous in determining the number of appropriate segments, which shows that unobserved heterogeneity is not a problem [110].

Table 7. Fit indices for the one to two segments solutions.

	Number of Segments	
	1	2
AIC	1142.702	**1130.045**
AIC$_3$	1151.702	**1149.045**
AIC$_4$	**1160.702**	1168.045
BIC	**1170.71**	1189.173
CAIC	**1179.71**	1208.173
HQ	1154.071	**1154.045**
MDL$_5$	**1354.741**	1577.684
LnL	−562.351	−546.022
EN	na	0.61
NFI	na	0.637
NEC	na	64.746

Note: AIC: Akaike's information criterion. AIC3: Modified AIC with factor 3. AIC4: Modified AIC with factor 4. BIC: Bayesian information criteria. CAIC: consistent AIC. HQ: Hannan Quinn criterion. MDL5: Minimum description length with factor 5. LnL: Log likelihood. EN: Entropy statistic. NFI: Non-fuzzy index. NEC: Normalized entropy criterion. na: not available. Numbers in bold indicate the best outcome per segment retention criterion.

4.5.3. Nonlinear Effects

We checked for the existence of nonlinear effects by following the instructions set out by Svenson [111]. First, we carried out the Ramsey RESET [112] using Stata V.16 with the values obtained for the latent variables. The findings in Table 8 show that none of the partial regressions are subject to nonlinearities. Second, we included interaction terms to represent the quadratic effects between the variables. The findings, with 10,000 samples, show that none of the nonlinear effects are significant. Therefore, in this model the linear effects are robust.

Table 8. Nonlinear effects.

Nonlinear Relationship	Coefficient	p Values	f²	Ramsey RESET	
CSR * CSR → Perf	0.012	0.403	0.480	F (3.159) = 1.21	p = 0.215
HRM * HRM → Perf	0.065	0.058	0.188		
Cus * Cus → Perf	0.062	0.098	0.267		
CSR * CSR → Cus	0.065	0.138	0.325	F (3.160) = 1.21	p = 0.219
HRM * HRM → Cus	0.095	0.076	0.177		

*: $p < 0.05$.

5. Discussions and Conclusions

Our study has found indications that CSR actions induce a positive influence on human resource management due to the generation of synergies and greater cohesion among employees [25], which allows companies to achieve their objectives more efficiently [27] and therefore improve their performance by observing the influence of human resources management on performance.

Similarly, a positive relationship has been observed between CSR and customer satisfaction as a consequence of increased customer loyalty and satisfaction through CSR practices. This is in line with previously published results [23]. This increase in customer satisfaction also provides companies with an important competitive advantage that influences performance.

On this basis, it has been possible to intuit an indirect influence of CSR on the performance of companies since, through increased customer satisfaction and appropriate human resources management, companies obtain a series of competitive advantages that allow them to increase their performance.

It is also interesting to note that the results obtained show us that human resource management directly and positively affects customer satisfaction. Through proper human resource management, it is possible to create a team with higher skills and capabilities [45], which impacts the service provided to customers and, therefore, improves their satisfaction [24]. Thus, it has been shown to achieve higher customer satisfaction by influencing human resource practices, as already stated [57].

Finally, an important finding of this research has been the observation of a sequential mediation of HRM and customer satisfaction in the relationship between CSR and performance. When companies carry out CSR practices aimed at satisfying the conditions of their employees, this has an impact on customer care by increasing customer loyalty and satisfaction, which enables companies to increase their performance.

With these results, this paper contributes to filling a gap related to the indirect effect of CSR through human resources management and customer satisfaction on performance.

Through a sample of 166 Spanish SMEs in the food and beverage manufacturing sector and using PLS-SEM, this research has focused on analyzing the effect of CSR practices on the performance of these companies. In addition, the mediating effect of human resource management and customer satisfaction on this relationship has been analyzed, which is a step further in relation to previous research.

The literature is divided on the effect of CSR practices on performance [5,6]. This paper contributes to evidence demonstrating no significant effect. However, when CSR is oriented towards customer and employee satisfaction, CSR practices have a significant indirect effect on firm performance and such strategies do allow firms to increase their chances of survival in the current uncertain environment.

From a theoretical point of view, this paper contributes to shedding light on the effect of CSR on the financial and non-financial performance of companies, integrating the role that human resource management and customer satisfaction play in this relationship, demonstrating that it can allow companies to obtain interesting competitive advantages, which is crucial for their growth. This is justified because the relationship between CSR and performance is fully mediated and the mediation is total (we have no evidence of direct relationship). Moreover, the VAF of the mediation is 97.83%, well above the acceptance threshold.

This research also has important implications for managers of SMEs in this sector and policymakers. From a practical point of view, it has shown how CSR practices aimed at improving customer and employee satisfaction not only contribute to creating a better society but also enable companies to improve their performance. In line with Yáñez-Araque et al. [113], the results obtained in this research show how, for these companies, the benefits of implementing CSR practices outweigh the costs, thereby increasing the profit obtained by companies. This should serve to encourage company managers to develop a CSR strategy that will bring them interesting competitive advantages. Regarding public policies, these results show how the establishment of awareness-raising campaigns and aid aimed at encouraging SMEs to develop CSR practices would produce significant benefits, both for society as a whole and these companies.

This article is not without limitations, which serve to establish future lines of research. This article is based on results obtained from a sample of Spanish companies only. For this reason, these results may not be extrapolated to other regions, as CSR depends on aspects such as culture, ethics, legislation, and the economic environment [114]. Therefore, future studies could use a larger sample size covering other regions. Likewise, this article has only used cross-sectional information, so these could change over time. For this reason, it would be interesting for future research to use longitudinal data in order to assess possible changes over time.

Author Contributions: Conceptualization, J.M.S.-J. and F.G.-A.; methodology, J.M.S.-J. and M.P.-M.; software, J.M.S.-J. and H.H.G.-S.; validation, J.M.S.-J., M.P.-M. and F.G.-A.; formal analysis, M.P.-M. and J.M.S.-J.; investigation, F.G.-A. and J.M.S.-J.; resources, M.P.-M. and H.H.G.-S.; data curation, J.M.S.-J. and F.G.-A.; writing-original draft preparation, J.M.S.-J. and F.G.-A.; writing-review and editing, M.P.-M. and H.H.G.-S.; visualization, M.P.-M.; supervision, J.M.S.-J.; project administration, H.H.G.-S.; funding acquisition, F.G.-A. All authors have read and agreed to the published version of the manuscript.

Funding: This research received no external funding.

Institutional Review Board Statement: Not applicable.

Informed Consent Statement: Not applicable.

Data Availability Statement: Not applicable.

Acknowledgments: This study has been supported by FAEDPYME (Fundación para el análisis económico de la Pyme).

Conflicts of Interest: The authors declare no conflict of interest.

Appendix A

Table A1. Survey questions used in the research.

CSR	
Regarding the CSR in your company, assess your level of conformity, from 1(absolutely disagree) to 5 (absolutely agree), with the following statements	
CSR.1	Is broadly understood by management and implemented in company.
CSR.2	Refers to achieve social and economic values.
CSR.3	The company performs its activities spending less energy and other resources.
CSR.4	The company implements effective recycling measures.
CSR.5	In recent years, transparency towards customers and suppliers has improved.

Table A1. Cont.

	Employee Satisfaction
Thinking of your employees as a whole, please indicate your level of agreement with the following statements: "In the last 2 years, the company … , from 1 (absolutely disagree) to 5 (absolutely agree)"	
EMP.1	Has assessed performance and given feedback on time.
EMP.2	Has ensured equal treatment in salaries.
EMP.3	Has allowed incentives based on results achieved.
EMP.4	Has improved career development.
EMP.5	Has provided opportunities to be involved in decision making.
EMP.6	Has applied accurately the requirements for each position in the recruitments.
EMP.7	Has invested enough time and money in training.
EMP.8	Has allowed successive training programmes.
	Customer satisfaction
Indicate your degree of agreement with the following statements: "In the last 2 years, the company ……. from 1(absolutely disagree) to 5 (absolutely agree)"	
CUS.1	The company has improved its corporate identity and reputation in the last years.
CUS.2	The quality of products and services has increased in the last years.
CUS.3	The company has enhanced the customer satisfaction in the last years.
	Performance
Indicate your degree of conformity with the following performance indicators of your company, from 1 (absolutely disagree) to 5 (absolutely agree)	
PERF.1	The company adapts earlier to changes in the market than competitors.
PERF.2	The company is growing more than competitors
PERF.3	The company is more profitable than competitors

Note: The research questions and their results were drawn from a broader study of companies in all sectors and in order to analyze many other variables. Only those that have been used in the research are reported here.

Table A2. Correlation matrix.

	1	2	3	4	5	6	7	8	9	10	11	12	13	14	15	16	17	18
CSR01	1																	
CSR02	0.698 **	1																
CSR03	0.480 **	0.547 **	1															
CSR04	0.330 **	0.487 **	0.598 **	1														
CSR05	0.437 **	0.529 **	0.541 **	0.522 **	1													
HRM01	0.235 **	0.203 **	0.127	0.208 **	0.300 **	1												
HRM02	0.253 **	0.281 **	0.187 *	0.259 **	0.261 **	0.398 **	1											
HRM03	0.172 *	0.252 **	0.267 **	0.294 **	0.298 **	0.385 **	0.424 **	1										
HRM04	0.251 **	0.314 **	0.237 **	0.289 **	0.412 **	0.529 **	0.458 **	0.569 **	1									
HRM05	0.335 **	0.405 **	0.372 **	0.336 **	0.374 **	0.565 **	0.452 **	0.658 **	0.572 **	1								
HRM06	0.295 **	0.333 **	0.332 **	0.314 **	0.253 **	0.353 **	0.379 **	0.518 **	0.495 **	0.629 **	1							
HRM07	0.226 **	0.278 **	0.282 **	0.321 **	0.441 **	0.530 **	0.365 **	0.560 **	0.484 **	0.598 **	0.898 **	1						
CUS01	0.331 **	0.358 **	0.247 **	0.260 **	0.218 **	0.269 **	0.263 **	0.298 **	0.367 **	0.480 **	0.465 **	0.223 **	1					
CUS02	0.235 **	0.272 **	0.361 **	0.173 *	0.330 **	0.242 **	0.275 **	0.233 **	0.286 **	0.282 **	0.227 **	0.154 *	.705 **	1				
CUS03	0.208 **	0.322 **	0.327 **	0.175 *	0.392 **	0.207 **	0.285 **	0.231 **	0.326 **	0.303 **	0.313 **	0.251 **	.687 **	0.596 **	1			
PERF01	0.359 **	0.333 **	0.392 **	0.365 **	0.266 **	0.252 **	0.272 **	0.385 **	0.277 **	0.362 **	0.412 **	0.392 **	.508 **	0.548 **	0.406 **	1		
PERF02	0.320 **	0.307 **	0.402 **	0.280 **	0.428 **	0.214 **	0.292 **	0.332 **	0.323 **	0.381 **	0.377 **	0.311 **	0.471 **	0.540 **	0.596 **	0.723 **	1	
PERF03	0.292 **	0.302 **	0.344 **	0.138	0.243 **	0.197 *	0.269 **	0.320 **	0.254 **	0.398 **	0.318 **	0.291 **	0.449 **	0.542 **	0.597 **	0.605 **	0.790 **	1

** $p < 0.01$; * $p < 0.05$.

References

1. Story, J.; Neves, P. When corporate social responsibility (CSR) increases performance: Exploring the role of intrinsic and extrinsic CSR attribution. *Bus. Ethics A Eur. Rev.* **2015**, *24*, 111–124. [CrossRef]
2. Surroca, J.; Tribó, J.A.; Waddock, S. Corporate responsibility and financial performance: The role of intangible resources. *Strateg. Manag. J.* **2010**, *31*, 463–490. [CrossRef]
3. Latif, K.F.; Sajjad, A. Measuring corporate social responsibility: A critical review of survey instruments. *Corp. Soc. Responsib. Environ. Manag.* **2018**, *25*, 1174–1197. [CrossRef]
4. Martinez-Conesa, I.; Soto-Acosta, P.; Palacios-Manzano, M. Corporate social responsibility and its effect on innovation and firm performance: An empirical research in SMEs. *J. Clean. Prod.* **2017**, *142*, 2374–2383. [CrossRef]
5. Tang, Z.; Hull, C.E.; Rothenberg, S. How Corporate Social Responsibility Engagement Strategy Moderates the CSR-Financial Performance Relationship. *J. Manag. Stud.* **2012**, *49*, 1274–1303. [CrossRef]
6. Bahta, D.; Yun, J.; Islam, M.R.; Bikanyi, K.J. How does CSR enhance the financial performance of SMEs? The mediating role of firm reputation. *Econ. Res.-Ekon. Istraživanja* **2020**, *34*, 1428–1451. [CrossRef]
7. Freeman, R.E. Stakeholder theory. In *Wiley Encyclopedia of Management, 12: Strategic Management*; John Wiley & Sons: Hoboken, NJ, USA, 2015; pp. 1–6.
8. Parmar, B.L.; Freeman, R.E.; Harrison, J.S.; Wicks, A.C.; Purnell, L.; De Colle, S. Stakeholder theory: The state of the art. *Acad. Manag. Ann.* **2010**, *4*, 403–445. [CrossRef]
9. Jamali, D. A stakeholder approach to corporate social responsibility: A fresh perspective into theory and practice. *J. Bus. Ethics* **2008**, *82*, 213–231. [CrossRef]
10. Torugsa, N.A.; O'Donohue, W.; Hecker, R. Capabilities, proactive CSR and financial performance in SMEs: Empirical evidence from an Australian manufacturing industry sector. *J. Bus. Ethics* **2012**, *109*, 483–500. [CrossRef]
11. Campbell, J.L. Why would corporations behave in socially responsible ways? An institutional theory of corporate social responsibility. *Acad. Manag. Rev.* **2007**, *32*, 946–967. [CrossRef]
12. Hao, X.; Tong, Y.; Hu, C. A Study on the Impact of Corporate Social Performance on Corporate Financial Performance: From a View of Social Capital. *Sci. Sci. Manag. S. T* **2011**, *10*.
13. Liping, Z.; Yan, C.; Yujian, J. An empirical study on the relationship between corporate social responsibility and financial performance—Based on the analysis and interpretation based on the perspective of corporate reputation. *Jiangsu Soc. Sci.* **2016**, *3*, 95–102.
14. Oeyono, J.; Samy, M.; Bampton, R. An examination of corporate social responsibility and financial performance: A study of the top 50 Indonesian listed corporations. *J. Glob. Responsib.* **2011**, *2*, 100–112. [CrossRef]
15. Tian, H. The correlation between CSR and corporate performance-cased on the empirical data in China's telecommunication industry. *Econ. Manag.* **2009**, *31*, 72–79.
16. Wagner, M.; Van Phu, N.; Azomahou, T.; Wehrmeyer, W. The relationship between the environmental and economic performance of firms: An empirical analysis of the European paper industry. *Corp. Soc. Responsib. Environ. Manag.* **2002**, *9*, 133–146. [CrossRef]
17. Margolis, J.D.; Walsh, J.P. Misery loves companies: Rethinking social initiatives by business. *Adm. Sci. Q.* **2003**, *48*, 268–305. [CrossRef]
18. Mahmood, F.; Qadeer, F.; Saleem, M.; Han, H.; Ariza-Montes, A. Corporate social responsibility and firms' financial performance: A multi-level serial analysis underpinning social identity theory. *Econ. Res.-Ekon. Istraživanja* **2021**, *12*, 1–39. [CrossRef]
19. Islam, T.; Islam, R.; Pitafi, A.H.; Xiaobei, L.; Rehmani, M.; Irfan, M.; Mubarak, M.S. The impact of corporate social responsibility on customer loyalty: The mediating role of corporate reputation, customer satisfaction, and trust. *Sustain. Prod. Consum.* **2021**, *25*, 123–135. [CrossRef]
20. Kramer, M.R.; Porter, M. *Creating Shared Value*; January–February 2011 Harvard Business Review; Harvard Business Publishing: Brighton, MA, USA, 2011; Volume 17.
21. Wang, Y.; Xu, S.; Wang, Y. The consequences of employees' perceived corporate social responsibility: A meta-analysis. *Bus. Ethics A Eur. Rev.* **2020**, *29*, 471–496. [CrossRef]
22. Flammer, C. Does corporate social responsibility lead to superior financial performance? A regression discontinuity approach. *Manag. Sci.* **2015**, *61*, 2549–2568. [CrossRef]
23. Chung, K.-H.; Yu, J.-E.; Choi, M.-G.; Shin, J.-I. The effects of CSR on customer satisfaction and loyalty in China: The moderating role of corporate image. *J. Econ. Bus. Manag.* **2015**, *3*, 542–547. [CrossRef]
24. Cantarello, S.; Filippini, R.; Nosella, A. Linking human resource management practices and customer satisfaction on product quality. *Int. J. Hum. Resour. Manag.* **2012**, *23*, 3906–3924. [CrossRef]
25. Chang, W.A.; Huang, T.C. Relationship between strategic human resource management and firm performance: A contingency perspective. *Int. J. Manpow.* **2005**, *26*, 434–449. [CrossRef]
26. Cheema, S.; Afsar, B.; Javed, F. Employees' corporate social responsibility perceptions and organizational citizenship behaviors for the environment: The mediating roles of organizational identification and environmental orientation fit. *Corp. Soc. Responsib. Environ. Manag.* **2020**, *27*, 9–21. [CrossRef]
27. Santana, M.; Morales-Sánchez, R.; Pasamar, S. Mapping the link between corporate social responsibility (CSR) and human resource management (HRM): How is this relationship measured? *Sustainability* **2020**, *12*, 1678. [CrossRef]

28. Shen, J.; Benson, J. When CSR is a social norm: How socially responsible human resource management affects employee work behavior. *J. Manag.* **2016**, *42*, 1723–1746. [CrossRef]
29. Wang, C.-C. Corporate social responsibility on customer behaviour: The mediating role of corporate image and customer satisfaction. *Total Qual. Manag. Bus. Excell.* **2020**, *31*, 742–760. [CrossRef]
30. Luo, X.; Bhattacharya, C.B. Corporate social responsibility, customer satisfaction, and market value. *J. Mark.* **2006**, *70*, 1–18. [CrossRef]
31. Saeidi, S.P.; Sofian, S.; Saeidi, P.; Saeidi, S.P.; Saaeidi, S.A. How does corporate social responsibility contribute to firm financial performance? The mediating role of competitive advantage, reputation, and customer satisfaction. *J. Bus. Res.* **2015**, *68*, 341–350. [CrossRef]
32. Zhang, Q.; Cao, M.; Zhang, F.; Liu, J.; Li, X. Effects of corporate social responsibility on customer satisfaction and organizational attractiveness: A signaling perspective. *Bus. Ethics A Eur. Rev.* **2020**, *29*, 20–34. [CrossRef]
33. Hair, J.F.; Sarstedt, M. Factors versus Composites: Guidelines for Choosing the Right Structural Equation Modeling Method. *Proj. Manag. J.* **2019**, *50*, 619–624. [CrossRef]
34. Segarra-Moliner, J.R.; Moliner-Tena, M.Á. Customer equity and CLV in Spanish telecommunication services. *J. Bus. Res.* **2016**, *69*, 4694–4705. [CrossRef]
35. Hannah, S.T.; Sayari, N.; Harris, F.H.d.B.; Cain, C.L. The Direct and Moderating Effects of Endogenous Corporate Social Responsibility on Firm Valuation: Theoretical and Empirical Evidence from the Global Financial Crisis. *J. Manag. Stud.* **2021**, *58*, 421–456. [CrossRef]
36. Freeman, R.E. *Strategic Management: A Stakeholder Approach*; Cambridge University Press: Cambridge, UK, 1984; ISBN 9780521151740.
37. Freeman, R.E. The politics of stakeholder theory: Some future directions. *Bus. Ethics Q.* **1994**, *4*, 409–421. [CrossRef]
38. Roberts, R.W. Determinants of corporate social responsibility disclosure: An application of stakeholder theory. *Account. Organ. Soc.* **1992**, *17*, 595–612. [CrossRef]
39. Yang, W.; Yang, S. An empirical study on the relationship between corporate social responsibility and financial performance under the Chinese context-based on the contrastive analysis of large, small and medium-size listed companies. *Chin. J. Manag. Sci.* **2016**, *24*, 143–150.
40. Barney, J. Firm resources and sustained competitive advantage. *J. Manag.* **1991**, *17*, 99–120. [CrossRef]
41. Sun, W.; Price, J.M. The impact of environmental uncertainty on increasing customer satisfaction through corporate social responsibility. *Eur. J. Mark.* **2016**, *50*, 1209–1238. [CrossRef]
42. Stavrou, E.T.; Brewster, C.; Charalambous, C. Human resource management and firm performance in Europe through the lens of business systems: Best fit, best practice or both? *Int. J. Hum. Resour. Manag.* **2010**, *21*, 933–962. [CrossRef]
43. Cho, S.J.; Chung, C.Y.; Young, J. Study on the Relationship between CSR and Financial Performance. *Sustainability* **2019**, *11*, 343. [CrossRef]
44. Matten, D.; Moon, J. "Implicit" and "explicit" CSR: A conceptual framework for a comparative understanding of corporate social responsibility. *Acad. Manag. Rev.* **2008**, *33*, 404–424. [CrossRef]
45. Zehir, C.; Gurol, Y.; Karaboga, T.; Kole, M. Strategic human resource management and firm performance: The mediating role of entrepreneurial orientation. *Procedia-Soc. Behav. Sci.* **2016**, *235*, 372–381. [CrossRef]
46. Sánchez, A.A.; Marín, G.S.; Morales, A.M. The mediating effect of strategic human resource practices on knowledge management and firm performance. *Rev. Eur. Dir. Econ. Empresa* **2015**, *24*, 138–148. [CrossRef]
47. Xie, X.; Jia, Y.; Meng, X.; Li, C. Corporate social responsibility, customer satisfaction, and financial performance: The moderating effect of the institutional environment in two transition economies. *J. Clean. Prod.* **2017**, *150*, 26–39. [CrossRef]
48. Sharma, N. An Examination of Customer Relationship Value in High vs. Low Technology Industries. *Acad. Mark. Stud. J.* **2020**, *24*, 1–24.
49. Høgevold, N.M.; Svensson, G.; Otero-Neira, C. Validating action and social alignment constituents of collaboration in business relationships: A sales perspective. *Mark. Intell. Plan.* **2019**, *37*, 721–740. [CrossRef]
50. Mpinganjira, M.; Roberts-Lombard, M.; Svensson, G. Validating the relationship between trust, commitment, economic and non-economic satisfaction in South African buyer-supplier relationships. *J. Bus. Ind. Mark.* **2017**, *32*, 421–431. [CrossRef]
51. Gandhi, S.K.; Sachdeva, A.; Gupta, A. Impact of service quality on satisfaction and loyalty at manufacturer-distributor dyad: Insights from Indian SMEs. *J. Adv. Manag. Res.* **2019**, *16*, 91–122. [CrossRef]
52. Ferro, C.; Padin, C.; Svensson, G.; Payan, J. Trust and commitment as mediators between economic and non-economic satisfaction in manufacturer-supplier relationships. *J. Bus. Ind. Mark.* **2016**, *31*, 13–23. [CrossRef]
53. Geyskens, I.; Steenkamp, J.-B.E.M.; Kumar, N. A meta-analysis of satisfaction in marketing channel relationships. *J. Mark. Res.* **1999**, *36*, 223–238. [CrossRef]
54. Johnston, W.J.; Le, A.N.H.; Cheng, J.M.-S. A meta-analytic review of influence strategies in marketing channel relationships. *J. Acad. Mark. Sci.* **2018**, *46*, 674–702. [CrossRef]
55. Geyskens, I.; Steenkamp, J.-B.E.M. Economic and social satisfaction: Measurement and relevance to marketing channel relationships. *J. Retail.* **2000**, *76*, 11–32. [CrossRef]
56. Chuang, C.; Liao, H.U.I. Strategic human resource management in service context: Taking care of business by taking care of employees and customers. *Pers. Psychol.* **2010**, *63*, 153–196. [CrossRef]

57. Chand, M. The impact of HRM practices on service quality, customer satisfaction and performance in the Indian hotel industry. *Int. J. Hum. Resour. Manag.* **2010**, *21*, 551–566. [CrossRef]
58. Phillips, S.; Thai, V.V.; Halim, Z. Airline Value Chain Capabilities and CSR Performance: The Connection Between CSR Leadership and CSR Culture with CSR Performance, Customer Satisfaction and Financial Performance. *Asian J. Shipp. Logist.* **2019**, *35*, 30–40. [CrossRef]
59. Jamali, D.R.; El Dirani, A.M.; Harwood, I.A. Exploring human resource management roles in corporate social responsibility: The CSR-HRM co-creation model. *Bus. Ethics A Eur. Rev.* **2015**, *24*, 125–143. [CrossRef]
60. Porter, M.E.; Kramer, M.R. Creating shared value: How to reinvent capitalism—And unleash a wave of innovation and growth. *Harv. Bus. Rev.* **2011**, *89*, 62–77.
61. Fisher, R.J. Social desirability bias and the validity of indirect questioning. *J. Consum. Res.* **1993**, *20*, 303–315. [CrossRef]
62. Cohen, J. *Statistical Power Analysis for the Behavioral Sciences*, 2nd ed.; Routledge: London, UK, 1988.
63. Podsakoff, P.M.; MacKenzie, S.B.; Lee, J.-Y.; Podsakoff, N.P. Common method biases in behavioral research: A critical review of the literature and recommended remedies. *J. Appl. Psychol.* **2003**, *88*, 879. [CrossRef]
64. Galbreath, J.; Shum, P. Do customer satisfaction and reputation mediate the CSR–FP link? Evidence from Australia. *Aust. J. Manag.* **2012**, *37*, 211–229. [CrossRef]
65. Adinata, G. CSR Expenditures, Financial Distress Prediction, and Firm Reputation: A Pathway Analysis. *Perspekt. Akunt.* **2019**, *2*, 1–18. [CrossRef]
66. Agyemang, O.S.; Ansong, A. Corporate social responsibility and firm performance of Ghanaian SMEs. *J. Glob. Responsib.* **2017**. [CrossRef]
67. Caro, N.; Salazar, I. La responsabilidad social y la competitividad de las MYPES de Tingo María. *Balance's* **2019**, *6*, 4–12.
68. Esparza Aguilar, J.L.; Reyes Fong, T. Practices of corporate social responsibility developed by mexican family businesses and their impact on competitive success and innovation. *Tec Empres.* **2019**, *13*, 45–57.
69. Ikram, M.; Sroufe, R.; Mohsin, M.; Solangi, Y.A.; Shah, S.Z.A.; Shahzad, F. Does CSR influence firm performance? A longitudinal study of SME sectors of Pakistan. *J. Glob. Responsib.* **2019**, *11*, 27–53. [CrossRef]
70. Liman, L.; Tarigan, J.; Jie, F. Corporate Social Responsibility, Financial Performance, and Risk in Indonesian Natural Resources Industry. *Soc. Responsib. J.* **2019**, *16*, 73–90.
71. Santos-Jaén, J.M.; Madrid-Guijarro, A.; García-Pérez-de-Lema, D. The impact of corporate social responsibility on innovation in small and medium-sized enterprises: The mediating role of debt terms and human capital. *Corp. Soc. Responsib. Environ. Manag.* **2021**, *28*, 1200–1215. [CrossRef]
72. Ozutku, H.; Ozturkler, H. The determinants of human resource practices: An empirical investigation in the Turkish manufacturing industry. *Ege Acad. Rev.* **2009**, *9*, 73–93.
73. Ngwenya, L.; Aigbavboa, C. Improvement of productivity and employee performance through an efficient human resource management practices. In *Advances in Human Factors, Business Management, Training and Education*; Springer: Berlin/Heidelberg, Germany, 2017; pp. 727–737.
74. Bombiak, E.; Marciniuk-Kluska, A. Socially responsible human resource management as a concept of fostering sustainable organization-building: Experiences of young polish companies. *Sustainability* **2019**, *11*, 1044. [CrossRef]
75. Wright, P.M.; Gardner, T.M.; Moynihan, L.M.; Allen, M.R. The relationship between HR practices and firm performance: Examining causal order. *Pers. Psychol.* **2005**, *58*, 409–446. [CrossRef]
76. Raineri, A. Linking human resources practices with performance: The simultaneous mediation of collective affective commitment and human capital. *Int. J. Hum. Resour. Manag.* **2017**, *28*, 3149–3178. [CrossRef]
77. Al-Hawary, S.I.S.; Shdefat, F.A. Impact of Human Resources Management Practices on Employees' Satisfaction A Field Study on the Rajhi Cement Factory. *Int. J. Acad. Res. Account. Financ. Manag. Sci.* **2016**, *6*, 274–286.
78. Ruiz-Palomo, D.; Diéguez-Soto, J.; Duréndez, A.; Santos, J.A.C. Family management and firm performance in family SMEs: The mediating roles of management control systems and technological innovation. *Sustainability* **2019**, *11*, 3805. [CrossRef]
79. Ubeda-García, M.; Claver-Cortés, E.; Marco-Lajara, B. Corporate social responsibility and firm performance in the hotel industry. The mediating role of green human resource management and environmental outcomes. *J. Bus. Res.* **2021**, *123*, 57–69. [CrossRef]
80. Rigdon, E.E. Choosing PLS path modeling as analytical method in European management research: A realist perspective. *Eur. Manag. J.* **2016**, *34*, 598–605. [CrossRef]
81. Henseler, J. Partial least squares path modeling: Quo vadis? *Qual. Quant.* **2018**, *52*, 1–8. [CrossRef]
82. Chin, W.W. The Partial Least Squares Approach to Structural Modeling. *Mod. Methods Bus. Res.* **1998**, *295*, 295–336.
83. Dijkstra, T.K.; Henseler, J. Linear indices in nonlinear structural equation models: Best fitting proper indices and other composites. *Qual. Quant.* **2011**, *45*, 1505–1518. [CrossRef]
84. Tenenhaus, M. Component-based structural equation modelling. *Total Qual. Manag. Bus. Excell.* **2008**, *19*, 871–886. [CrossRef]
85. Ringle, C.M.; Wende, S.; Becker, J.-M. *SmartPLS 3 2015*; SmartPLS GmbH: Boenningstedt, Germany, 2015.
86. Henseler, J.; Dijkstra, T.K.; Sarstedt, M.; Ringle, C.M.; Diamantopoulos, A.; Straub, D.W.; Ketchen, D.J.; Hair, J.F.; Hult, G.T.M.; Calantone, R.J. Common Beliefs and Reality About PLS: Comments on Rönkkö and Evermann (2013). *Organ. Res. Methods* **2014**, *17*, 182–209. [CrossRef]
87. Henseler, J.; Schuberth, F. Using confirmatory composite analysis to assess emergent variables in business research. *J. Bus. Res.* **2020**, *120*, 147–156. [CrossRef]

88. Hair, J.F.; Hult, G.T.M.; Ringle, C.; Sarstedt, M. *A Primer on Partial Least Squares Structural Equation Modeling (PLS-SEM)*; Sage Publications: Thousand Oaks, CA, USA, 2016; ISBN 1483377431.
89. Felipe, C.M.; Roldán, J.L.; Leal-Rodríguez, A.L. Impact of organizational culture values on organizational agility. *Sustainability* **2017**, *9*, 2354. [CrossRef]
90. Hair, J.F.; Howard, M.C.; Nitzl, C. Assessing measurement model quality in PLS-SEM using confirmatory composite analysis. *J. Bus. Res.* **2020**, *109*, 101–110. [CrossRef]
91. Fornell, C.; Larcker, D.F. Evaluating structural equation models with unobservable variables and measurement error. *J. Mark. Res.* **1981**, *18*, 39–50. [CrossRef]
92. Henseler, J.; Noonan, R. Partial Least Square Path Modeling: Basic Concepts issues and Application. *Adv. Methods Model. Mark. Int. Ser. Quant. Mark.* **2017**, 361–381. [CrossRef]
93. Hu, L.-T.; Bentler, P.M. Fit indices sensitivity to misspecification. *Psychol. Methods* **1998**, *3*, 424–453. [CrossRef]
94. Khan, G.F.; Sarstedt, M.; Shiau, W.L.; Hair, J.F.; Ringle, C.M.; Fritze, M.P. Methodological research on partial least squares structural equation modeling (PLS-SEM): An analysis based on social network approaches. *Internet Res.* **2019**, *29*, 407–429. [CrossRef]
95. Evermann, J.; Tate, M. Assessing the predictive performance of structural equation model estimators. *J. Bus. Res.* **2016**, *69*, 4565–4582. [CrossRef]
96. Kock, N. Common method bias in PLS-SEM: A full collinearity assessment approach. *Int. J. e-Collab.* **2015**, *11*, 1–10. [CrossRef]
97. Hair, J.F.; Sarstedt, M.; Ringle, C.M.; Gudergan, S.P. *Advanced Issues in Partial Least Squares Structural Equation Modeling*; Sage Publications: Los Angeles, CA, USA, 2017; ISBN 1483377385.
98. Falk, R.F.; Miller, N.B. *A Primer for Soft Modeling*; University of Akron Press: Akron, OH, USA, 1992; ISBN 0962262846.
99. Chin, W.W. How to write up and report PLS analyses. In *Handbook of Partial Least Squares*; Springer: Berlin/Heidelberg, Germany, 2010; pp. 655–690.
100. Hair, J.F.; Ringle, C.M.; Sarstedt, M. PLS-SEM: Indeed a silver bullet. *J. Mark. Theory Pract.* **2011**, *19*, 139–152. [CrossRef]
101. Shmueli, G. To explain or to predict? *Stat. Sci.* **2010**, *25*, 289–310. [CrossRef]
102. Straub, D.; Gefen, D. Validation Guidelines for IS Positivist Research. *Commun. Assoc. Inf. Syst.* **2004**, *13*, 380–427. [CrossRef]
103. Shmueli, G.; Sarstedt, M.; Hair, J.F.; Cheah, J.H.; Ting, H.; Vaithilingam, S.; Ringle, C.M. Predictive model assessment in PLS-SEM: Guidelines for using PLSpredict. *Eur. J. Mark.* **2019**, *53*, 2322–2347. [CrossRef]
104. Park, S.; Gupta, S. Handling endogenous regressors by joint estimation using copulas. *Mark. Sci.* **2012**, *31*, 567–586. [CrossRef]
105. García-pérez-de-lema, D.; Ruiz-palomo, D.; Diéguez-soto, J. Analysing the Roles of CEO' s financial literacy and financial constraints on Spanish SMEs technological innovation. *Technol. Soc.* **2021**, *64*, 1–21. [CrossRef]
106. Sarstedt, M.; Ringle, C.M.; Cheah, J.H.; Ting, H.; Moisescu, O.I.; Radomir, L. Structural model robustness checks in PLS-SEM. *Tour. Econ.* **2020**, *26*, 531–554. [CrossRef]
107. Hair, J.F.; Astrachan, C.B.; Moisescu, O.I.; Radomir, L.; Sarstedt, M.; Vaithilingam, S.; Ringle, C.M. Executing and interpreting applications of PLS-SEM: Updates for family business researchers. *J. Fam. Bus. Strateg.* **2021**, *12*, 100392. [CrossRef]
108. Antonakis, J.; Bendahan, S.; Jacquart, P.; Lalive, R. Causality and endogeneity: Problems and solutions. In *The Oxford Handbook of Leadership and Organizations*; Oxford University Press: New York, NY, USA, 2014; pp. 93–117. [CrossRef]
109. Huit, G.T.M.; Hair, J.F.; Proksch, D.; Sarstedt, M.; Pinkwart, A.; Ringle, C.M. Addressing endogeneity in international marketing applications of partial least squares structural equation modeling. *J. Int. Mark.* **2018**, *26*, 1–21. [CrossRef]
110. Ringle, C.M.; Sarstedt, M.; Mitchell, R.; Gudergan, S.P. Partial least squares structural equation modeling in HRM research. *Int. J. Hum. Resour. Manag.* **2020**, *31*, 1617–1643. [CrossRef]
111. Svensson, G.; Ferro, C.; Høgevold, N.; Padin, C.; Carlos Sosa Varela, J.; Sarstedt, M. Framing the triple bottom line approach: Direct and mediation effects between economic, social and environmental elements. *J. Clean. Prod.* **2018**, *197*, 972–991. [CrossRef]
112. Ramsey, J.B. Tests for specification errors in classical linear least-squares regression analysis. *J. R. Stat. Soc. Ser. B* **1969**, *31*, 350–371. [CrossRef]
113. Yáñez-Araque, B.; Sánchez-Infante Hernández, J.P.; Gutiérrez-Broncano, S.; Jiménez-Estévez, P. Corporate social responsibility in micro-, small- and medium-sized enterprises: Multigroup analysis of family vs. nonfamily firms. *J. Bus. Res.* **2020**, *124*, 581–592. [CrossRef]
114. García-Piqueres, G.; García-Ramos, R. Is the corporate social responsibility–innovation link homogeneous?: Looking for sustainable innovation in the Spanish context. *Corp. Soc. Responsib. Environ. Manag.* **2020**, *27*, 803–814. [CrossRef]

MDPI
St. Alban-Anlage 66
4052 Basel
Switzerland
Tel. +41 61 683 77 34
Fax +41 61 302 89 18
www.mdpi.com

Mathematics Editorial Office
E-mail: mathematics@mdpi.com
www.mdpi.com/journal/mathematics